Analogy Thinking for Photonics Innovation

For a complete listing of titles in the *Artech House Applied Photonics Library*, turn to the back of this book.

Analogy Thinking for Photonics Innovation

Khaled Ahmed

ARTECH HOUSE
BOSTON | LONDON
artechhouse.com

Library of Congress Cataloging-in-Publication Data
A catalog record for this book is available from the US Library of Congress.

British Library Cataloguing in Publication Data
A catalogue record for this book is available from the British Library.

Cover design by Iain Hill

ISBN 13: 978-1-68569-107-3

685 Canton Street
Norwood, MA 02062

10 9 8 7 6 5 4 3 2 1

Contents

Foreword

For more than 25 years, I have had the privilege of knowing the author of this remarkable book, an engineer whose curiosity has always extended well beyond the traditional boundaries of his discipline. While his technical work spans microelectronics, nanophotonics, solar energy harvesting, and advanced photonic devices, what has distinguished his career is not only its breadth but the lens through which he approaches innovation. As a virologist by training—one who has spent the majority of my professional life studying the behavior and evolution of bacteria and viruses, and later collaborating with engineers in both academic and industrial settings—I have come to recognize a familiar pattern in his approach: the instinct to draw meaning from analogy.

My own path began in the laboratories of the University of California, Irvine, where I completed my doctoral studies in virology before entering an engineering consulting firm for nearly a decade. These years working alongside mechanical, chemical, and electrical engineers taught me the value of cross-disciplinary translation. Biological systems rarely reveal themselves to a single vantage point; they reward integrative thinking. Later, during seven years in academic biotechnology research, I found that engineers, when immersed in the complexity of living systems, often perceive principles that biologists overlook. It is from this vantage—rooted in biology but strengthened through collaboration with engineering—that I have observed the author's career unfold.

The engineer who authored this book has made a practice of finding structural resonance across scientific worlds. He has worked in five distinct companies over three decades, each focused on a different domain, yet he learned to treat these differences not as obstacles but as opportunities for abstraction. Whether designing photonic crystals or advancing solar microstructures, he has continually reframed problems from one field using the conceptual scaffolding of another. This practice has contributed to a patent portfolio of more than two hundred inventions, each a testament to the generative power of analogy.

In biology, analogy is not merely a rhetorical device; it is a functional tool for understanding systems that defy direct observation. We compare viral replication to assembly lines, immune responses to communication networks, gene regulatory circuits to logic gates. These analogies are imperfect, of course, yet they illuminate patterns that would otherwise remain obscured. The author of this book has mastered a similar cognitive mode, applying it to the physics of light, the engineering of materials, and the challenges of photonic design.

This book makes explicit the mental process that has implicitly guided much of modern scientific discovery. As the early chapters reveal, analogy operates by drawing structured correspondences between a well-understood source domain and a less familiar target domain. The reader is taken through fundamental elements—mapping, abstraction, transfer, constraint matching—each illustrated through case studies that range from the steam engine to biomimetic optical systems. The content is broad in scientific scope yet coherently unified by a single theme: that innovation often arises not from acquiring more information but from perceiving existing information differently.

What resonates strongly for me, as a biologist, is how the book honors the long-standing tradition of borrowing insights from life. Biology has always been a prolific library of functional prototypes: the graded-index surface of a moth's eye that inspires antireflective coatings, the layered microstructures of butterfly wings that lead to vibrant structural colors, the wide-field vision of the lobster that informs next-generation X-ray imaging. This cross-domain borrowing, discussed richly in the chapters on biomimicry and photonic materials, mirrors the intellectual journey many of us take when grappling with complex biological phenomena. Nature often solves problems long before humans name them.

In virology, for instance, we routinely model replication dynamics using ecological analogies, or describe viral entry using chemical or mechanical metaphors. Such analogies do not trivialize biology; they scaffold our understanding. Likewise, the analogies presented in this book do not dilute photonics, they enrich it. They provide conceptual tools that help researchers navigate a field where intuition alone is rarely sufficient.

What is particularly valuable about this work is that it demystifies analogy thinking. The author makes clear that analogies are neither decorative nor incidental. They are deliberate, structured, and methodical. They demand a balance between imagination and discipline, a balance I have witnessed repeatedly in his professional collaborations. Engineers and biologists often speak different technical languages, yet analogy provides a shared grammar. It enables the collective exploration of ideas that cannot be confined to a single discipline.

Throughout my career—from molecular virology to engineering consulting to academic biotechnology—I have seen firsthand how innovation thrives at the boundaries between domains. The most transformative ideas often emerge when scientists allow themselves to look sideways, to juxtapose systems that appear unrelated. This book offers readers not only historical and scientific examples of that phenomenon but a framework for practicing it themselves.

In a time when disciplines are increasingly specialized, the author provides a refreshing and necessary counterpoint: a reminder that creative progress is built upon the willingness to bridge conceptual divides. His approach reflects a scientist's humility and an engineer's boldness, a combination that makes this book both intellectually rigorous and inspiring.

It is fitting that this work comes from someone who has spent 30 years navigating the interfaces between electronics, optics, materials, and biological inspiration. His career demonstrates that analogy thinking is not an abstract philosophy but a practical, repeatable method for generating insight. The chapters that follow will, I am certain, help engineers, physicists, biologists, and students of all backgrounds cultivate a mindset capable of discovery.

As someone who has spent a career investigating systems far more dynamic and unpredictable than any engineered device, I find deep resonance in the central thesis of this book: that understanding emerges when we dare to see one domain through the lens of another. The author has given readers a guide to doing just that—rigorously, creatively, and with an appreciation for the hidden structures that unite our scientific pursuits.

Leila A. Khatib
Professor of Virology
San Jose State University, California
April 2026

Preface

Innovation often begins with a question that breaks from tradition, a question that dares to reframe what is accepted, to dissolve boundaries between disciplines, and to draw meaning where others see only difference. At the heart of this leap is analogy thinking: the cognitive alchemy that transforms familiarity in one domain into insight in another. In the world of photonics, where waves dance at the speed of light and materials sculpt the invisible, this form of reasoning has repeatedly served as a catalyst for discovery. This book is not a catalog of devices, nor a textbook of optical equations. It is, instead, an exploration of how the mind navigates the unfamiliar by anchoring itself to the known. It is a tribute to the human instinct to find patterns across the chasm of complexity—a guidebook on how to think analogically in a field where intuition often falters. The book opens by grounding the reader in the nature and power of analogy. Before lasers and fibers, before interference and entanglement, comes the process of abstraction. We take a problem, often technical and jargon-laden, and we peel it back to its essence. We reword it in plain language, distilling it to a form that resonates with something else we understand well. This is more than simplification; it is the first spark of analogy. Reframing the problem allows the mind to wander into unexpected territories. In those moments, we are freed from the rigid heuristics of our training and able to look sideways, to borrow metaphors from nature, history, or even childhood memory. That is how Newton turned an apple into a metaphor for gravity, or how a dog's nose inspired chemical sensors. This approach is not limited to physics, and certainly not exclusive to photonics, but in this domain it has proven particularly fertile. Photonics, as a discipline, is inherently interdisciplinary. It draws from quantum mechanics, electromagnetism, solid-state physics, and increasingly, from biology and machine learning. Yet this wealth of interconnectedness also makes it intimidating. It becomes easy to hide behind symbols and equations, to speak in a dialect accessible only to the initiated. But innovation rarely thrives in cloisters. To develop analogies, one must first abandon the protective armor of jargon. One must describe a photonic bandgap not as a plot or a dispersion curve, but perhaps as a musical silence, a range of notes a violin string refuses to play. One must reimagine total internal reflection not as a Snell's law solution, but as a stubborn echo in a canyon of glass. The chapters that follow chart a progression. We begin with the earliest inklings of analogy thinking applied to problems of light. Long before the term "photonics" existed, light was already being compared to water, to particles, to waves in a pond. These analogies shaped the early understanding of reflection, refraction, and interference. As fiber optics emerged, engineers likened light propagation to fluid flow through pipes, and later to electrons moving in waveguides. The development of optical fibers was a triumph not only of materials science and manufacturing, but of the ability to reframe a stubborn problem—the high loss of light in glass—through the lens of both theory and metaphor. Charles Kao's vision of ultrapure glass waveguides was not purely empirical; it was built on the belief that the problem could be recast in a simpler, more abstract form. We then explore photonic crystals, and how the profound idea of a band gap—borrowed directly from semiconductor physics—transformed our ability to mold light itself. It was not enough to know that materials could be periodic. The insight came from recognizing that photons, like electrons, could be guided and forbidden by structure alone. The analogy to electronic band theory wasn't superficial; it was foundational. It enabled a generation of researchers to use a well-honed toolkit of solid-state physics and apply it to an entirely new medium. In this act of cognitive recycling, photons became electrons in disguise. The materials changed, the tools evolved, but the conceptual leap came from analogy. This way of thinking continues with devices like quantum dot lasers, which owe their very existence to the particle-in-a-box analogy. By confining charge carriers within nanostructures reminiscent of the discrete energy states of an atom, engineers were able to build devices with threshold behaviors and efficiencies unattainable in traditional bulk or even quantum well designs. The leap wasn't merely in fabrication. It was in seeing that a concept used to teach undergraduates quantum mechanics—one almost cartoonish in its simplicity—could be a design blueprint for a revolution in optoelectronics. The same theme appears in the story of antireflective coatings. What began as a curious observation about moth eyes—eyes that shimmered less, reflected less—became a roadmap for creating broadband antireflection films. The key insight was to abandon the planar thinking of multilayer stacks and embrace the graded-index surfaces found in nature. Once again, biology whispered a solution to a problem physics had been grappling with. The moth did not need to understand impedance matching or quarter-wave coatings. It only needed to survive. Engineers, listening to its whisper, learned to sculpt materials at the nanoscale, mimicking its corneal structures and achieving optical feats with simple geometrical elegance. This book also covers more recent stories: how lobster eyes inspired new types of wide-field X-ray telescopes, how butterflies taught us to fabricate vivid structural colors without dyes, and how the brain's network of neurons is being mirrored in optical neural networks with Mach-Zehnder interferometers standing in for synapses. These stories span centuries, continents, and disciplines, but they share a common spine. Someone, somewhere, asked, "What does this remind me of?" And from that moment, innovation began. Perhaps the most important goal of this book is not to inform but to equip. The reader should leave not just with knowledge of past analogies, but with a skillset for finding new ones. This is where the art blends with the science. Learning to describe a complex system in plain language is not easy. It requires the humility to accept that expertise can become a barrier to clarity. It means discarding the comfort of mathematical formality, if only briefly, to test whether the core idea still holds when translated into the language of a child or even a dog. This is not an insult to complexity, but an homage to insight. The act of explaining to someone without technical training is often the moment when new connections become visible, when the fog lifts and the heart of the problem is revealed. We include examples from multiple domains not to distract from photonics, but to provide practice in cross-domain thinking. The steam engine, the artificial kidney, the bionic limb—each was born from a willingness to see the familiar in the unfamiliar. These examples show that analogy is not a trick or shortcut. It is a deeply rooted cognitive mechanism, one that science has too often ignored in its rush to formalism. But innovation, the kind that upends industries and changes paradigms, is rarely formal at its birth. It is playful, intuitive, and often born of metaphor. If there is a single thesis that runs through these pages, it is this: analogy is not the enemy of rigor. It is the companion of discovery. The ability to reason analogically, to bridge domains, to describe complexity in plain language—these are not merely soft skills. They are the architecture of breakthrough thinking. I wrote this book for the engineer trying to see the problem differently. For the student who feels lost in a forest of equations. For the researcher who senses that the next idea might come from a field they have never studied. And for the curious mind who wonders how a moth's eye, a butterfly wing, or a lobster's vision could possibly inform the future of photonics. May this book help you see more clearly, think more creatively, and invent more courageously. Not because you mastered the details, but because you dared to draw connections that others overlooked. In the end, that is the power of analogy. It lights the way when the path is uncertain. It builds bridges where none exist. It transforms how we see the world—and how we shape its future.

1

Analogy Thinking

After reading this chapter you should be able to:

- Understand elements of analogy thinking;
- Prepare to practice analogy thinking;
- Create habits that put you in a position to master analogy thinking.

1.1 Introduction

The process of innovation often hinges not on discovering entirely new elements, but on seeing familiar ones through a different lens. Analogy thinking is one of the most enduring and powerful cognitive tools for achieving this. By transferring knowledge across seemingly unrelated domains, it allows individuals to form new connections, ask better questions, and sometimes even reshape entire fields. This chapter explores the structure of analogy thinking, breaking it down into a sequence of distinct yet deeply interwoven elements that enable the transformation of insight from one context to another. The journey begins with the identification of the source domain, which is the place where knowledge already exists in an organized and often intuitive form. This is where the analogy starts, whether it be in nature, physics, history, or a man-made system. The choice of source domain is foundational, because the depth and internal logic of this domain will determine the richness of what can be transferred. While the initial appeal may stem from visual or structural similarity, the most fruitful analogies usually come from recognizing deeper functional parallels that operate under comparable constraints or objectives. Once the source is selected, attention shifts to the target domain. This is the problem space that demands rethinking, often because conventional approaches have reached a plateau. The target may be new, complex, or ambiguous. It may suffer from inefficiencies or design challenges. Whatever the case, the target domain presents a need that cannot easily be addressed within its own conceptual framework. It is this tension between a well-understood source and a challenging target that sets the stage for meaningful analogical exploration. The heart of analogy thinking lies in the act of mapping. This is where the conceptual alignment between the two domains begins to take shape. Mapping is not about copying but about translating roles, functions, relationships, and mechanisms from one domain into the other. It requires a sensitivity to structure, a feel for correspondence, and a willingness to overlook superficial mismatches in order to uncover meaningful parallels. During mapping, one begins to test whether the analogy has legs—whether it can carry over enough substance to spark useful insights in the new domain. But mapping only becomes truly effective when it is supported by abstraction. Abstraction allows the thinker to lift out the core idea from the specific details of the source domain. It removes clutter, revealing the skeleton of the concept that might be applicable elsewhere. This can involve distilling geometric relationships, dynamic interactions, or governing principles that exist independently of context. Abstraction is what makes analogy scalable. It allows a localized observation to be converted into a generalized insight.

Once the abstraction is in hand, the next step is transfer. This is the moment when the insights and patterns drawn from the source are put to use within the target domain. The process of transfer can be straightforward or deeply iterative. Sometimes it produces a direct implementation; at other times it prompts a shift in perspective that suggests entirely new ways of approaching the problem.

Transfer is where the analogy is tested in action. It is where thought becomes design, and metaphor becomes mechanism. No analogy can succeed without careful attention to constraints. What works in one domain may falter in another due to differences in scale, materials, environment, or objectives. Constraint matching serves as a filter that checks the realism of the analogy and adapts it as needed. It helps avoid forced fits by revealing mismatches early and prompting refinements to the original mapping or abstraction. Often, this phase determines whether an analogy remains a clever idea or becomes a viable solution. All of these elements—source, target, mapping, abstraction, transfer, and constraint matching—culminate in insight generation. Insight may come as a sudden leap or an incremental realization, but in both cases it marks a turning point. It signals that the analogy has yielded something new: a design breakthrough, a conceptual shift, a functional improvement. Insight is not the end of analogy thinking but rather its most valuable byproduct, and one that often sets in motion further cycles of analogical reasoning. This chapter aims to make each of these elements explicit and tangible. Through detailed discussion and real-world examples, it shows how analogy thinking can be broken down, understood, and deliberately practiced—not as a mystical flash of genius, but as a repeatable framework for generating innovation. As shown in Figure 1.1, the analogy thinking process operates as a dynamic and recursive learning framework rather than a linear sequence. Each connection between blocks serves a distinct cognitive function that collectively makes the framework resilient and generative. The forward progression from source domain through transfer represents the movement of conceptual structure—from what is known to what is uncertain. However, it is the feedback loops that make the process powerful.

Validation connects back to both transfer and the source domain, ensuring that each iteration incorporates critical assessment and refinement. When

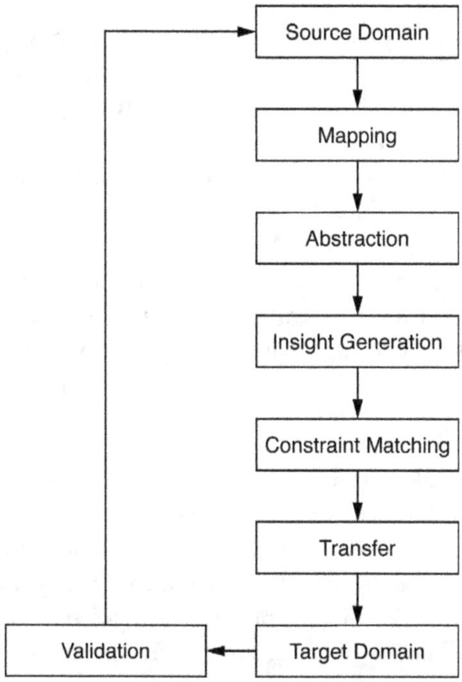

Figure 1.1 Process flow of analogy thinking. The diagram illustrates the sequential and cyclic structure underlying the analogy thinking framework described in the text. The process begins in the source domain, where a well-understood system or concept is identified. Through mapping, relationships and structural correspondences are drawn between the source and a new context. The abstraction stage extracts core principles stripped from domain-specific details, enabling the transfer of these generalized insights into the target domain. In the target space, new configurations or perspectives emerge, giving rise to insight generation. These insights are subsequently subjected to validation, where empirical or conceptual testing determines their soundness. The feedback loops—from validation back to transfer and source—illustrate the iterative nature of analogy thinking, where each cycle refines mappings, corrects mismatches, and deepens understanding across domains.

a proposed mapping fails empirical scrutiny, the loop forces reevaluation of assumptions, driving a return to earlier stages for recalibration. The linkage between transfer and insight generation underscores the transition from analogy as a mapping exercise to analogy as a tool for creative discovery. By grounding insights in validated mappings, the process avoids superficial similarities and instead cultivates deep structural reasoning—the essence of innovative thought in photonics and beyond. The final connection from insight generation back to source domain symbolizes intellectual growth: the source itself evolves, enriched by feedback from its application, enabling more sophisticated analogies in subsequent cycles.

1.2 Analogy Thinking Elements

Before examining the individual components of analogy thinking, it is useful to understand how these elements collectively support the creative process. Analogical reasoning operates by drawing structured relationships between a familiar domain and a novel problem space, enabling innovators to reframe challenges, uncover hidden patterns, and discover unconventional solutions. The subsections that follow outline the major stages involved in this cognitive workflow—from identifying an appropriate source of inspiration, to abstracting its underlying principles, to transferring and refining those principles within new constraints. Together, these elements form a systematic foundation for generating insight across diverse technical and conceptual domains.

1.2.1 Source Domain

The foundation of analogy thinking begins with the identification of the source domain. This is the familiar, often well-understood area from which knowledge is drawn. It may come from personal experience, established scientific understanding, historical precedent, or even a completely unrelated discipline. The power of a source domain lies not in its surface similarity to the problem at hand, but in the depth of its internal logic, structure, and behavior that can be mapped onto a new context. For example, consider the development of the quantum dot laser. Its conceptual origins were inspired by the analogy of electrons confined in nanostructures behaving like particles trapped in a three-dimensional (3D) potential well, akin to particles in a box in classical mechanics. This analogy provided a powerful mental model for visualizing quantum confinement and led to the refinement of the bandgap engineering in semiconductor nanostructures [1]. The idea of drawing from quantum mechanical thought experiments, such as the particle-in-a-box, illustrates how a well-understood theoretical source can drive technological invention in a different domain. Another compelling example is the use of the eye of a moth to inspire antireflective coatings on solar panels and camera lenses. The microscopic structure of a moth's eye minimizes reflection in low-light environments to avoid detection by predators. Engineers used this biological source domain to design nanostructured surfaces that reduce reflection across a broad spectrum, leading to efficiency improvements in photovoltaic and imaging systems [2]. This bioinspired analogy demonstrates how nature can serve as a deep reservoir of functional templates that can be applied to technological challenges. Selecting the right source domain is not about superficial resemblance but rather about uncovering systems that behave in structurally similar ways. The creative act of analogy thinking often begins by noticing something compelling in a familiar system and wondering: Could this work here, too?

1.2.2 Target Domain

While the source domain is the starting point, the target domain is the destination: the new, less-understood problem space where innovation is needed. In analogy thinking, the goal is not just to understand the target domain better but to reimagine it using insights and mechanisms borrowed from the source domain. The target domain is often a technical challenge or a system in need of improvement, redesign, or explanation. It could be a novel optical device, a software architecture problem, or even a business model. What makes a good target domain for analogical reasoning is that it contains unknowns, inefficiencies, or emerging questions that are ripe for conceptual reframing. An illustrative case comes from optical interconnects in data centers. Engineers faced increasing limitations in data throughput and energy efficiency using traditional electrical interconnects. Looking for a new framework, they turned to biological systems like the human nervous system as a source domain. Neurons use spatially parallel, low-power signal transmission. Mapping this onto the problem of interconnects led to architectures that use parallel optical channels, such as those employing arrays of microLEDs and photodetectors [3]. The nervous system thus served as a structural template to rethink bandwidth scalability and parallelism in optical communication. In another example, the design of soft robotics often uses animal locomotion as the target domain. Traditional rigid robots struggle in dynamic, unstructured environments. By analogizing the behavior of octopus arms or elephant trunks, both highly dexterous and flexible structures, engineers designed soft actuators that better conform to their environment and perform nuanced tasks that rigid robots cannot [4]. Here, the analogy helps redefine the mechanical design problem by setting a new target for behavior and adaptability. Crucially, the target domain must be flexible enough to accommodate novel frameworks. It's in this space where analogy thinking proves its worth, not by forcing a solution to fit, but by redefining the contours of the problem itself through cross-domain insight.

1.2.3 Target Mapping

Mapping is the cognitive act of establishing correspondence between elements of the source domain and those of the target domain. It's where the real engine of analogy resides. Effective mapping goes beyond surface-level similarities and instead identifies deeper relational or functional parallels between systems. This process often involves abstracting away superficial details and focusing on the roles and relationships of system components. In the case of the lobster-eye-inspired X-ray imaging system, the analogy was not based on the material or color of the lobster eye but on the function and geometry of the reflective tubes that capture light. Scientists replicated the square, mirror-lined channels of the crustacean's eye in microchannel plates to reflect incoming X-rays to a common focal point [5]. The mapping here centered on geometric optics and reflection paths rather than biological function. A historical example comes from the Wright brothers, who drew inspiration from birds. They did not try to mimic the exact flapping motion of wings, but instead mapped the principles of lift and control through adjustable surfaces. The elevator and rudder of their aircraft were functionally analogous to a bird's tail and wings in flight maneuvering. This mapping was conceptual and functional rather than literal, which is often the most powerful kind. Mapping is also a test of creativity and technical skill. It requires understanding both domains deeply enough to spot useful structural parallels, even when the contexts differ significantly. This skill can be cultivated through practice, interdisciplinary exposure, and deliberate exercises in comparison.

1.2.4 Abstraction

Abstraction is the process of stripping away domain-specific details to isolate the underlying structure or principle that might transfer across domains. This element of analogy thinking allows innovators to see through the noise and uncover generalizable patterns or mechanisms. For instance, in the design of photonic crystals, engineers abstracted from the periodic structure of electronic bandgap materials to consider the effect of periodic dielectric constants on photons instead of electrons. The underlying idea—that periodicity could lead to forbidden energy bands—was translated from solid-state physics into optics by abstracting away the specific particles involved [6, 7]. This enabled a whole class of devices that manipulate light propagation with precision. Another good example comes from cloud computing. The abstraction of "utility" from the domain of electricity enabled thinkers to conceptualize computing resources as metered, scalable, and centrally managed infrastructure. This conceptual reframing catalyzed the development of the modern cloud ecosystem. The ability to abstract away from the wires and servers to focus on the service delivery model made the analogy between electricity and computing truly transformative. Abstraction acts as a bridge between the source and target. It provides the common ground that allows for meaningful mapping. Without abstraction, analogies often remain stuck at the metaphorical or poetic level. With it, they become engineering blueprints and business models.

1.2.5 Transfer

Transfer is where the rubber meets the road: the act of applying insights or mechanisms from the source domain into the target domain to develop a solution, framework, or prototype. This element operationalizes the analogy. It is inherently experimental, requiring validation and iteration. In the case of the Mirasol display, engineers transferred the principle of interference seen in butterfly wings—where nanostructures reflect certain wavelengths—to create display pixels that produce vivid color without pigments [8]. The design involved building MEMS-based reflectors whose spacing could be tuned to interfere constructively for specific wavelengths. The concept transfer was elegant, but its commercial realization proved challenging, reminding us that successful analogy thinking must also consider manufacturability and system integration. Another transfer example is found in the use of bee swarm behavior to design decentralized algorithms for autonomous drone fleets. The idea of local sensing and collective decision-making was translated from natural behavior into a robust engineering framework for distributed control. Each drone, like each bee, acts on local information, but the system as a whole behaves in a coordinated manner. Transfer is both the culmination and the true test of analogy thinking. It challenges inventors to recontextualize insights and often reveals practical obstacles that need further mapping, abstraction, or constraint adjustment.

1.2.6 Constraint Matching

Even the most compelling analogies can fail if the constraints in the source domain do not align with those in the target. Constraint matching ensures that transferred concepts obey the physical, functional, or contextual boundaries of the new domain. For instance, while birds provided early inspiration for human flight, the analogy broke down when engineers tried to build flapping-wing aircraft at human scale. The mechanical and power constraints were too different. Only when designers adjusted the analogy to fixed-wing principles, while retaining control surfaces, did successful aircraft emerge. Constraint mismatches can derail analogy-driven innovation unless they are identified and adapted for. In photonic device design, the use of graphene was initially inspired by its exceptional electrical properties. However, attempts to replicate its performance in optical modulators ran into challenges due to weak light-matter interaction. Engineers had to adjust the analogy, coupling graphene with resonant cavities or plasmonic structures to meet performance constraints [9]. Constraint matching often requires iteration and recalibration. It's where the creative flexibility of analogy thinking is tested against the hard edges of engineering reality. Successful innovators use constraints not as blockers but as refinement tools that sharpen the analogy into a usable solution.

1.2.7 Insight Generation

All previous elements build up to this: the generation of new insights. Whether it's a novel architecture, a reframed question, or an inventive prototype, insight is the ultimate product of analogy thinking. This insight can emerge suddenly or grow incrementally as mappings and transfers are refined. One of the most well-known examples is the invention of the Velcro fastener. Swiss engineer George de Mestral observed how burrs stuck to his dog's fur and realized the hook-and-loop structure could be replicated in synthetic materials to create a fastening system. The insight came not just from noticing the burrs but from analyzing their function and mapping it onto a human need [10, 11]. In high-tech fields, insight generation might be more abstract. The leap from early neural network analogies to modern AI architectures illustrates how initial metaphors, once tested and refined, can scale into revolutionary technologies. What begins as an analogy becomes a fully fledged paradigm. Insight is not just the endpoint but also a feedback mechanism. Successful insights validate the analogy; failed insights prompt revisiting the mapping, abstraction, or constraints. As such, insight is both an outcome and a tool within the analogy thinking cycle.

1.3 Practicing for Analogy Thinking

In the pages that follow, we move from understanding analogy thinking as a concept to engaging with it as a deliberate, trainable practice. While analogies often seem to emerge in moments of inspiration or serendipitous insight, they are more than mental accidents. They are the result of patterns noticed, structures recognized, and connections made, sometimes consciously, often through disciplined exposure to unfamiliar material. But like any cognitive skill, analogy thinking sharpens with use. It can be stretched, flexed, and strengthened through habit and intention.

What makes analogy thinking particularly rich is that it operates at the intersection of curiosity and structure. It requires both openness to the strange, and sensitivity to the deep logic of systems. The following sections outline practical behaviors that help cultivate this dual capacity. They are not formulas. Rather, they are habits of attention and framing—ways of moving through the world that leave the mind stocked with potential source domains, primed to reframe problems, and alert to the structural harmonies between otherwise unrelated ideas.

Taken together, these behaviors form a toolkit, not for forcing analogies, but for creating the conditions under which useful ones can surface, be tested, and ultimately transformed into innovative action.

1.3.1 Observe Across Domains

One of the most fruitful habits an innovator can develop is a curiosity that roams freely across disciplines. The practice of observing systems beyond one's core field opens a mental reservoir of analogical material. Whether drawn from the natural world, cultural patterns, mechanical designs, or abstract mathematics, each domain holds structural models and functional insights waiting to be reused in a novel context. This behavior sharpens the ability to identify compelling source domains, which is the foundational starting point for analogy thinking. To strengthen this habit, one might start a daily or weekly practice of intentional exposure. Spend time reading outside your field. Watch documentaries that explain how ant colonies organize, how medieval builders constructed arches, or how Japanese carpentry evolved without nails. Sketch out the operating logic of systems you encounter: how a seed disperses, how crowds flow through a train station, or how stories propagate in folklore. By routinely breaking the silos of expertise, innovators can stock their minds with well-formed systems that may one day become the surprising key to solving problems in unfamiliar territory. An example comes from the field of architecture. The Eastgate Centre in Harare, Zimbabwe, is an office complex that uses natural ventilation principles observed in termite mounds to regulate temperature [12–16]. Architect Mick Pearce and engineer Arup applied the termite mound's system of chimneys and air circulation as a biological source domain to redesign building airflow without traditional HVAC systems. This behavior doesn't only strengthen analogy thinking; it rewires the lens through which the world is seen. The observer begins to view functional forms, not static things. They look at a tree and see hydraulic networks. They look at coral and see distributed fabrication. By practicing this observation, the mind becomes more nimble in spotting analogical structures embedded in daily life.

1.3.2 Frame Problems Broadly

Defining a challenge too narrowly risks trapping the mind within conventional boundaries. Innovation demands that problems be framed in ways that leave room for unexpected associations. When an innovator reframes a situation in functional, purpose-driven terms, rather than technical specifics, they expand the space in which analogies can form. This behavior refines the ability to articulate target domains in a way that makes them more receptive to foreign but structurally relevant insights. A practical method for broadening a problem frame is to describe what needs to happen without referencing how it has traditionally been done. For instance, instead of asking, "How can we build a lighter electric motor?" one might ask, "How can rotational force be generated without mass or contact?" That reframing opens the door to exploring acoustic, electrostatic, or even biological solutions. In industrial design, the OXO Good Grips kitchen tools emerged not from thinking about redesigning utensils, but from reframing the challenge as "creating tools that are comfortable for people with arthritis" [17–20]. That broader problem definition invited ergonomic and materials-focused thinking from domains as diverse as surgical instrument grips and bicycle handles. This approach expands the analogical canvas. A well-framed problem invites fresh metaphors. It gives analogies room to land. Most importantly, it prevents the target domain from becoming a cul-de-sac. Framing broadly means the innovator is not only solving a problem, they are designing a question that welcomes unexpected answers.

1.3.3 Train Structural Sensitivity

At the heart of effective analogy thinking lies a kind of vision: not a visual one, but a structural one. This is the capacity to see through surface differences and grasp the bones of a system—its interdependencies, feedback loops, and causal chains. Training structural sensitivity directly strengthens the mapping element of analogy thinking, where correspondences between source and target must be drawn with precision. This behavior can be strengthened by a simple yet powerful exercise: for any system encountered, try to redraw it using entirely different objects but the same relationships. Imagine replacing the nervous system with subway lines, or describing a thermostat as a legal negotiation between heat input and environmental resistance. These practices force the mind to lift away from appearances and focus on interactions and functions. The invention of the computer mouse, as documented in early work by Douglas Engelbart, stemmed not from input devices per se, but from understanding how people physically interact with spatial elements in their environment [21]. Engelbart studied how pilots navigate and how surgeons operate before drawing a map between human hand motion and on-screen navigation. This was not a design copied from existing tools but a mapping of function to new form. Developing structural sensitivity also means revisiting old analogies and questioning what made them work. Was it the control mechanism? The feedback loop? The distributed architecture? The goal is to become fluent in the language of systems, so that structural resonance can be recognized quickly, even when cloaked in unfamiliar materials or scales.

1.3.4 Practice Abstraction

Abstraction is not reduction. It is transformation. Practicing abstraction means stripping away context-specific noise while retaining the essence of what makes a system work. This behavior targets the abstraction element of analogy thinking and builds a mind capable of extracting the portable core from a situation. Start with small exercises. Describe how a vending machine works—but without mentioning coins, snacks, or electronics. Try to define a traffic light system without using the words light, red, or green. These constraints reveal the core logic underneath the implementation: a vending machine is a conditional release mechanism; a traffic system is a signal-controlled priority allocator.

One striking example of abstraction in practice is the analogy that led to Transmission Control Protocol/Internet Protocol (TCP/IP). Early pioneers in computer networking abstracted away the complexity of delivery routes and postal systems and instead focused on the concept of layered protocols and data packets—analogous in some ways to how cargo containers function in international shipping. The structural abstraction made it possible to build protocols that didn't rely on hardware compatibility but on functional handoffs [22]. This habit can be further reinforced by keeping a journal of abstracted concepts from the day's experiences. Whether it's observing a recipe, a negotiation, or a classroom, one can practice distilling the structure: What is being gated? What triggers change? What kind of flow is being managed? Over time, the mind starts seeing patterns as transferable templates rather than isolated events.

1.3.5 Prototype to Learn

To transfer an analogy from idea to implementation requires experimentation. Prototyping is how we explore whether what works in one domain still functions in another. This behavior sharpens the transfer element of analogy thinking and builds a bias toward action without premature commitment. Unlike final products, prototypes are disposable hypotheses. Their value lies in their ability to test mappings and abstractions quickly. An innovator might build a rough paper model, simulate the behavior in code, or even just walk through a physical space imagining how the system would operate if transformed. Each of these is a form of embodied analogy testing. The development of the Dyson vacuum was a result of prototyping based on an analogy to industrial cyclone separators. James Dyson built over 5,000 prototypes using cardboard and duct tape before refining a consumer-grade product. The key wasn't knowing the answer in advance—it was treating the analogy as a live experiment [23]. Practicing this behavior requires giving oneself permission to fail repeatedly. It means decoupling identity from outcome and investing in learning through building. Each prototype sharpens the analogy itself by forcing specifics to the surface: where the abstraction holds, where it breaks, and what must be adapted.

1.3.6 Interrogate Mismatches

Failures in analogical thinking are often rich in insight if one knows how to mine them. Practicing the habit of interrogating mismatches hones the constraint matching element. It helps innovators refine analogies instead of abandoning them, adjusting for fit rather than discarding the whole frame. When a transfer breaks down, ask: Is the scale wrong? Is the material inappropriate? Are temporal dynamics incompatible? For example, early attempts at solar-powered flight failed not because the idea was wrong, but because energy storage and material weight weren't yet aligned. Over time, with new battery chemistries and ultralight materials, the analogy became viable, as seen in aircraft like Solar Impulse [24]. Another example: engineers tried applying hydraulic logic to neural circuits, inspired by fluid motion [25, 26]. The analogy failed because delay and nonlinearity in biological systems are far more complex than in fluid mechanics. However, the exploration led to better understanding of synaptic delays and eventually informed neuromorphic circuit design. By embracing what doesn't work and systematically investigating why, innovators learn to tune their analogies. This is the difference between analogy as metaphor and analogy as engineering. Practicing this behavior builds resilience and precision, traits that are essential when working at the frontier of disciplines.

1.3.7 Reflect on Insights

Moments of insight are often treated as magical flashes. But in analogy thinking, they are clues, evidence that something aligned across domains. Reflecting on these moments, tracing them backward, and dissecting how they emerged, strengthens the insight generation element of the process. A useful practice is keeping an insight log, where unexpected solutions or conceptual leaps are recorded along with the mental steps that led there. What was the trigger? What mapping clicked? What prior observation suddenly made sense? This metacognitive practice turns intuition into a learnable skill. The invention of the sewing machine needle, often credited to Elias Howe, reportedly came from a dream in which warriors carried spears with holes at the tips, not the ends, leading to the insight of positioning the eye of the needle differently. Whether apocryphal or not, the story reflects how reflective thinking on imagery, structure, and function can lead to real-world breakthroughs [27]. This behavior reinforces a feedback loop. Insights become not just outcomes but teachers. Each one enriches the innovator's repository of functional correspondences, preparing them for future analogies. It transforms innovation from isolated sparks to an evolving process of pattern mastery.

1.4 Little Nudges

Innovation doesn't always arrive in loud, dramatic moments. Often, it sneaks in sideways, nudged into existence by subtle adjustments to the way we move through our days. While analogy thinking can be sharpened by deliberate cognitive exercises, it can also be gently fostered through habits that at first glance may seem unconnected to problem-solving. These small behaviors don't demand grand changes; they work quietly, cultivating flexibility, mental elasticity, and openness—all critical traits for the analogical mind.

1.4.1 Take a Different Route to Work

Routine is comforting. But in excess, it hardens thought patterns. Something as simple as changing your daily commute—choosing an unfamiliar road, taking a scenic detour, or even walking part of the way—can loosen cognitive rigidity. When the environment shifts, the brain's prediction engine kicks in with greater intensity. This increased alertness makes one more aware of patterns, more sensitive to deviations, and more curious about alternate configurations. On a different street, you may notice a structure you've never seen before or a cluster of activities unfolding in a novel way. These moments accumulate into a more adaptable mental posture, encouraging your brain to rehearse the same flexibility it will need when transferring ideas from one domain to another. That's the hidden gift: by choosing unpredictability, you prepare yourself to embrace surprise, not just in traffic, but in thinking.

1.4.2 Read Widely and Outside Your Discipline

Technical depth builds mastery, but analogy thinking thrives on breadth. Reading about topics far removed from one's field—such as marine biology, medieval history, architecture, psychology—feeds the mind with source domains you didn't know you needed. It isn't just about collecting facts; it's about observing how other disciplines frame problems, define systems, and pursue explanations. For instance, a materials scientist reading about coral reefs might absorb new ideas about structural growth or environmental adaptation without even intending to. Later, when facing a challenge in nanostructured coatings, those seemingly irrelevant impressions might whisper their way into relevance. Over time, this habit of diverse reading forms a kind of cognitive compost that is rich, fertile ground for analogies to take root.

1.4.3 Spend Time in Restorative Environments

There's something profoundly generative about the beach, a forest path, or even a quiet room with sunlight slanting through the blinds. Not because inspiration strikes like lightning, but because these moments allow mental noise to fade. When the mind stops bracing against pressure, it starts to wander. And wandering, unpressured thought is where analogy often begins. A walk by the ocean can trigger reflections on wave dynamics, rhythm, persistence. Without effort, metaphors begin to unfold, some of which may later crystalize into insight. These restorative settings don't deliver solutions on demand. What they do is reset the mental circuitry, allowing new connections to arise without force.

1.4.4 Be Curious About Odd Behaviors

When someone behaves in a way you find strange, do not rush to judgment. Pause. Ask yourself what unseen framework might explain this behavior. Practicing this kind of interpretive empathy—imagining the mental model someone else might be working from—builds the very muscles required for analogical mapping. This nudge expands the range of how you understand systems. People, after all, are complex systems themselves. What you observe in them—feedback loops, incentive structures, adaptive behaviors—are transferable conceptual tools. That person who always questions authority may be modeling distributed control. The one who hoards notes may be reflecting on long-term information resilience. You're not just learning about people; you're training your mind to look beneath the surface, to infer hidden logic.

1.4.5 Invent "What If" Worlds for Fun

While waiting in line or doing the dishes, invent alternate worlds. What if trees could migrate? What if elevators were living organisms? These little daydreams are not escapist distractions, they are low-stakes simulations for analogical play. In these imagined spaces, you practice crossing conceptual boundaries, unconsciously training your ability to map unfamiliar structures to known ideas. The beauty of these thought experiments is that they are playful yet potent. They keep your mind agile, primed for insight, and more capable of flipping between domains when the real-world challenge arrives.

1.5 Reflections

While analogy thinking is a powerful cognitive engine for discovery, it is not without limits. Analogies offer structured correspondences between domains, but they are not the same as innovation itself. Innovation requires not only seeing connections but also testing, refining, and often transcending them. Analogical reasoning may spark an idea, yet the journey from analogy to innovation involves validation, constraint matching, and practical realization. Inspiration and creativity may emerge from the recognition of resemblance, but innovation demands translation into a working solution that changes how a system performs or how knowledge is applied. In this sense, analogy thinking should be seen as a fertile starting ground—not a guarantee of novelty. A metaphor may illuminate; only experimentation and iteration can transform it into invention.

The limits of analogy thinking often appear when resemblance replaces reasoning. Superficial parallels—those based on shape, appearance, or metaphor—can mislead. An analogy drawn from nature or another field is valuable only if its underlying structure or governing mechanism is faithfully transferable to the target domain. Without careful abstraction and constraint matching, analogies risk becoming decorative rather than generative. The photonic engineer who borrows from biology, for instance, must recognize that the moth's eye or the butterfly's wing is not a ready-made solution, but a prompt for exploring how function arises from structure under specific physical constraints. Analogy thinking becomes productive when it moves beyond resemblance toward re-creation of principle.

Equally important is clarifying that analogy thinking is distinct from mere inspiration or general creativity. Inspiration is affective—it arises from wonder or aesthetic resonance. Creativity is generative—it produces multiple possibilities. Analogy thinking, by contrast, is structural—it identifies and manipulates relational patterns that can be mapped across domains. The process is disciplined rather than purely spontaneous, built on a deliberate sequence of mapping, abstraction, transfer, and validation. It is not a substitute for experimentation or analysis but a complement that expands the conceptual space in which both can operate. Thus, not every creative leap is analogical, and not every analogy is creative in itself.

It should not be inferred that all moments of insight arise exclusively from analogy thinking. While analogies often catalyze insight by connecting the known to the unknown, insight can also arise from other cognitive processes, such as pattern recognition within a single domain, serendipitous observation, or deductive reconfiguration of existing principles. Analogy thinking represents one of several routes to insight, distinguished by its cross-domain nature. Other insights emerge from iterative refinement, empirical anomaly, or even random association later rationalized into coherence. The reflective practitioner recognizes that insight is multifaceted: sometimes analogical, sometimes inferential, sometimes emergent from direct engagement with materials or data. Ultimately, analogy thinking should be understood as both a method and a mindset. Its value lies not in claiming every insight as its own, but in reminding innovators that the world's structures, whether biological, mechanical, or conceptual, can mirror one another in ways that illuminate. Reflecting on where analogies succeed and where they fail deepens this awareness, turning analogy from a spark of intuition into a disciplined pathway for sustained innovation.

1.6 Conclusion

The act of thinking through analogy is more than a clever mental shortcut—it is a disciplined process that, when understood and practiced, becomes an enduring pathway to original ideas. What begins as a curious comparison between two seemingly unrelated things can evolve into a methodical unraveling of innovation. But this transformation doesn't occur all at once. It comes from repeatedly engaging with the building blocks: choosing rich source domains, defining meaningful target problems, drawing mappings with precision, abstracting core mechanisms, testing transfers, and respecting real-world constraints. Each element serves its purpose, but their strength lies in how they interlace. Together, they form a repeatable structure that mirrors the way some of the most transformative discoveries have come to life. As with any creative discipline, the key lies in practice. The more one moves through this sequence—intentionally, reflectively—the more fluent and intuitive it becomes. Over time, patterns that once seemed hidden begin to stand out. Connections across disciplines emerge naturally. Instead of reaching for analogies, they start to present themselves, almost as if the mind is primed to think that way. This is the real reward of mastering analogy thinking. It's not just about solving one problem differently, but about developing a mental habit that expands what's possible. The process doesn't end with insight; in many cases, that insight simply becomes the new starting point for the next analogy. In this way, analogy thinking is less a destination and more a way of seeing—one that reshapes how challenges are approached, how knowledge is applied, and how innovation is born.

References

[1] Arakawa, Y., and H. Sakaki, "Multidimensional Quantum Well Laser and Temperature Dependence of its Threshold Current," *Applied Physics Letters*, Vol. 40, No. 11, 1982, pp. 939–941, https://doi.org/10.1063/1.92959.

[2] Phillips, B. M., and P. Jiang, "Biomimetic Antireflection Surfaces," in *Engineered Biomimicry*, New York: Elsevier, Ch. 12, 2013.

[3] Ahmed, Khaled, *MicroLED Devices and Systems*, Norwood, MA: Artech House, 2024.

[4] Majidi, C., "Soft Robotics: A Perspective—Current Trends and Prospects for the Future," *Soft Robotics*, Vol. 1, No. 1, 2014, pp. 5–11, https://doi.org/10.1089/soro.2013.0001.

[5] Kaaret, P. E., and P. Geissbuehler, "Lobster-Eye X-Ray Optics using Microchannel Plates," in *SPIE Proceedings*, Vol. 1546, 1992, pp. 82–90, https://doi.org/10.1117/12.51261.

[6] Yablonovitch, E., "Inhibited Spontaneous Emission in Solid-State Physics and Electronics," *Physical Review Letters*, Vol. 58, No. 20, 1987, pp. 2059–2062, https://doi.org/10.1103/PhyiguresRevLett.58.2059.

[7] John, S., "Strong Localization of Photons in Certain Disordered Dielectric Superlattices," *Physical Review Letters*, Vol. 23, 1987, p. 2486, https://doi.org/10.1103/PhysRevLett.58.2486.

[8] Tsai, H.-K., 2003, *Optical Interference Display Panel*, US Patent 7978396B2.

[9] Liu, M., et al., "A Graphene-Based Broadband Optical Modulator," *Nature*, Vol. 474, No. 7349, 2011, pp. 64–67, https://doi.org/10.1038/nature10067.

[10] Velcro Companies, *History of Velcro® Brand*, https://www.velcro.com/about-us/history/.

[11] Hwang, J., et al., "Biomimetics: Forecasting the Future of Science, Engineering, and Medicine," *International Journal of Nanomedicine*, Vol. 10, No. 1, 2015, pp. 5701–5713, https://doi.org/10.2147/IJN.S83642.

[12] King, H., et al., "Termite Mounds Harness Diurnal Temperature Oscillations for Ventilation," *Proceedings of the National Academy of Sciences*, Vol. 112, No. 7, https://doi.org/10.48550/arXiv.1703.08067.

[13] Abdullah, A., et al., "Cooling Strategies in the Biological Systems and Termite Mound: The Potential of Emulating them to Sustainable Architecture and Bionic Engineering," *ARPN Journal Of Engineering and Applied Sciences*, Vol. 13, No. 8, 2018, 2018, https://www.arpnjournals.org/jeas/research papers/rp 2018/jeas 1018 7313.pdf.

[14] Vesala, R., et al., "Termite Mound Architecture Regulates Nest Temperature and Correlates with Species Identities of Symbiotic Fungi," *PeerJ*, Vol. 6, 2019, p. e6237, https://doi.org/10.7717/peerj.6237.

[15] Nasir, O., and M. A. Kamal, "Inspiration from Nature: Biomimicry as a Paradigm for Architectural and Environmental Design," *American Journal of Civil Engineering and Architecture*, Vol. 10, No. 3, 2022, pp. 126–136, https://doi.org/10.12691/ajcea-10-3-3.

[16] Wei, Y., et al., "Simulation and Optimization Study on the Ventilation Performance of High-Rise Buildings Inspired by the White Termite Mound Chamber Structure," *Biomimetics*, Vol. 8, No. 8, 2023. https://doi.org/10.3390/biomimetics8080607.

[17] Lidwell, W., and G. Manacsa, *Deconstructing Product Design: Exploring the Form, Function, Usability, Sustainability, and Commercial Success of 100 Amazing Products*, Gloucester, MA: Rockport Publishers, 2009.

[18] Kelsheimer, H. L., et al., "Older Adult Women Find Food Preparation Easier with Specialized Kitchen Tools," *Journal of the American Dietetic Association*, Vol. 100, No. 8, 2000, pp. 950–952, https://doi.org/10.1016/S0002-8223(00)00273-X.

[19] McDonagh, D., "Needs Beyond the Functional," *The Routledge Companion to Design Studies*, 2016, p. 282.

[20] Story, M. F., and J. L. Mueller, "Universal Design of Products," *Universal Design Handbook*, 2001.

[21] Engelbart, D. C., et al., "A Research Center for Augmenting Human Intellect," in *Proceedings of the AFIPS '68 Fall Joint Computer Conference, Part 1*, Vol. 33, 1968, pp. 395–410, https://doi.org/10.1145/1476589.1476645.

[22] Cerf, V., and R. Kahn, "A Protocol for Packet Network Intercommunication," *IEEE Transactions on Communications*, Vol. 22, No. 5, 1974, pp. 637–648, https://doi.org/10.1109/TCOM.1974.1092259.

[23] Dyson, J., *Against the Odds: An Autobiography*, Cheshire, UK: Texere, 2002.

[24] G. Abbe et al., "Technological Development Trends in Solar-Powered Aircraft Systems," *Renew. Sustain. Energy Rev.*, Vol. 60, 2016, pp. 770–783, https://doi.org/10.1016/j.rser.2016.01.053.

[25] Rohner, P., *Fluid Power Logic Circuit Design: Analysis, Design Methods and Worked Examples*, Springer, 1979, https://doi.org/10.1007/978-1-349-16126-3.

[26] Vacca, A., and G. Franzoni, *Hydraulic Fluid Power: Fundamentals, Applications, and Circuit Design*, John Wiley & Sons, 2021.

[27] McNeil, I., *An Encyclopaedia of the History of Technology*, UK: Routledge, 1990.

2

Case Studies in Analogy Thinking

After reading this chapter, you should be able to understand:

- The ways in which analogy thinking drives innovation across diverse fields;
- Why biomimicry represents a powerful application of analogy thinking;
- How cross-domain analogies fuel the development of emerging technologies.

2.1 The Steam Engine

The development of the steam engine is a quintessential example of analogy-based innovation, demonstrating how the application of analogical reasoning can lead to groundbreaking technological advancements. The steam engine, which played a pivotal role in the Industrial Revolution, was the result of engineers and inventors drawing upon existing knowledge and applying it in novel ways to solve emerging problems. This process of innovation was heavily influenced by analogies with natural processes and earlier mechanical devices, which together laid the groundwork for the steam engine's design and subsequent impact on society. One of the key figures in the development of the steam engine was Thomas Newcomen, who in 1712 built the first practical steam engine, known as the Newcomen engine [1, 2]. Newcomen's invention was directly inspired by the work of earlier inventors, such as Thomas Savery, who had developed a primitive steam pump for removing water from mines [3].

Savery's device, in turn, was based on analogies between the behavior of steam and natural forces like atmospheric pressure. By observing how steam could create a vacuum when condensed, Savery realized that this principle could be harnessed to lift water. Newcomen expanded on this idea by incorporating mechanical elements such as a piston and cylinder, which were analogous to the mechanisms used in water pumps of the time. The result was a more efficient and powerful engine that could be used not only for pumping water but also for other industrial applications.

James Watt, another key figure in the history of the steam engine, further advanced the technology by applying analogical reasoning to improve the efficiency of Newcomen's design. Watt, a skilled instrument maker and engineer, recognized that the inefficiency of the Newcomen engine was due in part to the repeated heating and cooling of the cylinder, which wasted a significant amount of energy. Drawing on his knowledge of thermodynamics and heat transfer, Watt devised an analogy between the steam engine and natural systems, particularly the idea of conserving energy. He introduced a separate condenser to cool the steam without cooling the cylinder, which greatly improved the engine's efficiency. This innovation, based on Watt's analogical thinking, transformed the steam engine into a versatile and powerful tool that could drive machinery in factories, ships, and locomotives.

The steam engine also serves as an example of how analogies between different fields can spur innovation. For instance, the concepts of pressure and vacuum that were crucial to the development of the steam engine were originally studied in the context of atmospheric science. Engineers like Newcomen and Watt adapted these ideas to the mechanical domain, demonstrating how knowledge from one field can be effectively applied to solve problems in another. This cross-disciplinary application of analogies not only led to the refinement of the steam engine but also contributed to the broader understanding of physics and engineering principles. Moreover, the success of the steam engine highlights the importance of iterative design and continuous improvement in innovation, processes that are often driven by analogical reasoning. As engineers and inventors encountered challenges in the operation and application of steam engines, they drew upon analogies with existing technologies and natural processes to develop solutions. For example, the analogy between the human body's circulatory system and the movement of steam within the engine informed the design of more efficient engines that mimicked the continuous flow of blood. This iterative process of analogy-based innovation allowed the steam engine to evolve over time, becoming more reliable, powerful, and versatile.

As illustrated in Figure 2.1, Watt's steam engine operates on the principle of converting thermal energy from high-pressure steam into mechanical work. The high-pressure steam enters the cylinder through the valve slide, pushing the piston linearly within the cylinder. This reciprocating motion is transmitted

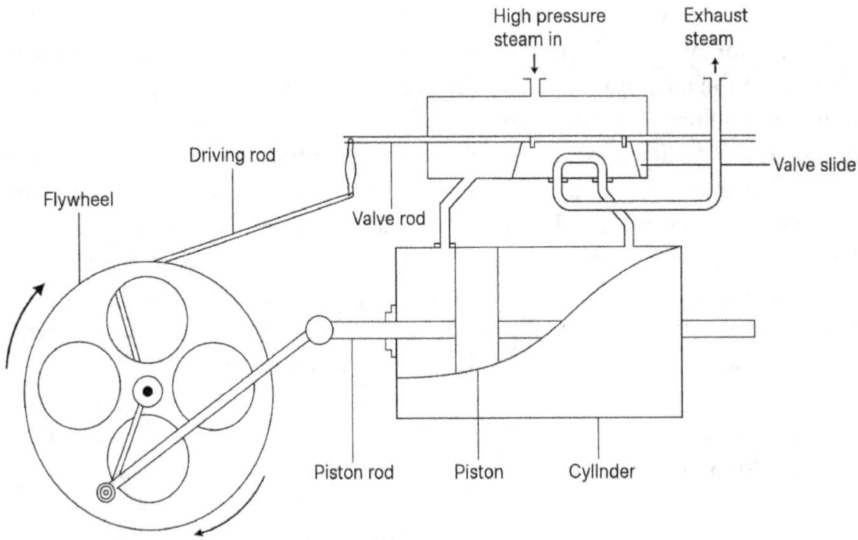

Figure 2.1 Schematic representation of Watt's steam engine showing key components such as the flywheel, piston, valve system, and connecting rods. The diagram illustrates how high-pressure steam drives the piston, converting reciprocating motion into rotary motion via the flywheel mechanism.

to the flywheel via the piston rod and driving rod, where it is converted into rotary motion. The flywheel ensures a uniform rotational speed by storing kinetic energy.

The valve mechanism, controlled by the valve rod, regulates the flow of steam into and out of the cylinder, alternating between the high-pressure inlet and the exhaust outlet. When steam expands inside the cylinder, it pushes the piston, and once the stroke is complete, the exhaust steam is released, allowing the piston to return for the next cycle. This controlled sequence of intake and exhaust ensures a continuous and smooth conversion of steam energy into rotational motion. The coordination between the valve slide and the piston movement is a key innovation that distinguished Watt's design from earlier steam engines, significantly improving efficiency and mechanical output.

2.2 Human Physiology

Human physiology has served as a rich source of inspiration for many groundbreaking innovations across various fields, thanks to the powerful tool of analogy-based thinking. By drawing parallels between the complex systems and processes found in the human body and those in engineering, technology, and medicine, innovators have been able to develop solutions that address a wide range of challenges. These case studies in analogy-based innovation highlight how understanding and mimicking the functions of the human body can lead to the creation of technologies that revolutionize industries and improve human life. One of the most notable examples of analogy-based innovation inspired by human physiology is the development of artificial organs and prosthetics. The human body's ability to perform highly specialized functions, such as pumping blood, filtering waste, and processing nutrients, has provided engineers and medical researchers with a blueprint for creating artificial systems that replicate these processes. For instance, the artificial heart is a direct result of analogical reasoning, where the mechanical function of a pump is used to mimic the heart's ability to circulate blood throughout the body. Early designs of artificial hearts, such as those developed by Dr. Paul Winchell and later refined by Dr. Robert Jarvik, drew heavily on analogies with the body's natural cardiovascular system [4–6]. These devices replicated the heart's pumping action through mechanical means, helping to save the lives of patients with severe heart conditions.

As illustrated in Figure 2.2, the total artificial heart replicates the anatomical configuration and hemodynamic function of the natural human heart. The artificial heart replaces both ventricles and the four native valves with mechanical counterparts that sustain pulmonary and systemic circulation. Blood enters the right and left atria through the venae cavae and pulmonary veins, respectively, before being actively pumped into the pulmonary artery and aorta

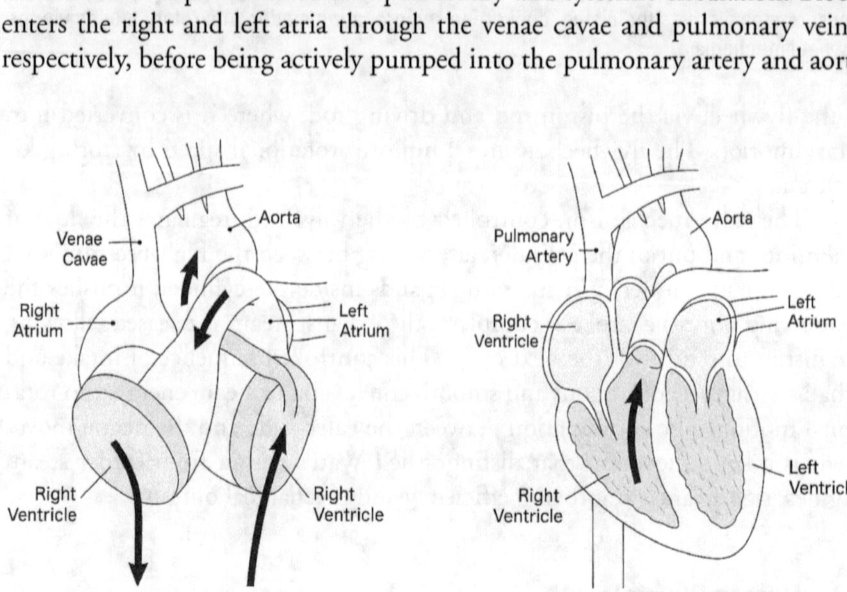

Figure 2.2 Comparison between a total artificial heart and a human heart. The schematic highlights the analogous structure and function of both systems. In the human heart (right), muscular ventricles generate pressure to circulate blood through the lungs and body, while in the total artificial heart (left), mechanical pumps perform the same task using synthetic ventricles and valves. Both maintain systemic and pulmonary circulation by ensuring continuous flow between the venae cavae, atria, ventricles, pulmonary artery, and aorta.

by the artificial ventricles. This mechanical system mimics the natural heart's rhythmic contractions, ensuring continuous oxygenation and nutrient delivery throughout the body. Although the materials and energy sources differ—biological muscle versus mechanical pumps—the operational principles of flow regulation and pressure generation remain fundamentally similar.

Another example is the development of dialysis machines, which were designed to perform the function of the kidneys in patients with renal failure. The kidneys are responsible for filtering blood and removing waste products from the body, a process that involves complex chemical and physical interactions [7]. By studying these interactions and drawing analogies between the natural filtration process and mechanical filtration systems, researchers were able to create dialysis machines that mimic the kidney's function [8, 9]. These machines work by passing the patient's blood through a series of membranes that filter out waste products, much like the glomeruli in the kidneys. This analogy between biological and mechanical filtration has allowed countless patients to manage their kidney disease effectively, improving their quality of life and extending their lifespans.

As illustrated in Figure 2.3, the dialysis process serves as an artificial replacement for kidney function in patients with renal failure. Blood is withdrawn from an artery and circulated through a pump that regulates flow into the dialysis unit. Within this unit, a semipermeable membrane separates the patient's blood from the dialysate solution. This membrane allows the selective diffusion of metabolic waste products such as urea and creatinine, as well as

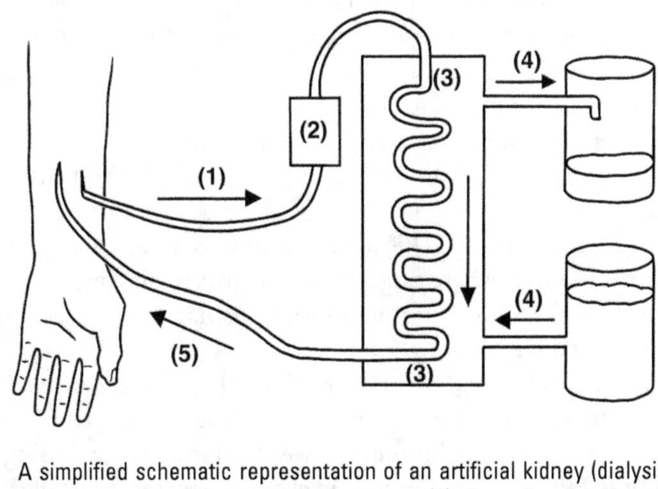

Figure 2.3 A simplified schematic representation of an artificial kidney (dialysis machine). Blood is drawn from an artery (1), and driven by a pump (2), through a dialysis unit (3), which contains a semipermeable membrane. This allows waste products and excess ions to diffuse into the dialysate while retaining essential blood components. Fresh dialysate (4), enters the dialysis unit and exits as used dialysate, while the cleaned blood (5), is returned to the patient's vein.

excess salts, into the dialysate, while preventing the loss of larger molecules like proteins and blood cells.

The dialysate, which continuously flows in the opposite direction to the blood, maintains a steep concentration gradient to maximize the efficiency of diffusion. As the dialysate becomes saturated with waste materials, it is replaced by a fresh solution to sustain purification. The cleansed blood is then safely returned to the patient's venous system, ensuring that electrolyte balance and toxin removal mimic the physiological role of the kidneys.

Human physiology has also inspired innovations in the field of robotics, particularly in the development of bionic limbs and exoskeletons [10–12]. These technologies are designed to restore mobility and functionality to individuals who have lost limbs or suffer from physical disabilities. Bionic limbs, such as those developed by companies like Ossur and DEKA Research, use analogies between the body's musculoskeletal system and mechanical systems to create prosthetics that closely mimic the natural movement of human limbs [13–15].

These devices incorporate sensors, actuators, and advanced materials that replicate the function of muscles, tendons, and bones, allowing users to perform tasks with a high degree of dexterity and control. The development of these prosthetics was made possible by a deep understanding of human physiology and the application of that knowledge to engineering, demonstrating the power of analogy-based innovation in enhancing human capabilities.

Another remarkable case of analogy-based innovation inspired by human physiology is the development of drug delivery systems. The human body's circulatory system is highly efficient at transporting nutrients, oxygen, and other essential substances to various tissues and organs [16–19]. By drawing analogies between this natural distribution system and the need to deliver medications to specific parts of the body, researchers have developed targeted drug delivery technologies that improve the efficacy and reduce the side effects of treatments. For example, nanoparticles can be engineered to mimic the behavior of red blood cells, allowing them to travel through the bloodstream and release drugs directly at the site of disease. This approach has been particularly effective in the treatment of cancer, where targeted drug delivery can concentrate chemotherapy agents at the tumor site, minimizing damage to healthy tissues and improving patient outcomes.

As illustrated in Figure 2.4, the drug delivery process can be interpreted as a molecular communication (MC) system, where drug particles serve as information carriers. The injection point functions as the transmitter, releasing particles into the bloodstream, the communication channel represented by the cardiovascular system. Through the natural flow of blood, these particles are transported to the intended delivery site (receiver), such as a specific organ or tissue region. This abstraction enables the application of communication-theoretic tools to

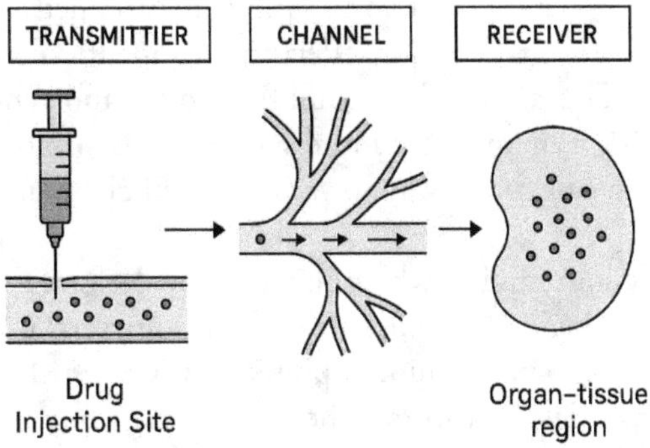

Figure 2.4 Schematic representation of the drug delivery system (DDS) modeled through MC. The transmitter corresponds to the drug injection, while the receiver corresponds to the DDS. The cardiovascular system acts as the communication channel through which drug particles propagate, enabling the analysis of the pharmacokinetics and targeting efficiency using communication-theoretic frameworks.

analyze and optimize the propagation, delay, and concentration of therapeutic agents. By modeling the dynamics of particulate motion within biological fluids, it becomes possible to improve targeting precision, reduce systemic side effects, and enhance the overall efficiency of the drug delivery system.

As shown in Figure 2.5, targeted drug delivery systems (TDDSs) represent a multifaceted therapeutic approach for pulmonary fibrosis (PF), utilizing biomaterials to enhance pharmacological precision and minimize systemic toxicity. The schematic outlines the integration of drugs within nanocapsules, which form nanoparticles that are later embedded in hydrogel matrices. This hierarchical structure allows gradual and localized drug release directly within fibrotic lung tissues, reducing off-target effects. Hydrogels, owing to their

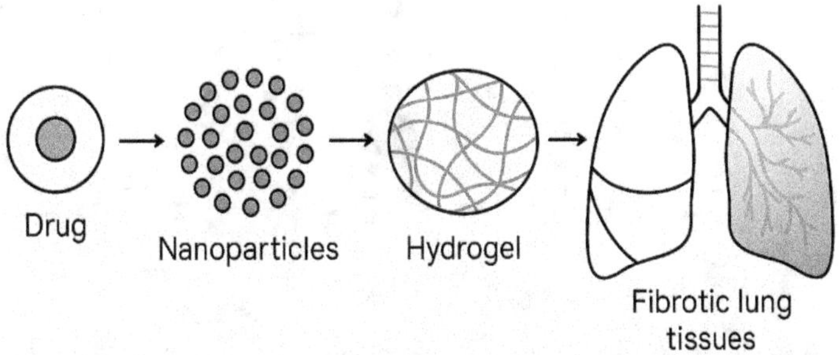

Figure 2.5 Schematic illustration of a targeted drug delivery system designed for pulmonary fibrosis therapy. The process involves encapsulating therapeutic drugs within nanocapsules, forming nanoparticles that are subsequently incorporated into hydrogels. These hydrogels enable controlled and sustained release at fibrotic sites in the lungs, promoting efficient drug penetration and localized therapeutic effects.

three-dimensional polymeric networks, provide sustained drug diffusion at the inflammation site, maintaining effective therapeutic concentrations over extended periods. The approach exemplifies how nanotechnology and biomaterial engineering synergize to address limitations of conventional antifibrotic therapies, offering a pathway toward more efficient and personalized PF management.

The field of biomimetics, which involves the design of materials and devices inspired by biological systems, also offers numerous examples of analogy-based innovation rooted in human physiology. One such innovation is the development of materials that mimic the properties of human skin, such as its ability to heal itself, sense pressure, and regulate temperature. Researchers have created synthetic skins that incorporate these features, using analogies with the body's natural processes to design materials that can be used in a variety of applications, from medical devices to wearable electronics [20–24]. For instance, self-healing polymers, inspired by the body's ability to repair damaged tissue, are being developed for use in coatings, adhesives, and electronics. These materials can automatically repair small cracks and tears, extending the life of products and reducing the need for maintenance.

As depicted in Figure 2.6, human skin is a highly complex, multifunctional organ composed of three primary layers—the epidermis, dermis, and

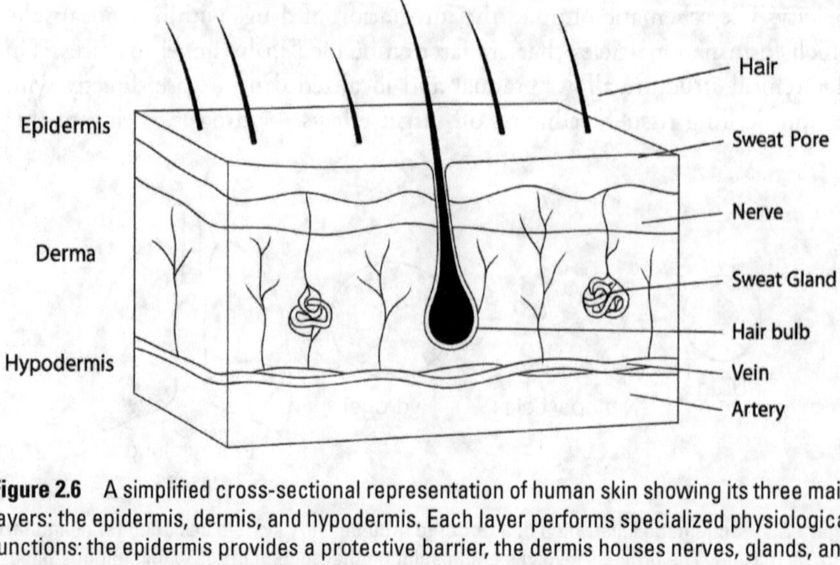

Figure 2.6 A simplified cross-sectional representation of human skin showing its three main layers: the epidermis, dermis, and hypodermis. Each layer performs specialized physiological functions: the epidermis provides a protective barrier, the dermis houses nerves, glands, and hair follicles responsible for sensation and thermoregulation, and the hypodermis contains adipose tissue for insulation and energy storage. This natural structure has inspired biomimetic innovations in synthetic materials that emulate the skin's abilities to sense pressure, heal damage, and regulate temperature.

hypodermis—each serving distinct physiological roles. The epidermis forms the outermost protective barrier that prevents dehydration and pathogen entry, while the dermis beneath it contains an intricate network of blood vessels, nerves, sweat glands, and hair follicles that collectively regulate temperature, enable tactile sensing, and maintain homeostasis. The innermost hypodermis, rich in adipose tissue, serves as both an energy reserve and a thermal insulator.

This layered biological architecture has become a foundational model in the field of biomimetics, inspiring the creation of synthetic materials that replicate the skin's multifunctionality. Researchers have developed artificial skins capable of self-healing, pressure sensing, and thermoregulation by drawing direct analogies from these natural processes. For example, self-healing polymers mimic the tissue-repair mechanisms of the dermis, allowing materials to autonomously repair microcracks [25, 26]. Likewise, flexible electronic skins integrate conductive nanomaterials to simulate the skin's nerve response, providing feedback in wearable devices and prosthetics. Such bioinspired innovations demonstrate how the intricate design of human skin continues to influence advancements in material science, robotics, and medical engineering.

Moreover, human physiology has inspired innovations in environmental engineering, particularly in the design of systems that manage and purify water. The human body's ability to process and excrete waste through the urinary and digestive systems has served as a model for creating more efficient and sustainable water treatment technologies. Engineers have developed wastewater treatment systems that mimic the body's natural filtration processes, using membranes and biological agents to remove contaminants and recycle water for reuse. These systems draw on analogies between the body's excretory functions and the need to manage waste in urban environments, leading to more effective and environmentally friendly solutions for water management [27–31].

As shown in Figure 2.7, the human body's internal systems provide a remarkable model for environmental engineering innovations. The liver, kidneys, large intestine, and bladder collaboratively perform complex filtration, detoxification, and excretion processes that maintain homeostasis and water balance. Environmental engineers have drawn direct inspiration from these natural mechanisms to design wastewater treatment technologies that replicate similar principles of biological filtration and selective absorption. By mimicking the body's ability to separate, neutralize, and expel waste, these engineered systems achieve sustainable water purification and reuse—offering an efficient, biomimetic approach to managing waste in urban environments. This convergence of human physiology and environmental technology underscores how biological insights can shape more sustainable engineering solutions.

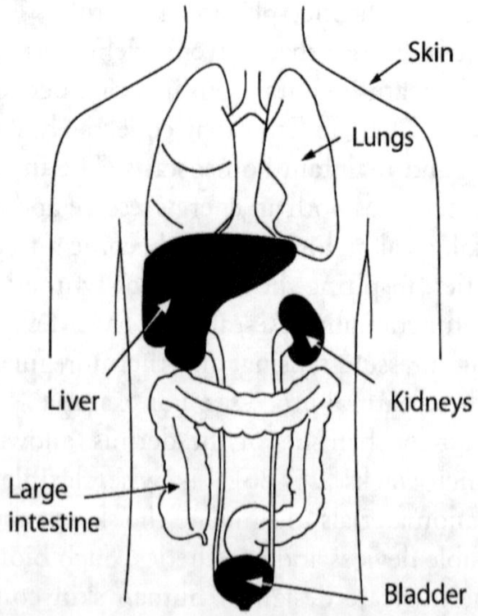

Figure 2.7 Illustration of the human body's waste management systems, highlighting the major organs involved in processing and excreting waste. The diagram depicts the liver, kidneys, large intestine, and bladder, which collectively demonstrate the body's natural filtration and purification mechanisms. These biological systems have inspired environmental engineers to design advanced water treatment processes that mimic similar physiological functions for efficient waste removal and water recycling.

2.3 Aerodynamics Inspired by Bird Flights

The field of aerodynamics is one of the most profound examples of analogy-based innovation, where the principles of bird flight inspired the development of aircraft and significantly advanced human engineering [32, 33]. The study of how birds fly, with their ability to navigate complex air currents, achieve lift, and maneuver with precision, provided the foundational understanding for the science of aerodynamics. This understanding, when applied through analogical reasoning, led to innovations that have revolutionized transportation, exploration, and even military strategy.

As depicted in Figure 2.8, a gliding bird experiences a complex interplay of aerodynamic and gravitational forces that determine its equilibrium and flight stability. The lift force acts perpendicular to the direction of motion, counterbalancing the bird's weight, while drag opposes forward motion along the x-axis. The side force component, aligned with the y-axis, contributes to lateral stability, particularly in crosswind conditions or during coordinated turns. All of these forces converge at the aerodynamic center, where the net moments—pitch, roll, and yaw—are generated around their respective axes. The bird's ability to modulate wing camber and angle of attack allows fine control of these moments, maintaining steady gliding flight without excessive energy

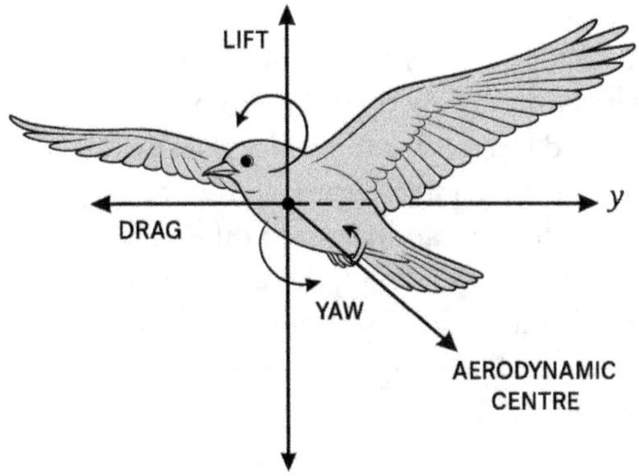

Figure 2.8 Free-body diagram of a gliding bird illustrating the primary aerodynamic forces and moments acting on a gull during steady flight. The diagram identifies the lift, drag, side force, and weight vectors, all applied at the aerodynamic center. Additionally, the rotational motions of roll, pitch, and yaw about the respective axes (x, y, and z) are indicated, showing how these interactions govern the bird's stability and control during gliding.

expenditure. Understanding this force distribution provides valuable insights into avian aerodynamics and has informed biomimetic approaches to aircraft design, emphasizing efficient control mechanisms and natural stability.

From the earliest days of human curiosity about flight, birds have been a natural source of inspiration. Ancient civilizations observed birds and wondered how their lightweight structures and wing movements allowed them to soar and glide effortlessly. However, it wasn't until the Renaissance that serious attempts were made to scientifically understand and replicate bird flight. Leonardo da Vinci, one of the earliest pioneers in this area, meticulously studied birds and produced detailed sketches of their anatomy, particularly their wings [34, 35]. He observed the curvature and the movement of wings during flight, which led him to conceptualize the first flying machines, such as the ornithopter, which mimicked the flapping motion of bird wings. Although da Vinci's designs were never built, his work laid the groundwork for the analogical thinking that would later fuel the development of aerodynamics.

The breakthrough in understanding the mechanics of bird flight came with the development of the principles of lift and drag, which are critical components of aerodynamics. In the eighteenth century, scientists like Sir George Cayley, often considered the father of aerodynamics, began to formalize the study of flight by drawing analogies between the forces acting on birds and those that could be applied to man-made machines [36, 37]. Cayley identified the four forces of flight—lift, drag, thrust, and weight—and demonstrated how the curvature of a bird's wing, known as the airfoil shape, creates lift by generating a pressure difference between the upper and lower surfaces of the wing. This concept of lift was directly inspired by observations of bird wings, where the curved shape and angle of attack allowed birds to rise into the air and stay aloft.

As shown in Figure 2.9, an aircraft in flight is subjected to four principal aerodynamic forces that govern its motion and stability. The lift force acts vertically upward, generated primarily by the wings as air flows over their surfaces, opposing the downward weight due to gravity. Meanwhile, the thrust produced by the engines propels the aircraft forward, counteracting the resistive drag arising from aerodynamic friction and pressure differences. In steady, level flight, these opposing forces are balanced such that lift equals weight and thrust equals drag, allowing the plane to maintain a constant altitude and velocity. The diagram also emphasizes the alignment of the motion vector with the thrust direction, clearly distinguishing between the propulsive and resistive forces influencing aircraft dynamics.

Cayley's work led to the construction of the first gliders, which mimicked the gliding flight of birds. These early gliders had fixed wings with airfoil shapes designed to replicate the lift generated by bird wings. The success of these gliders validated the analogy between bird flight and human flight, proving that the principles observed in nature could be applied to create machines capable of sustained flight. This analogical reasoning was critical in moving from the dream of flight to the reality of piloted aircraft. The Wright brothers, Orville and Wilbur, are perhaps the most famous figures in the history of aviation, and their success can also be attributed to their careful study of birds [38, 39]. The brothers conducted extensive research on bird flight, focusing particularly on how birds achieve balance and control in the air. They observed that birds tilt their wings to adjust their flight path and maintain stability. This observation led to the development of wing warping, a method the Wright brothers used to control the roll of their aircraft. Wing warping involved twisting the wings of the aircraft in flight, analogous to how birds adjust their wing surfaces to steer and stabilize themselves. This innovation was crucial to the Wright brothers'

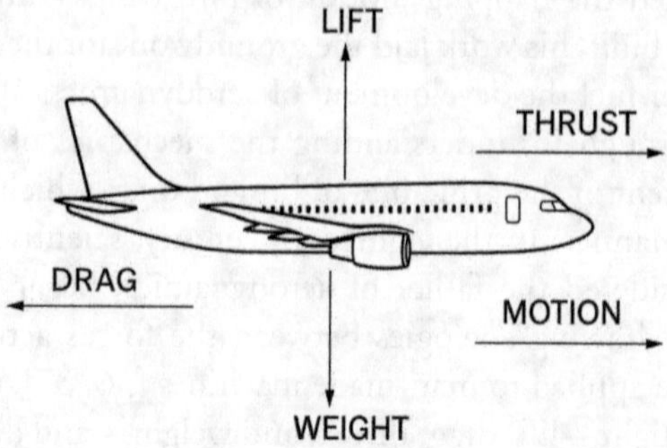

Figure 2.9 Illustration of the four fundamental aerodynamic forces acting on an aircraft in steady-level flight: lift, weight (gravity), thrust, and drag. Lift acts upward to counteract weight, while thrust propels the aircraft forward against the resistive drag. The diagram shows their relative directions with respect to the aircraft's motion.

success, as it allowed them to achieve controlled, powered flight for the first time in 1903.

The analogy between bird flight and human flight did not stop at the basic mechanics of lift and control; it also extended to the materials and structures used in aircraft design. Birds have lightweight, yet strong, skeletons made of hollow bones, which are crucial for efficient flight. This inspired early aircraft designers to use lightweight materials like wood and canvas to construct their planes. The biplane design, with its double-wing structure, was particularly influenced by the natural design of bird wings, providing the necessary strength and lift while maintaining a light overall weight. As materials science advanced, engineers continued to draw on this analogy, eventually developing stronger yet lighter materials like aluminum alloys and composite materials used in modern aircraft, which maintain the balance of strength and lightness observed in bird physiology.

Another important area of analogy-based innovation in aerodynamics inspired by bird flight is the study of turbulence and flow control. Birds are masters of navigating turbulent air and exploiting air currents to their advantage. Engineers have studied the wingtips of birds, particularly those of large soaring birds like eagles and albatrosses, which have specialized feathers that spread apart in flight. These feathers reduce drag and improve lift efficiency by minimizing the formation of vortices at the wingtips, which are a source of energy loss. This observation led to the development of winglets—upward curved extensions at the tips of airplane wings [40–42]. Winglets reduce drag by controlling the formation of vortices, similar to how a bird's feathers function, and have become a standard feature in modern aircraft design, significantly improving fuel efficiency.

The concept of biomimicry, where engineers and designers look to nature for solutions to human problems, is deeply rooted in the analogy between bird flight and aerodynamics. For instance, the study of bird flight dynamics has also inspired advancements in unmanned aerial vehicles (UAVs) or drones. Engineers have developed drones with flapping wings that mimic the motion of bird wings, allowing for more agile and efficient flight in environments where traditional fixed-wing or rotary-wing designs might be less effective [43]. These bioinspired drones can navigate tight spaces, perch on ledges, and hover with precision, capabilities that are directly borrowed from the observations of birds in nature.

As illustrated in Figure 2.10, a flapping-wing micro air vehicle (MAV) employs a biomimetic mechanism in which aerodynamic forces are produced through oscillatory wing motions rather than rotating propellers [44–46]. The figure highlights the essential components: a lightweight fuselage housing the actuation and control modules, a pair of primary wings that generate lift and thrust via controlled flapping, and a tail assembly that aids in stability and

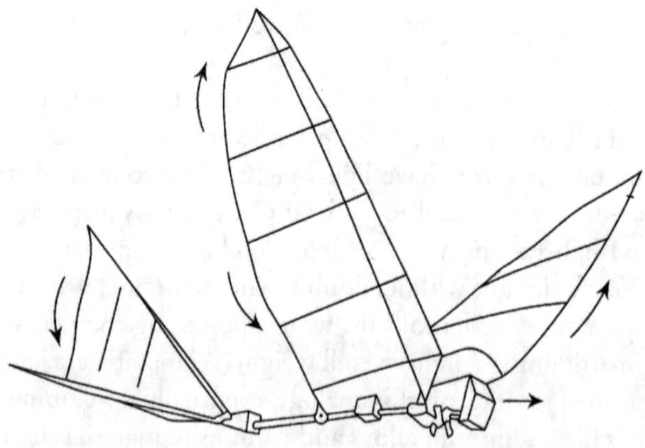

Figure 2.10 Schematic representation of a flapping-wing micro air vehicle (ornithopter). The design illustrates a bioinspired configuration where lift and thrust are generated through the periodic flapping of flexible wings. The primary wings, actuated by a compact mechanical drive system, provide both propulsion and control, while the tail surfaces contribute to stability and maneuverability. Such architectures enable bird-like flight characteristics including hovering, agile maneuvers, and low-noise operation.

directional control. Unlike conventional multirotor configurations, such systems can modulate wing kinematics (frequency, amplitude, and twist) to adapt efficiently across diverse flight regimes—from hovering to forward flight—thereby achieving enhanced maneuverability at low Reynolds numbers. This design paradigm not only improves aerodynamic performance for small-scale vehicles but also allows for quieter, more natural flight dynamics suitable for operations in human-proximal or sensitive environments.

Moreover, the analogy between bird flight and aerodynamics extends to more advanced concepts like morphing wings, which are inspired by the way birds can adjust the shape and structure of their wings in response to changing flight conditions. In recent years, aerospace engineers have been developing morphing wing technologies that can change their shape in real-time to optimize flight performance. This concept, which is being explored for future aircraft, allows for greater flexibility and adaptability during flight, much like how birds adjust their wings to maximize efficiency, speed, or maneuverability depending on their needs.

As shown in Figure 2.11, morphing-wing UAVs demonstrate how adaptive geometry can enhance aerodynamic efficiency across diverse flight regimes. In this schematic, the wing transitions from a rigid, fixed-span configuration to a morphing geometry capable of spanwise extension and camber deformation. Such mechanisms, as detailed by Chu et al. [47], allow the UAV to dynamically adjust its aerodynamic properties in response to varying mission requirements and airflow conditions. By increasing wingspan, the aircraft improves

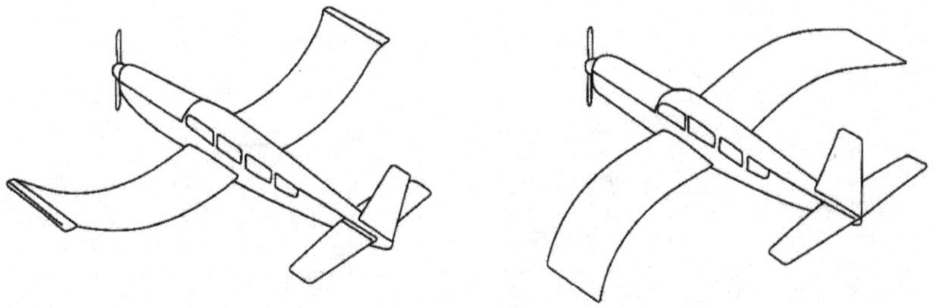

Figure 2.11 Schematic illustration of a variable-span and cambered-span morphing wing UAV. The left configuration shows the aircraft in its undeformed state, while the right configuration depicts the morphing condition where the wing exhibits adaptive spanwise and camber deformation. Such morphing enables dynamic alteration of lift, drag, and stability characteristics during flight, reflecting the fundamental design principle of bioinspired and flexible-wing UAVs for improved aerodynamic efficiency and maneuverability [47].

lift-to-drag ratio during cruise, while downward camber deformation enhances maneuverability and stability during low-speed or gusty conditions.

The morphing mechanism also alleviates the structural and aerodynamic limitations of conventional fixed-wing designs by redistributing aerodynamic loads and delaying stall onset. This dual-mode adaptation—spanning and camber morphing—embodies the design philosophy of biomimetic aircraft, where flexible structures and active materials enable seamless shape modulation without discrete control surfaces. Such designs directly contribute to the ongoing evolution of intelligent morphing UAVs capable of energy-efficient, multienvironment operation.

Even the study of flight formation, where birds like geese fly in a V-formation, has inspired innovations in aerodynamics. Researchers have found that this formation allows birds to conserve energy by taking advantage of the upwash created by the wings of the bird in front. This principle has been tested in aviation, where aircraft flying in close formation can similarly reduce fuel consumption by exploiting the aerodynamic benefits of flying in the wake of another aircraft. This insight from bird flight is being considered in the design of future flight strategies to improve fuel efficiency and reduce the environmental impact of aviation.

As illustrated in Figure 2.12, geese instinctively organize themselves into a characteristic V-formation during flight. This pattern is not merely aesthetic—it plays a critical role in reducing aerodynamic drag and improving overall flight efficiency. Each bird flies slightly above the one in front of it, benefiting from the upwash created by the preceding bird's wingtip vortices. This cooperative strategy allows the flock to conserve a significant amount of energy compared to flying alone. Moreover, the formation enables effective visual communication and coordination among the members of the group, ensuring stability and cohesion during long migratory journeys.

Figure 2.12 Geese flying in a V-formation, a natural aerodynamic arrangement that minimizes air resistance and allows individuals to conserve energy during long migratory flights.

In conclusion, the analogy between bird flight and aerodynamics has been a driving force behind some of the most significant innovations in the history of aviation, and continues to inspire advancements in modern aerospace engineering. From the earliest gliders to the development of advanced materials, wing structures, and even UAV technology, the principles derived from observing and understanding bird flight have provided engineers with the insights needed to overcome the challenges of human flight. This case study in analogy-based innovation highlights the profound impact that nature can have on technological development and demonstrates the enduring value of looking to the natural world for inspiration in solving human challenges. As aerospace technology continues to evolve, the lessons learned from bird flight will likely continue to play a crucial role in shaping the future of flight.

2.4 More Examples of Biomimicry

Biomimicry, a field of innovation that draws inspiration from nature's time-tested patterns and strategies, represents one of the most profound applications of analogy-based thinking. The concept of biomimicry is rooted in the idea that nature, through billions of years of evolution, has already solved many of the challenges humans face today. By studying and mimicking the designs, processes, and systems found in the natural world, engineers and designers can develop solutions that are not only innovative but also sustainable, efficient, and resilient.

This approach to problem-solving has led to numerous breakthroughs across various industries, where the application of natural principles has inspired the creation of advanced materials, efficient structures, and sustainable technologies. In this extensive exploration, we will delve into several case studies that highlight how biomimicry has influenced engineering solutions, demonstrating the power of nature-inspired innovation.

The case studies discussed here demonstrate the immense potential of biomimicry as a source of inspiration for engineering solutions. By studying and emulating the designs and processes found in nature, engineers and designers have developed technologies that are not only innovative but also highly efficient, sustainable, and effective. From Velcro to wind turbines, and from water-harvesting surfaces to high-performance materials, biomimicry has led to breakthroughs across a wide range of industries. The principles observed in nature, honed over billions of years of evolution, provide a rich repository of knowledge that humans can tap into to solve some of the most pressing challenges of our time. As we continue to face global challenges such as climate change, resource depletion, and environmental degradation, the importance of sustainable innovation cannot be overstated. Biomimicry offers a path forward by encouraging us to look to nature not just for inspiration but for solutions that are inherently aligned with the principles of sustainability. By adopting nature-inspired designs and processes, we can develop technologies that meet human needs while minimizing environmental impact, promoting resilience, and fostering a deeper connection with the natural world. In the future, as technology continues to advance, the potential for biomimicry to drive innovation will only grow.

Advances in fields such as materials science, genetic engineering, and nanotechnology will enable even more precise and sophisticated emulation of natural systems. This will open up new possibilities for creating materials, devices, and structures that are lighter, stronger, more efficient, and more sustainable than ever before. Moreover, the interdisciplinary nature of biomimicry, which brings together biology, engineering, design, and environmental science, will continue to foster collaboration and creativity, leading to solutions that are as diverse as they are impactful.

2.4.1 Velcro: Inspired by Burrs

One of the most famous examples of biomimicry is Velcro, a technology inspired by the simple yet ingenious design of burrs—those sticky seed pods that cling to clothing and animal fur. The story of Velcro began in 1941 when Swiss engineer Georges de Mestral went for a hike in the Alps. Upon returning, he noticed that his clothes and his dog were covered in burrs. Curious about how these seeds managed to stick so firmly, he examined them under a microscope and discovered that they were covered in tiny hooks that latched onto the loops in fabric and fur. This observation led de Mestral to develop a two-part fastening system composed of hooks and loops, which he named Velcro—a combination of the words "velvet" and "crochet" [48, 49]. Velcro has since become a ubiquitous fastening solution, used in everything from clothing and footwear to aerospace and medical devices. This innovation is a perfect example of how a simple natural mechanism can inspire a transformative engineering solution.

As illustrated in Figure 2.13, the concept behind Velcro emerged from the natural design of burrs, which possess numerous tiny hooks on their surface [48, 49]. When these burrs come into contact with fibrous materials, such as animal fur or fabric, the hooks attach firmly by catching onto loops or threads. This biological mechanism of temporary adhesion inspired the creation of a synthetic analog, Velcro, which replicates this hook-and-loop system. By engineering one surface with flexible hooks and another with soft loops, Velcro achieves a reusable fastening solution that is both strong and easily detachable. This figure exemplifies the essence of biomimicry, demonstrating how studying natural phenomena can lead to innovative technological advancements in material science and product design.

2.4.2 The Kingfisher and the Bullet Train

Another remarkable case of biomimicry in engineering can be found in the design of Japan's Shinkansen bullet trains [50, 51]. In the late 1980s, engineers faced a significant challenge with these high-speed trains: when the trains exited tunnels at speeds exceeding 200 miles per hour, the sudden change in air pressure caused a loud sonic boom, disturbing residents nearby and potentially causing structural damage to the tunnels. To solve this problem, engineers looked to nature for inspiration. The lead engineer, Eiji Nakatsu, an avid birdwatcher, observed that the kingfisher bird could dive into water at high speed without making a splash. The bird's beak, long and pointed, allowed it to smoothly transition between air and water, minimizing resistance. Nakatsu applied this analogy to the bullet train, redesigning the train's nose to resemble the kingfisher's beak. This biomimetic design not only eliminated the sonic boom but also reduced air resistance, leading to increased speed and energy efficiency.

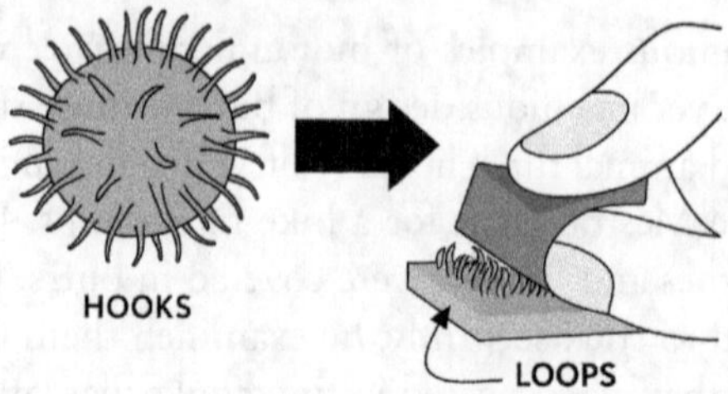

Figure 2.13 Illustration showing how the invention of Velcro was inspired by the hooklike structures of burrs. The burr's surface features microscopic hooks that latch onto loops in fabric or animal fur, a natural mechanism mimicked in Velcro's design [48, 49].

The success of the Shinkansen redesign showcases how biomimicry can solve complex engineering problems while also improving performance.

As illustrated in Figure 2.14, the sleek and tapered beak of the kingfisher bird served as the biological inspiration for the aerodynamic design of modern bullet trains. Engineers observed that the kingfisher can dive into water at high speed with minimal splash due to its elongated and streamlined beak shape, which efficiently manages sudden pressure changes. By mimicking this natural design, the redesigned bullet train nose effectively reduces aerodynamic drag, noise, and the problematic sonic booms that occurred when trains exited tunnels at high speed. This adaptation not only improved passenger comfort but also enhanced energy efficiency and operational stability. The example beautifully illustrates the principle of biomimicry—how insights from nature can lead to transformative innovations in engineering and sustainable design.

2.4.3 Termite Mounds and Sustainable Architecture

The architectural design of buildings has also been profoundly influenced by nature, particularly in the development of sustainable and energy-efficient structures. One of the most notable examples is the Eastgate Centre in Harare, Zimbabwe, a commercial building inspired by the natural ventilation system of termite mounds [52–56]. Termites in Africa construct large, complex mounds that maintain a constant internal temperature despite significant fluctuations in external temperatures. They achieve this by creating a network of tunnels that allow for passive air circulation, which regulates heat and moisture. Architect Mick Pearce, in collaboration with engineers from Arup, applied this natural principle to the design of the Eastgate Centre. The building utilizes a passive cooling system that mimics the termite mounds, with a series of chimneys and vents that facilitate natural air circulation. This design eliminates the need for

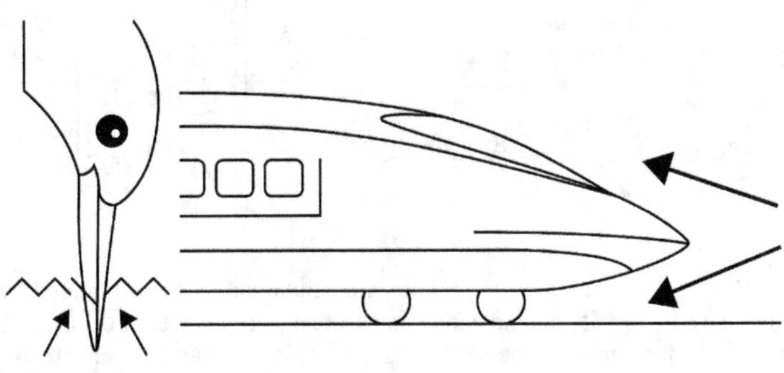

Figure 2.14 Illustration showing how the streamlined beak of the kingfisher bird inspired the aerodynamic redesign of the bullet train's nose, significantly reducing air resistance and tunnel sonic booms.

conventional air conditioning, reducing energy consumption by up to 90% compared to similar buildings. The Eastgate Centre stands as a pioneering example of how biomimicry can lead to sustainable architectural solutions that are both efficient and environmentally friendly.

As illustrated in Figure 2.15, the termite mound-inspired ventilation system integrates a natural convective airflow pattern that can be adapted to high-rise building design under passive ventilation conditions. In part (a), the termite mound structure demonstrates a balance between incoming fresh air at lower temperatures and the exhaust of warm air through upper channels, maintaining a nearly constant internal temperature around 28°C. This biological strategy effectively utilizes temperature gradients and surface porosity to regulate internal air movement. Part (b) extends this principle to a high-rise building model, where vertical core ventilation shafts act analogously to the termite mound's main chimney, promoting upward airflow driven by buoyancy and pressure differentials. Fresh air is drawn in through openings at lower elevations (*y*-axis direction) and distributed through functional floors before being expelled as hot air through upper vents along the *z*-axis. This configuration ensures stable and energy-efficient airflow circulation without relying on mechanical ventilation, thereby embodying a biomimetic approach to sustainable building design.

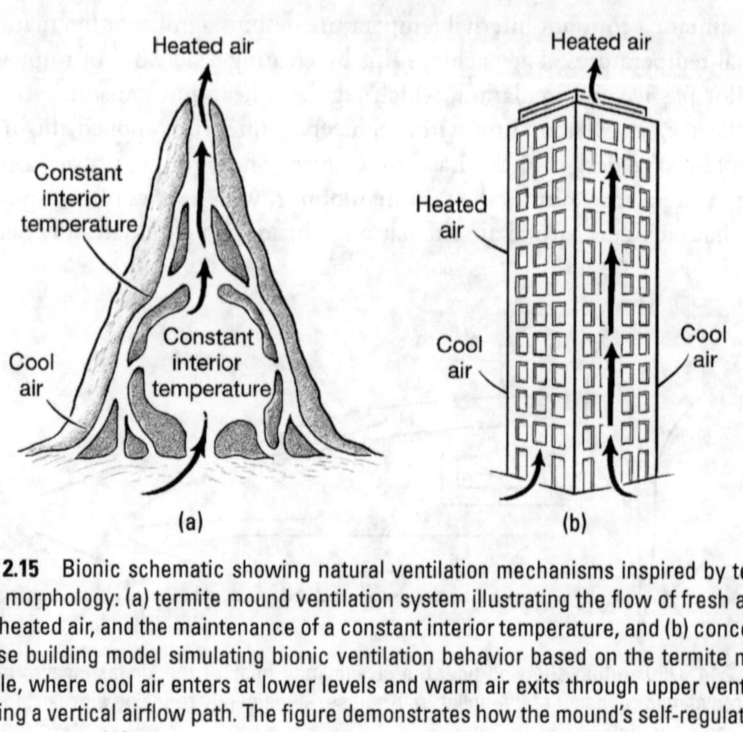

Figure 2.15 Bionic schematic showing natural ventilation mechanisms inspired by termite mound morphology: (a) termite mound ventilation system illustrating the flow of fresh air, the rise of heated air, and the maintenance of a constant interior temperature, and (b) conceptual high-rise building model simulating bionic ventilation behavior based on the termite mound principle, where cool air enters at lower levels and warm air exits through upper vents, establishing a vertical airflow path. The figure demonstrates how the mound's self-regulated air exchange system informs sustainable architectural ventilation strategies.

2.4.4 The Humpback Whale and Wind Turbine Blades

In the quest for renewable energy, biomimicry has played a crucial role in enhancing the efficiency of wind turbines. Engineers seeking to improve the performance of wind turbine blades turned to the ocean for inspiration, specifically to the humpback whale. Humpback whales are known for their remarkable agility in the water, despite their massive size. This agility is partly due to the unique structure of their flippers, which have large, irregular bumps called tubercles along their leading edges [57–60]. These tubercles disrupt the flow of water over the flippers, reducing drag and increasing lift, allowing the whale to make sharp turns and move with surprising speed. Researchers at the WhalePower Corporation recognized the potential of this natural design for wind turbines. By incorporating similar bumps along the leading edges of turbine blades, they were able to reduce drag and increase lift, resulting in blades that are more efficient and capable of capturing more energy from the wind. This biomimetic innovation not only improves the performance of wind turbines but also contributes to the broader goal of sustainable energy production.

As depicted in Figure 2.16, the tubercle system found on the leading edge of the humpback whale's pectoral fin serves as a remarkable example of nature-inspired aerodynamic optimization. These rounded protrusions, or tubercles, modulate fluid flow along the fin's surface, delaying stall and enhancing lift under varying flow conditions. Engineers have successfully mimicked this mechanism to improve performance in several technological applications

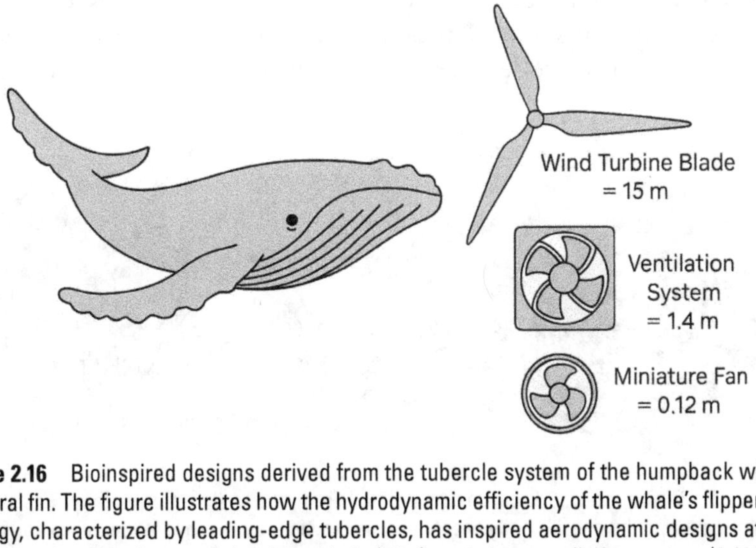

Figure 2.16 Bioinspired designs derived from the tubercle system of the humpback whale's pectoral fin. The figure illustrates how the hydrodynamic efficiency of the whale's flipper morphology, characterized by leading-edge tubercles, has inspired aerodynamic designs at various dimensional scales: a wind turbine blade (15m), an indoor ventilation system (1.4m), and a compact fan (0.12m). These adaptations showcase the translation of biological flow-control mechanisms into engineered technologies [61].

across scales—from large wind turbine blades measuring approximately 15m, to midscale indoor ventilation systems of about 1.4m, and even miniature fans with diameters near 0.12m. In each case, the tubercle-inspired leading-edge geometry reduces drag and increases efficiency by controlling vortex formation, thereby translating a naturally evolved hydrodynamic strategy into a robust engineering solution for air and fluid flow management.

2.4.5 Shark Skin and Antifouling Technologies

The study of shark skin has led to significant advancements in materials science, particularly in the development of antifouling technologies. Sharks are known for their ability to swim swiftly through the water, a capability attributed in part to their unique skin structure. Shark skin is covered with microscopic scales called dermal denticles, which reduce drag and prevent the accumulation of algae and barnacles [62–67]. Inspired by this natural design, engineers developed a synthetic material that mimics the structure of shark skin. This material, known as Sharklet, has been used to create surfaces that resist bacterial growth and biofouling. Applications of Sharklet technology include medical devices, such as catheters and surgical instruments, as well as ship hulls and water filtration systems. By reducing the need for chemical cleaners and coatings, this biomimetic approach not only enhances the performance of products but also promotes environmental sustainability by reducing the release of harmful substances into the environment.

As illustrated in Figure 2.17, the structure of shark skin has served as a remarkable model for bioinspired surface engineering. Sharks possess microscopic, toothlike scales known as dermal denticles that are aligned in a distinct riblet pattern. These denticles streamline fluid flow along the shark's body,

Figure 2.17 Bioinspired surface design derived from shark skin morphology. The dermal denticles covering the shark's body create microgrooved textures that reduce drag and inhibit microbial attachment. This natural architecture inspired the development of the Sharklet surface material, which replicates these micropatterns to resist bacterial growth and biofouling without the use of chemical agents.

minimizing drag while simultaneously discouraging the accumulation of microorganisms and other biofouling agents. By emulating this microtopography, engineers developed the Sharklet material—a synthetic surface featuring repeating micropatterns that mimic the denticle arrangement. This design effectively reduces bacterial colonization by disrupting microbial adhesion and growth, providing a chemical-free means of maintaining cleanliness. Such biomimetic innovations demonstrate how evolutionary adaptations in marine life can directly inspire sustainable technological solutions in medicine, marine engineering, and public health applications.

2.4.6 The Lotus Effect and Self-Cleaning Surfaces

Nature has also inspired innovations in surface materials, particularly those that are self-cleaning and water-repellent. The lotus flower, which grows in muddy waters, remains remarkably clean because of the microscopic structure of its leaves [68–70]. These leaves have tiny bumps covered with a waxy coating, which causes water droplets to bead up and roll off, taking dirt and contaminants with them. This phenomenon, known as the Lotus Effect, has been replicated in various man-made materials. Engineers have developed self-cleaning surfaces that mimic the microstructure of lotus leaves, leading to the creation of products such as self-cleaning glass, textiles, and paints. These materials are used in a wide range of applications, from building facades to solar panels, reducing maintenance costs and extending the lifespan of the products. The Lotus Effect is a prime example of how understanding and replicating natural processes can lead to practical solutions that are both efficient and environmentally friendly. As shown in Figure 2.18, the self-cleaning ability of the lotus leaf arises from its hierarchical surface structure composed of micropapillae and nanoscopic wax crystals. When a water droplet contacts this surface, air pockets form between the rough structures, significantly reducing the solid–liquid contact area and minimizing adhesive forces. Consequently, the droplet assumes an almost perfect spherical shape and rolls off the surface, capturing and removing dust particles in its path. In contrast, a droplet on a smooth solid surface (Figure 2.18(b)) exhibits greater contact area and adhesion, causing it to slide slowly or remain pinned, leaving behind contaminants. This distinct behavior demonstrates how the lotus leaf's micro/nanostructural features are key to achieving superhydrophobicity and self-cleaning performance. Such an understanding has inspired the development of artificial surfaces that mimic the lotus effect for applications in coatings, textiles, and microfluidic devices.

2.4.7 Gecko Feet and Adhesive Technologies

Geckos are known for their remarkable ability to climb smooth surfaces, such as glass, with ease. This ability is due to the unique structure of their feet, which

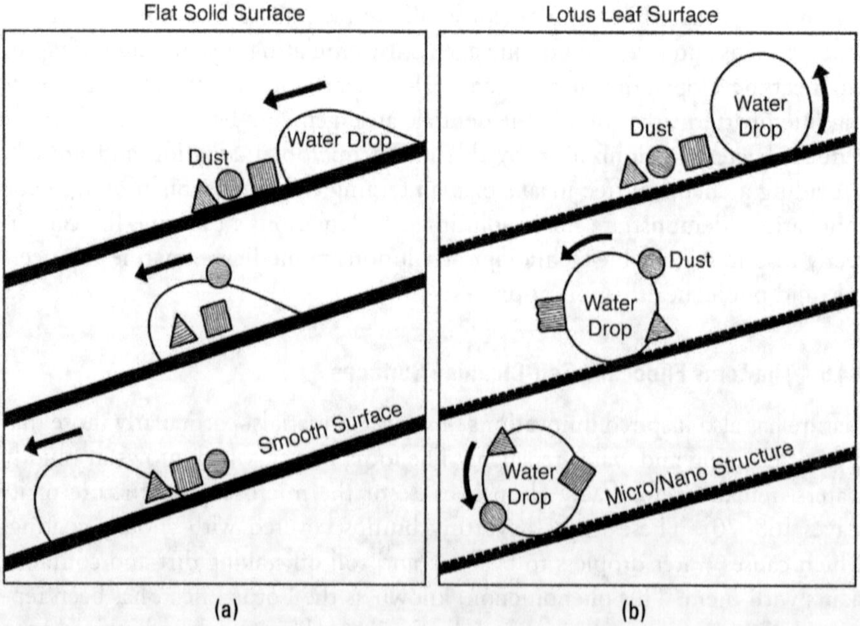

Figure 2.18 Schematic representation of the lotus effect mechanism illustrating the self-cleaning property of superhydrophobic surfaces: (a) water droplets rolling over a smooth solid surface, where the absence of surface roughness and trapped air pockets increases adhesion and prevents dust removal; (b) water droplets rolling over a lotus leaf surface with hierarchical microstructures and nanostructures that trap air pockets, resulting in low adhesion and effective removal of dust particles.

are covered in millions of tiny hairlike structures called setae [71–73]. These setae interact with surfaces at the molecular level through Van der Waals forces, allowing geckos to adhere to surfaces without the need for liquids or suction. This natural adhesive mechanism has inspired the development of synthetic materials that mimic the properties of gecko feet. Researchers have created gecko-inspired adhesives that are incredibly strong, reusable, and leave no residue. These adhesives have potential applications in a variety of fields, including robotics, where they can be used to create robots capable of climbing walls and ceilings, and in the medical field, where they could be used in surgical tapes and wound dressings that adhere gently to the skin. The study of gecko feet demonstrates how biomimicry can lead to the development of new materials with properties that are difficult to achieve through conventional manufacturing processes.

The remarkable climbing ability of geckos originates from the intricate hierarchical design of their toe pads (Figure 2.19). At the macroscopic level, each toe consists of lamellae, which are flattened ridges that increase the surface area available for adhesion. On a finer scale, these lamellae are covered with arrays of microscopic keratinous filaments called setae. Each seta then branches into hundreds of even smaller spatula-shaped termini, whose nanometric

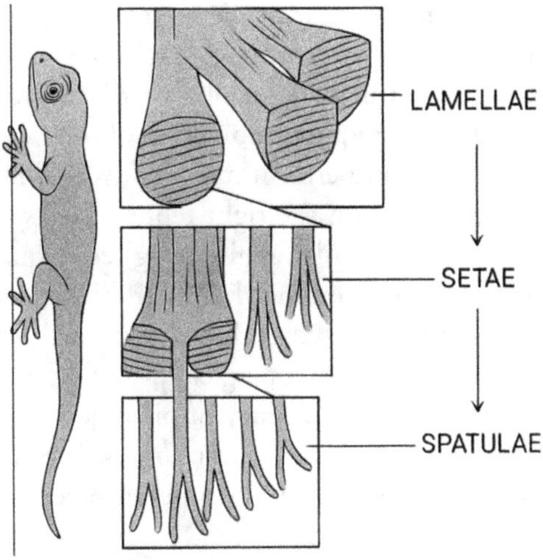

Figure 2.19 Schematic representation of the hierarchical structure in a gecko's adhesive system. The gecko foot features lamellae composed of densely packed keratinous filaments known as setae. Each seta further branches into numerous terminal spatulae that maximize surface contact with the substrate, allowing strong Van der Waals interactions responsible for the animal's exceptional climbing ability [74].

dimensions enable close contact with the substrate surface. This multiscale arrangement, extending from lamellae to spatulae, optimizes the number and strength of Van der Waals interactions, allowing geckos to adhere to and detach from surfaces efficiently. Such a biologically inspired design continues to inform synthetic adhesive development in biomimetics.

2.4.8 Spider Silk and High-Performance Fibers

Spider silk, known for its incredible strength and elasticity, has long fascinated scientists and engineers. Despite being thinner than a human hair, spider silk is stronger than steel and tougher than Kevlar, making it one of the most remarkable materials found in nature. Engineers have sought to replicate the properties of spider silk in synthetic materials, leading to the development of high-performance fibers with a wide range of applications [75–77]. Biomimetic approaches have involved both genetic engineering and chemical synthesis. In one approach, scientists have inserted the genes responsible for silk production into bacteria, plants, and even goats, enabling these organisms to produce silk proteins that can be harvested and spun into fibers. These bioengineered silks have potential applications in medical sutures, bulletproof clothing, and lightweight, high-strength composites used in aerospace engineering. The development of spider silk-inspired materials highlights how biomimicry can lead to the creation of new, advanced materials that push the boundaries of what is possible in engineering.

As shown in Figure 2.20, advances in biomimetic engineering have enabled scientists to replicate the extraordinary properties of natural spider silk within synthetic materials. Natural spider silk, produced by specialized glands in spiders, is renowned for its unique combination of tensile strength, elasticity, and lightness—qualities that outperform many conventional industrial fibers. The engineered fiber illustrated on the right represents a synthetic analog designed through genetic and molecular engineering techniques, where key protein motifs from spider silk are reproduced in microbial or polymer systems.

This visual comparison underscores the achievement of translating a complex natural process into scalable technological production. While the spider's web epitomizes the precision and efficiency of biological fabrication, the engineered fiber roll symbolizes human innovation in transforming these biological insights into materials suitable for medical sutures, protective fabrics, and high-performance composites.

2.4.9 The *Stenocara gracilipes*, a Namib Desert Beetle, and Water Harvesting Technologies

In arid regions where water is scarce, the ability to harvest moisture from the air can be a life-saving technology. The *Stenocara gracilipes*, a Namib Desert beetle, which lives in one of the driest places on Earth, has evolved a unique mechanism for collecting water from fog. The beetle's back is covered with a pattern of hydrophilic (water-attracting) and hydrophobic (water-repelling) bumps. As fog condenses on its back, water droplets form on the hydrophilic bumps and then roll off into the beetle's mouth, allowing it to drink. Inspired by this natural water-harvesting technique, engineers have developed materials and surfaces that can collect water from the air in a similar manner. These technologies have been applied to create fog-harvesting devices that provide fresh water in dry

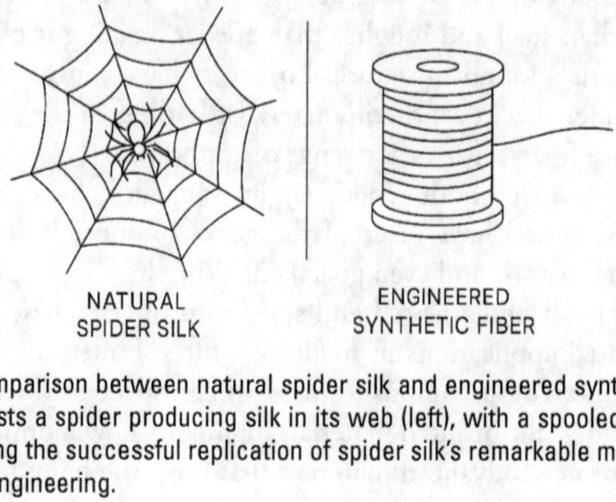

Figure 2.20 Comparison between natural spider silk and engineered synthetic fiber. The illustration contrasts a spider producing silk in its web (left), with a spooled engineered fiber (right), symbolizing the successful replication of spider silk's remarkable mechanical properties through bioengineering.

regions, as well as to design surfaces for buildings and vehicles that can self-clean and reduce water usage. The *S. gracilipes* serves as a powerful example of how studying natural adaptations can lead to innovative solutions for resource management in challenging environments [78–82].

Figure 2.21 illustrates the ingenious water-collection mechanism of the Namib Desert beetle, *S. gracilipes.* Living in one of the driest regions on Earth, this beetle has evolved a remarkable adaptation that allows it to survive in an environment where liquid water is scarce. Its dorsal surface exhibits a unique microstructure consisting of hydrophilic (water-attracting) peaks interspersed with hydrophobic (water-repelling) valleys. As fog-laden air passes over the beetle's body, moisture condenses on the hydrophilic spots to form droplets, which then roll off the hydrophobic regions and coalesce near the beetle's mouth for ingestion.

This natural condensation system represents a perfect example of how biological evolution optimizes material properties for environmental survival. Engineers and materials scientists have replicated this strategy to design advanced fog-harvesting meshes and coatings that efficiently capture water in arid climates. The *S. gracilipes* beetle thus stands as a model organism for sustainable technology inspired by the intersection of surface chemistry and functional design.

2.4.10 The Boxfish and Automotive Design

The study of animal shapes and structures has also influenced the design of more efficient and environmentally friendly vehicles. One such example is the boxfish, a marine creature with a unique, boxlike shape that is both hydrody-

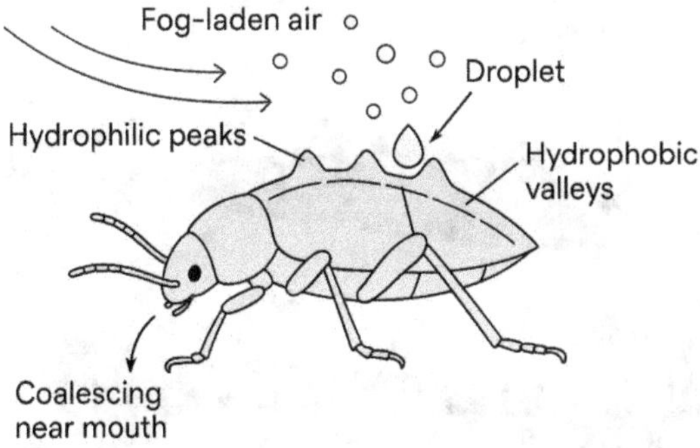

Figure 2.21 The Namib Desert beetle *S. gracilipes* harvesting water from fog. The beetle's back is patterned with alternating hydrophilic bumps and hydrophobic valleys, which enable it to condense microscopic fog droplets and channel them toward its mouth. This natural mechanism has inspired biomimetic designs for water-collection surfaces.

namic and stable in the water [83]. Mercedes-Benz engineers drew inspiration from the boxfish when designing the Mercedes-Benz Bionic concept car [49, 84, 85]. Despite its boxy appearance, the boxfish is highly streamlined, with a low drag coefficient that allows it to move efficiently through water. The engineers applied this analogy to the design of the car, creating a vehicle that is both aerodynamic and fuel-efficient. The Bionic concept car demonstrated that by mimicking the shapes found in nature, it is possible to achieve high levels of efficiency and performance while also reducing environmental impact. This case study in biomimicry illustrates how nature's designs can inspire not only functional improvements but also advancements in sustainable engineering.

As shown in Figure 2.22, the Mercedes-Benz Bionic concept car draws direct inspiration from the natural morphology of the boxfish. The boxfish exhibits a remarkably efficient hydrodynamic shape, characterized by a smooth, rounded body with subtle hexagonal surface patterns that minimize drag and enhance structural rigidity. Engineers translated these biological design principles into the Bionic car's streamlined geometry, resulting in reduced aerodynamic resistance and improved stability at high speeds. This biomimetic approach demonstrates how evolutionarily optimized natural forms can inform innovative automotive design, merging aesthetics with functional efficiency.

Figure 2.22 Comparison between the boxfish (top) and the Mercedes-Benz Bionic concept car (bottom). The streamlined body of the boxfish inspired the aerodynamic and structurally efficient design of the Bionic car, combining biological optimization with engineering innovation [49, 84, 85].

2.4.11 Batteries and the Electric Fish

In the final years of the eighteenth century, a quiet revolution sparked not with fireworks or fanfare, but through the humblest of creatures—the electric fish. Among the scholars of that age, few possessed the imaginative boldness of Alessandro Volta, whose mind moved freely between disciplines. It was through analogy thinking, a mental bridge from nature's mysteries to human invention, that Volta created the world's first battery, setting in motion the era of electrical science. Volta lived in a time when the very notion of electricity remained cloaked in mystery. Experiments by scientists such as Luigi Galvani had shown that a frog's leg could twitch when touched by metal probes, leading Galvani to propose a kind of animal electricity intrinsic to living tissue. Yet Volta, observing the same phenomena, drew a very different conclusion. Rather than see life as the source of the spark, Volta wondered: could it be the metals themselves that, when brought into contact across a moist conductor, generated electrical force? Here is where his analogy thinking truly shined. Volta turned to the electric fish—the torpedo ray and the electric eel—creatures that had fascinated naturalists for centuries. These animals could deliver strong shocks, and scholars speculated about the source of their power. Volta reasoned that if nature could stack biological cells inside a fish to produce electricity, then perhaps humans could construct an artificial stack—layer upon layer of materials designed to mimic this natural design. He imagined metals and moist membranes as stand-ins for the biological tissues of the fish. Copper and zinc could be the new muscle plates, and saltwater-soaked cloths would serve as the conductive fluids. In the first months of 1800, Volta built what would later be called the voltaic pile: alternating discs of copper and zinc, separated by brine-soaked pads, stacked high like a column of coins [86, 87]. When he connected the ends with a wire, he observed, for the first time in history, a continuous flow of electric current—no sparks, no fleeting shocks, but a steady, silent force. Without the fish as a mental model, the voltaic pile may never have been conceived. Volta's genius was not just his technical ability but his willingness to see across boundaries, to look at the anatomy of an eel and envision an entirely new kind of machine. In the voltaic pile, he built an artificial organ of sorts, a man-made creature capable of producing its own invisible energy. The impact of this moment cannot be overstated. Volta's invention tore open the doors of possibility. It allowed chemists to break water into hydrogen and oxygen, leading eventually to entire fields of electrochemistry. It paved the way for electromagnetism, electric motors, communications, and modern electronics. Yet at its core, the voltaic pile was born not purely from careful measurement or mechanical tinkering but from analogy—seeing echoes of nature in the design of machines. What makes Volta's achievement so remarkable is that it did not simply imitate nature's form. Instead, it captured nature's function, abstracted into a new medium. He did not need to grow tissues or breed fish; he only needed to understand the underlying pattern: layers, differences, and the movement of charge. That ability to distill the essence of a phenomenon and translate it into a new domain is the beating heart of innovation. Today, when we flip a switch, charge a phone, or send a signal across the world, we draw on a lineage that traces directly back to a man who looked at an electric fish and saw a battery. Volta's voltaic pile remains one of the most luminous examples of analogy thinking shaping the course of human history. It reminds us that invention often begins not with the cold march of reason alone, but with the wild leap of imagination—finding hidden kinships between the living world and the world we seek to build.

As shown in Figure 2.23, the electrocyte stacks of the electric eel extend nearly the entire length of the body and are anatomically partitioned into three principal electric organs [88, 89]. Hunter's organ occupies the anterior portion of the tail region and initiates high-voltage discharges through densely stacked, serially aligned electrocytes. The main organ comprises the largest longitudinal segment and provides the bulk of the eel's high-power output, enabling rapid and forceful electric pulses for predation and defense. In contrast, Sach's organ occupies the posterior region and is composed of electrocytes specialized for low-voltage, high-frequency discharges that support electrolocation and communication. Together, these subdivisions form an integrated bioelectrical system capable of producing a wide dynamic range of electric signals for behavioral modulation and environmental sensing.

The torpedo ray's electrogenic system represents one of the most sophisticated biological mechanisms for high-power electric discharge found in vertebrates [90–92]. As illustrated in Figure 2.24, the electric organs consist of large

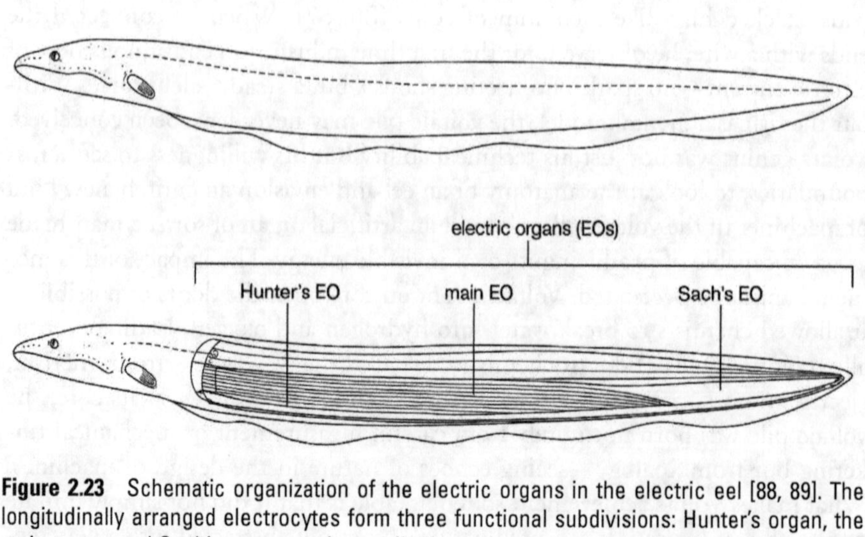

Figure 2.23 Schematic organization of the electric organs in the electric eel [88, 89]. The longitudinally arranged electrocytes form three functional subdivisions: Hunter's organ, the main organ, and Sach's organ, each contributing uniquely to electrogenesis and associated behavioral outputs.

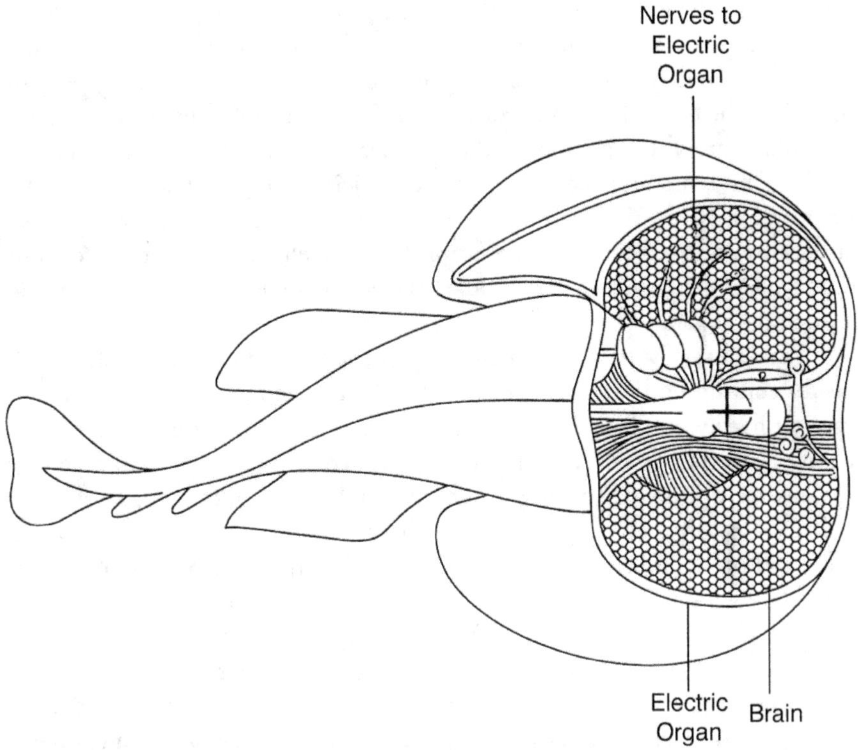

Figure 2.24 Anatomical organization of the electric organ of the torpedo ray. The paired electric organs, composed of densely packed electrocyte columns arranged in a hexagonal lattice, occupy much of the pectoral disc [90–92]. These columns are innervated by specialized cranial nerves that originate from the brain and fan out to provide highly synchronized activation required for generating strong electric discharges.

arrays of electrocytes arranged in stacked, columnar units that fill a substantial portion of the pectoral disc. Each electrocyte is a modified muscle-derived cell specialized for unidirectional ion flux and voltage generation. When activated, the electrocytes in a single column behave as a series of voltage sources, enabling the organ to achieve voltages far exceeding those produced by any individual cell.

The power and reliability of the torpedo's discharge depend critically on the precision of its neural control. The brain sends coordinated action potentials through specialized cranial and spinal nerves that branch extensively within the organ, as shown in Figure 2.24. These nerves synchronize the depolarization of thousands of electrocytes with millisecond accuracy. Because the electrocytes act effectively as capacitive plates whose output must sum constructively, any temporal dispersion in activation would dramatically reduce the net voltage and current. The nervous system therefore functions not only as an initiator of discharge but also as an enforcing mechanism for coherence across the organ.

Electrogenesis in the torpedo ray is also shaped by the organ's geometric regularity. The nearly hexagonal packing of electrocyte columns creates a dense, space-filling array that maximizes the number of units contributing to the discharge. This structural regularity ensures efficient longitudinal conductance and minimizes resistive losses between columns. The physical arrangement thus serves as an evolutionary compromise between packing density, cooling capacity, and the need to accommodate the elaborate nerve plexus that threads through the organ.

Unlike weakly electric fish, which generate continuous low-voltage signals for communication or electrolocation, the torpedo ray employs its electric organ primarily as a high-power, intermittent weapon used for predation and defense. This functional specialization is reflected in its electrocyte physiology: the cells are optimized for large, transient ionic fluxes rather than sustained oscillatory activity. The organ therefore behaves as a pulse-generating system with rapid onset and rapid recovery, allowing the ray to incapacitate prey with short, intense bursts.

In summary, the torpedo ray's electric organ demonstrates how cellular specialization, geometric organization, and tightly orchestrated neural control can be integrated to produce a robust biological high-voltage system. The structures depicted in Figure 2.24 reveal how evolution has engineered a compact, massively parallel electrogenic array capable of generating shocks powerful enough to deter predators and stun prey.

2.5 Lessons from Aerospace to Automotive

Analogy-based innovation has been a powerful driver of progress in both the automotive and aerospace industries, where lessons learned in one field have often been adapted to solve challenges in the other. The transfer of knowledge and ideas between these two industries highlights the value of cross-domain thinking, where concepts, designs, and technologies developed for one application are reimagined and repurposed for another. By examining case studies that illustrate this transfer, we can better understand how analogy-based innovation fosters creativity and leads to breakthroughs that benefit multiple sectors.

One of the most prominent examples of analogy-based innovation between the automotive and aerospace industries is the development of lightweight materials [93]. The aerospace industry, with its emphasis on fuel efficiency and performance, has long prioritized the use of lightweight materials such as aluminum alloys, composites, and carbon fiber. These materials provide the strength and durability needed to withstand the stresses of flight while minimizing the weight of the aircraft, which is crucial for fuel efficiency. Recognizing the potential benefits of these materials for the automotive industry, car manufacturers began to adopt them in the design and production of vehicles. The result has been the development of lighter, more fuel-efficient cars that maintain the safety and performance standards expected by consumers. For example, the widespread use of aluminum in car frames and the introduction of carbon fiber components in high-performance vehicles are direct outcomes of the automotive industry learning from aerospace engineering. This cross-industry application of lightweight materials has led to significant improvements in vehicle fuel economy, reduced emissions, and enhanced driving dynamics.

Another key area where the automotive industry has drawn lessons from aerospace is in the field of aerodynamics. The principles of aerodynamics, which are critical for the design of aircraft, have been increasingly applied to the design of cars, particularly as manufacturers strive to improve fuel efficiency and performance. In the 1980s, car manufacturers began to recognize that reducing drag could lead to significant gains in fuel economy. Inspired by the streamlined shapes of aircraft, automotive designers started to incorporate more aerodynamic features into their vehicles, such as smoother body lines, underbody panels, and rear spoilers. One notable example is the design of the Audi 100, introduced in the early 1980s, which featured a drag coefficient of just 0.30—remarkably low for a production car at the time. This design was heavily influenced by aerodynamic principles borrowed from aerospace, demonstrating how lessons from one industry can lead to innovation in another [93]. The success of the Audi 100 paved the way for other manufacturers to prioritize aerodynamics in their vehicle designs, leading to a new generation of cars that are both more efficient and visually distinctive.

The concept of active suspension systems provides another example of analogy-based innovation between the automotive and aerospace industries. Active suspension systems, which adjust a vehicle's suspension in real-time to optimize handling and comfort, were initially developed for aerospace applications, where they were used to improve the stability and performance of aircraft. These systems, which rely on sensors and computer-controlled actuators to adjust suspension settings dynamically, were later adapted for use in high-end sports cars and luxury vehicles. For instance, the Mercedes-Benz S-Class introduced an active suspension system known as Active Body Control (ABC) in the late 1990s. This system, which continuously adjusts the suspension based on driving conditions, was inspired by similar technologies used in aircraft to maintain stability during flight. By transferring this technology from aerospace to automotive, manufacturers were able to offer vehicles with unprecedented levels of comfort and handling, enhancing the driving experience and setting new standards for vehicle performance.

The development of safety features is another area where the automotive industry has benefited from aerospace innovations. The concept of crumple zones in cars, which absorb the energy of an impact to protect occupants, was inspired by the principles of energy absorption used in aircraft design. Aircraft are designed to withstand and dissipate the forces experienced during crashes or hard landings, and engineers recognized that similar principles could be applied to cars to improve passenger safety. The introduction of crumple zones, reinforced passenger cabins, and other safety features in cars has led to significant reductions in fatalities and injuries in road accidents. Additionally, the use of advanced materials like high-strength steel, which was originally developed for aerospace applications, has further enhanced vehicle safety by providing greater protection without adding excessive weight. Moreover, the adoption of fly-by-wire technology, which replaces traditional mechanical controls with electronic systems, is a clear example of how aerospace innovations have influenced the automotive industry. Fly-by wire systems, first used in military and commercial aircraft, allow for more precise control of the vehicle's systems by replacing bulky mechanical linkages with lightweight electronic signals. This technology has been adapted for use in modern vehicles, particularly in the development of electronic power steering, throttle control, and brake-by-wire systems. These systems offer improved responsiveness, reduced weight, and the ability to integrate advanced driver assistance features, such as adaptive cruise control and lane-keeping assist. The transition from mechanical to electronic control systems in cars is a direct result of the automotive industry adopting and adapting technology from aerospace, leading to more sophisticated and capable vehicles.

In conclusion, the exchange of ideas and technologies between the automotive and aerospace industries through analogy-based innovation has led to significant advancements in both fields. By applying lessons learned in aerospace to automotive engineering, and vice versa, engineers and designers have been able to create vehicles that are safer, more efficient, and more enjoyable to drive. The use of lightweight materials, aerodynamic designs, active suspension systems, and advanced safety features are just a few examples of how cross-domain thinking can lead to breakthroughs that benefit multiple industries. As technology continues to evolve, the ongoing collaboration and transfer of knowledge between these two sectors will undoubtedly lead to further innovations that enhance the capabilities of both cars and aircraft, pushing the boundaries of what is possible in transportation and engineering.

2.6 The Role of Analogy Thinking on the Field of AI and Machine Learning

Analogy thinking is pivotal in artificial intelligence (AI) and machine learning, where it enables the application of concepts and strategies from one domain to another, driving innovation and enhancing problem-solving capabilities. One prime example is the development of neural networks, inspired by the human brain's structure and function. By drawing an analogy between neurons in the brain and nodes in a network, researchers created algorithms capable of learning and processing information in ways that mimic human cognition.

Another example is transfer learning, where a model trained on a large dataset in one domain, like image recognition, is adapted to perform a different but related task, such as medical imaging analysis, with minimal additional training. This analogy between tasks allows AI to apply knowledge gained from one context to another, improving efficiency and effectiveness.

Lastly, reinforcement learning, which is modeled after behavioral psychology, uses analogies from the way humans and animals learn through rewards and punishments to train AI systems in decision-making processes, such as in autonomous vehicles navigating complex environments. These examples illustrate how analogy thinking in AI and machine learning leads to the development of adaptable, powerful technologies that can solve a wide range of challenges.

References

[1] Kanefsky, J., et al., "Steam Engines in 18th-Century Britain: A Quantitative Assessment," *Technology and Culture*, Vol. 21, No. 2, 1980, pp. 161–186, https://doi.org/10.2307/3103337.

[2] Frenken, K., et al., "The Early Development of the Steam Engine: An Evolutionary Interpretation using Complexity Theory," *Industrial and Corporate Change*, Vol. 13, No. 2, p. 419–450, https://doi.org/10.1093/icc/dth017.

[3] Bailey, J., "The Steam Age—Evolution of Steam Engines and the 1st Steam Pumping Devices," In: *Inventive Geniuses Who Changed the World*, Springer, 2021, pp. 23–36, https://doi.org/10.1007/978-3-030-81381-93.

[4] Khan, S., et al., "Evolution of Artificial Hearts: An Overview and History," *Cardiology Research*, Vol. 5, No. 5, 2014, pp. 121–125, https://doi.org/10.14740/cr354w.

[5] Cohn, W., et al., "Total Artificial Hearts: Past, Present, and Future," *Nature Reviews Cardiology*, Vol. 12, 2015, pp. 609–617, https://doi.org/10.1038/nrcardio.2015.79.

[6] Masarone, D., et al., "The Total Artificial Heart: A Historical Perspective," *Journal of Clinical Medicine*, Vol. 14, No. 17, 2025, p. 6290, https://doi.org/10.3390/jcm14176290.

[7] Cusumano, A. M., et al., "The Glomerular Filtration Rate: From the Diagnosis of Kidney Function to a Public Health Tool," *Front. Med.*, Vol. 8, 2021, https://doi.org/10.3389/fmed.2021.769335.

[8] Eknoyan, G., et al., "The History of Dialysis: The Wonderful Apparatus of John Jacob Abel Called the Artificial Kidney," *Seminars in Dialysis*, Vol. 22, No. 3, May/June 2009, pp. 287–296, https://doi.org/10.1111/j.1525-139X.2009.00527.x.

[9] van Noordwijk, J., *Dialysing for Life: The Development of the Artificial Kidney*, Springer Science & Business Media, 2012.

[10] Firouzi, V., et al., "Biomechanical Models in the Lower-Limb Exoskeletons Development: A Review," *Journal of NeuroEngineering and Rehabilitation*, Vol. 22, No. 12, 2025, https://doi.org/10.1186/s12984-025-01556-5.

[11] Zakaryan, N., et al., "Bio-Inspired Conceptual Mechanical Design and Control of a New Human Upper Limb Exoskeleton," *Robotics*, Vol. 10, No. 4, 2021, p. 123, https://doi.org/10.3390/robotics10040123.

[12] Yang, H., et al., "Compliant Actuators that Mimic Biological Muscle Performance with Applications in a Highly Biomimetic Robotic Arm," *Advanced Bionics*, Vol. 1, No. 1, 2025, pp. 2–18, https://doi.org/10.1016/j.abs.2024.12.001.

[13] Farina, D., et al., "Toward Higher-Performance Bionic Limbs for Wider Clinical Use," *Nature Biomedical Engineering*, Vol. 7, 2023, pp. 473–485, https://doi.org/10.1038/s41551-021-00732-x.

[14] Li, B., et al., "Intelligent Ankle–Foot Prosthesis Based on Human Structure and Motion Bionics," *Journal of NeuroEngineering and Rehabilitation*, Vol. 21, 2024, p. 119, https://doi.org/10.1186/s12984-024-01414-w.

[15] Varaganti, P., et al., "Recent Advances in Biomimetics for the Development of Bio-Inspired Prosthetic Limbs," *Biomimetics*, Vol. 9, No. 5, 2024, p. 273, https://doi.org/10.3390/biomimetics9050273.

[16] Sabu, C., et al., "Bioinspired and Biomimetic Systems for Advanced Drug and Gene Delivery," *Journal of Controlled Release*, Vol. 286, 2018, pp. 201–216, https://doi.org/10.1016/j.jconrel.2018.08.033.

[17] Zinger, A., et al., "Biomimetic Nano Drug Delivery Carriers for Treating Cardiovascular Diseases," *Nanomedicine: Nanotechnology, Biology and Medicine*, Vol. 33, 2021, p. 102360, https://doi.org/10.1016/j.naNo.2020.102360.

[18] Gu, Y., et al., "Biomembrane-Modified Biomimetic Nanodrug Delivery Systems: Frontier Platforms for Cardiovascular Disease Treatment," *Biomolecules*, Vol. 14, 2024, p. 960, https://doi.org/10.3390/biom14080960.

[19] Kang, W., et al., "Advances in Biomimetic Nanomaterial Delivery Systems: Harnessing Nature's Inspiration for Targeted Drug Delivery," *Journal of Materials Chemistry B*, Vol. 12, 2024, pp. 7001–7019, https://doi.org/10.1039/d4tb00565a.

[20] Root, S. E., et al., "A Damage-Perceptive, Self-Healing Electronic Skin with Millimeter Resolution," *Device*, 2025, https://doi.org/10.1016/j.device.2025.100802.

[21] Geng, Y., et al., "A Skin-Inspired Self-Adaptive System for Temperature Control During Dynamic Wound Healing," *Nano-Micro Letters*, Vol. 16, 2024, p. 152, https://doi.org/10.1007/s40820-024-01345-0.

[22] Li, G., et al., "Recent Progress in Self-Healing Triboelectric Nanogenerators for Artificial Skins," *Biosensors*, Vol. 15, No. 1, 2025, p. 37, https://doi.org/10.3390/bios15010037.

[23] Chen, J., et al., "Self-Healing Materials-Based Electronic Skin: Mechanism, Development and Applications," *Gels*, Vol. 8, No. 6, 2022, p. 356, https://doi.org/10.3390/gels8060356.

[24] Benight, S. J., et al., "Stretchable and Self-Healing Polymers and Devices for Electronic Skin," *Progress in Polymer Science*, Vol. 38, No. 12, 2013, pp. 1961–1977, https: //doi.org/10.1016/j.progpolymsci.2013.08.001.

[25] White, S. R., et al., "Autonomic Healing of Polymer Composites," *Nature*, Vol. 409, No. 6822, 2001, pp. 794–797, https://doi.org/10.1038/35057232.

[26] Ayeleru, O. O., et al., "Concept of Self-Healing in Polymeric Materials," Ma*terials Today: Proceedings*, Vol. 62, Supplement 1, 2022, pp. S158–S162, https://doi.org/10.1016/j.matpr.2022.04.811.

[27] Fonoll, X., et al., "Simulating Rumen Conditions using an Anaerobic Dynamic Membrane Bioreactor to Enhance Hydrolysis of Lignocellulosic Biomass," *Environmental Science & Technology*, 2024, https://doi.org/10.1021/acs.est.3c06478.

[28] Li, N., et al., "Hydrolytic Acidification via Rumen Microorganisms and Aerobic MBR to Reduce Contaminants in Pulping Midcourse Wastewater," *Desalination and Water Treatment*, 2016, https://doi.org/10.1080/19443994.2014.996772.

[29] Deng, R., et al., "Investigation on a Soft Tubular Model Reactor Based on Bionics of Small Intestine–Residence Time Distribution," *International Journal of Food Engineering*, 2014, https://doi.org/10.1515/ijfe-2014-0048.

[30] Zhang, S., et al., "A Soft Tubular Model Reactor Based on the Bionics of a Small Intestine: Anti-Particulate Fouling by Peristalsis," *Brazilian Journal of Chemical Engineering*, 2021, https://doi.org/10.1007/s43153-021-00196-1.

[31] Vorlander, D., et al., "PETR: A Novel Peristaltic Mixed Tubular Bioreactor Simulating Human Colonic Conditions," *Biotechnology and Bioengineering*, 2024, https://doi.org/10.1002/bit. 28636.

[32] Harvey, C., et al., "Lessons from Natural Flight for Aviation: Then, Now and Tomorrow," *Journal of Experimental Biology*, Vol. 226, 2023, p. jeb245409, https://doi.org/10.1242/jeb.245409.

[33] Han, J., et al., "Review on Bio-Inspired Flight Systems and Bionic Aerodynamics," *Chinese Journal of Aeronautics*, Vol. 34, No. 7, 2021, p. 170–186, https://doi.org/10.1016/j.cja.2020.03.036.

[34] Richardson, Philip L., "Leonardo da Vinci's Discovery of the Dynamic Soaring by Birds in Wind Shear," *Notes and Records: The Royal Society Journal of the History of Science*, Vol. 73, No. 3, 2018, pp. 285–301, https://doi.org/10.1098/rsnr.2018.0024.

[35] Liu, T., et al., "Engineering Perspective on Bird Flight: Scaling, Geometry, Kinematics and Aerodynamics," *Progress in Aerospace Sciences*, New York: Elsevier, 2023.

[36] Ackroyd, J. A. D., "Sir George Cayley, the Father of Aeronautics: Part 1–The Invention of the Aeroplane," *Notes and Records of the Royal Society of London*, Vol. 56, No. 2, 2002, pp. 167–181, https://doi.org/10.1098/rsnr.2002.0176.

[37] Ackroyd, J. A. D., "Sir George Cayley, the Father of Aeronautics: Part 2–The Subsequent Development of His Thinking," *Notes and Records of the Royal Society of London*, Vol. 56, No. 3, 2002, pp. 333–348, https://doi.org/10.1098/rsnr.2002.0186.

[38] Padfield, G. D., and B. Lawrence, "The Birth of Flight Control: An Engineering Analysis of the Wright Brothers' 1902 Glider," *The Aeronautical Journal*, Vol. 107, No. 1078, 2003, pp. 697–718, https://doi.org/10.1017/S0001924000013464.

[39] Velazquez, J., et al., "The Contribution of the Wright Brothers in Airplane Development: An Investigative Report," *International Journal of Professional Aviation Training & Testing Research*, Vol. 8, No. 1, 2016, https://ojs.library.okstate.edu/osu/index.php/IJAR/article/download/8116/7462/17043.

[40] Tucker, V. A., "Gliding Birds: Reduction of Induced Drag by Wing Tip Slots Between the Primary Feathers," *Journal of Experimental Biology*, Vol. 180, No. 1, 1993, pp. 285–310, https://doi.org/10.1242/jeb.180.1.285.

[41] Tucker, V. A., "Drag Reduction by Wing Tip Slots in a Gliding Harris' Hawk, *Parabuteo unicinctus*," *Journal of Experimental Biology*, Vol. 198, No. 3, 1995, pp. 775–781, https://doi.org/10.1242/jeb.198.3.775.

[42] Liu, D., et al., "A Brief Review on Aerodynamic Performance of Wingtip Slots and Research Prospect," *Journal of Bionic Engineering*, Vol. 18, No. 6, 2021, pp. 1255–1279, https://doi.org/10.1007/s42235-021-00116-6.

[43] Shyy, W., *An Introduction to Flapping Wing Aerodynamics*, Cambridge, UK: Cambridge University Press, 2013, https://doi.org/10.1017/ CBO9781139583916.

[44] Abas, M. F. B., et al., "Flapping Wing Micro-Aerial-Vehicle: Kinematics, Membranes, and Flapping Mechanisms of Ornithopter and Insect Flight," *Chinese Journal of Aeronautics*, Vol. 29, No. 5, 2016, pp. 1159–1177, https://doi.org/10.1016/j.cja.2016.08.003.

[45] Yang, W., et al., "Dove: A Biomimetic Flapping-Wing Micro Air Vehicle," *International Journal of Micro Air Vehicles*, Vol. 10, No. 1, 2018, pp. 70–84, https://doi.org/10.1177/1756829317734837.

[46] Jiang, S., et al., "Design and Analysis of an Innovative Flapping Wing Micro Aerial Vehicle with a Figure Eight Wingtip Trajectory," *Mechanical Sciences*, Vol. 12, 2021, pp. 603–613, https://doi.org/10.5194/ms-12-603-2021.

[47] Chu, L., et al., "Design, Modeling, and Control of Morphing Aircraft: A Review," *Chinese Journal of Aeronautics*, Vol. 35, No. 5, 2022, pp. 220–246, https://doi.org/10.1016/j.cja.2021.09.013.

[48] Velcro Companies, *History of Velcro® Brand*, https://www.velcro.com/about-us/history/.

[49] Hwang, J., et al., "Biomimetics: Forecasting the Future of Science, Engineering, and Medicine," *International Journal of Nanomedicine*, Vol. 10, No. 1, 2015, pp. 5701–5713, https://doi.org/10.2147/IJN.S83642.

[50] Kobayashi, K., et al., "Shinkansen Technology Learned From an Owl? The Story of Eiji Nakatsu," in *Technologies Learned from Living Things: Concepts and Examples–Front Line Reports. Japan for Sustainability (JFS) Newsletter*, 2005, http://www.japanfs.org/en/news/archives/news id027795.html.

[51] Linic, S., et al., "Experimental and Numerical Methods for Concept Design and Flow Transition Prediction on the Example of the Bionic High-Speed Train," in *Lecture Notes in Networks and Systems (LNNS)*, Vol. 153, *International Conference of Experimental and Numerical Investigations and New Technologies* (CNNTech), 2020, pp. 65–82, https://doi.org/10.1007/978-3-030-58362-0 5.

[52] King, H., et al., "Termite Mounds Harness Diurnal Temperature Oscillations for Ventilation," *Proceedings of the National Academy of Sciences*, 2015, arXiv preprint arXiv:1703.08067, https://doi.org/10.48550/arXiv.1703.08067.

[53] Abdullah, A., et al., "Cooling Strategies in the Biological Systems and Termite Mound: The Potential of Emulating them to Sustainable Architecture and Bionic Engineering," *ARPN Journal of Engineering and Applied Sciences*, Vol. 13, No. 8, 2018, https://www.arpnjournals.org/jeas/research papers/rp 2018/jeas 1018 7313.pdf.

[54] Vesala, R., et al., "Termite Mound Architecture Regulates Nest Temperature and Correlates with Species Identities of Symbiotic Fungi," *PeerJ*, Vol. 6, 2019, p. e6237, https://doi.org/10.7717/peerj.6237.

[55] Nasir, O., and M. A. Kamal, "Inspiration from Nature: Biomimicry as a Paradigm for Architectural and Environmental Design," *American Journal of Civil Engineering and Architecture*, Vol. 10, No. 3, 2022, pp. 126–136, https://doi.org/10.12691/ajcea-10-3-3.

[56] Wei, Y. et al., "Simulation and Optimization Study on the Ventilation Performance of High-Rise Buildings Inspired by the White Termite Mound Chamber Structure," *Biomimetics*, Vol. 8, No. 8, 2023, https://doi.org/10.3390/biomimetics8080607.

[57] Hansen, K. L., et al., "Performance Variations of Leading-Edge Tubercles for Distinct Airfoil Profiles," *AIAA Journal*, Vol. 49, No. 1, 2011, pp. 185–194, https://doi.org/10.2514/1.J050631.

[58] Chen, S., et al., "Numerical Investigation of Turbine Blades with Leading-Edge Tubercles in Uniform Current," *Water*, Vol. 13, No. 16, 2021, p. 2205, https://doi.org/10.3390/w13162205.

[59] Yirijor, J., "Design and Optimisation of Horizontal Axis Wind Turbine Blades Using Biomimicry of Whale Tubercles," *Journal of Engineering Research and Reports*, Vol. 25, No. 5, 2023, pp. 100–112, https://doi.org/10.9734/JERR/2023/v25i5915.

[60] Shi, W., et al., "Humpback Whale Inspired Design for Tidal Turbine Blades," *Proceedings of the 5th International Symposium on Marine Propulsion, 2017*, https://doi.org/10.13140/RG.2.2.21227.80160.

[61] Perricone, V., et al., "Organismal Design and Biomimetics: A Problem of Scale," *Biomimetics*, Vol. 6, No. 4, 2021, p. 56, https://doi.org/10.3390/biomimetics6040056.

[62] Pu, X., et al., "Preparation, Anti-Biofouling and Drag-Reduction Properties of a Biomimetic Shark Skin Surface," *Biology Open*, Vol. 5, No. 4, 2016, pp. 389–396, https://doi.org/10.1242/bio. 016899.

[63] Lang, T. G. P., et al., "Biomimetic Shark Skin: Design, Fabrication and Hydrodynamic Function," *Journal of Experimental Biology*, Vol. 217, No. 10, 2014, pp. 1656–1666, https://doi.org/10.1242/jeb.097097.

[64] Feld, K., et al., "Dermal Denticles of Three Slowly Swimming Shark Species: Microscopy and Flow Visualisation," *Biomimetics*, Vol. 4, No. 2, 2019, p. 38, https://doi.org/10.3390/biomimetics4020038.

[65] Mann, E. E., et al., "Surface Micropattern Limits Bacterial Contamination," *Antimicrobial Resistance & Infection Control*, Vol. 3, 2014, p. 28, https://doi.org/10.1186/2047-2994-3-28.

[66] May, R. M., et al., "Micro-Patterned Surfaces Reduce Bacterial Colonization and Biofilm Formation *In Vitro*: Potential for Enhancing Endotracheal Tube Designs," *Clinical & Translational Medicine*, Vol. 3, 2014, p. 8, https://doi.org/10.1186/2001-1326-3-8.

[67] Gabler-Smith, M. K., and George V. Lauder, "Ridges and Riblets: Shark Skin Surfaces Versus Biomimetic Models," *Frontiers in Marine Science*, Vol. 9, 2022, p. 975062, https://doi.org/10.3389/fmars.2022.975062.

[68] Barthlott, W., et al., "Purity of The Sacred Lotus, or Escape from Contamination in Biological Surfaces," *Planta*, Vol. 202, No. 1, 1997, pp. 1–8, https://doi.org/10.1007/s004250050096.

[69] Neinhuis, C., et al., "Characterization and Distribution of Water-Repellent, Self-Cleaning Plant Surfaces," *Annals of Botany*, Vol. 79, No. 6, 1997, pp. 667–677, https://doi.org/10.1006/anbo.1997.0400.

[70] Wang, L., et al., "A Critical Review on Robust Self-Cleaning Properties of Lotus Leaf," *Soft Matter*, Vol. 19, No. 6, 2023, pp. 1058–1075, https://doi.org/10.1039/D2SM01521H.

[71] Wang, W., et al., "Gecko-Like Dry Adhesive Surfaces and Their Applications: A Review," *Journal Bionic Eng*, Vol. 18, 2021, pp. 1011–1044, https://doi.org/10.1007/s42235-021-00088-7.

[72] Autumn, K., et al., "Frictional Adhesion: A New Angle on Gecko Attachment," *Journal of Experimental Biology*, Vol. 209, No. 18, 2006, pp. 3569–3579, https://doi.org/10.1242/jeb.02486.

[73] Autumn, K., et al., "Evidence for van der Waals Adhesion in Gecko Setae," *Proceedings of the National Academy of Sciences*, Vol. 99, No. 19, 2002, pp. 1252–1256, https://doi.org/10.1073/pnas.192252799.

[74] Heim, M., et al., "Hierarchical Structures Made of Proteins: The Complex Architecture of Spider Webs and their Constituent Silk Proteins," *Chemical Society Reviews*, 2009, https://doi.org/10.1039/b813273a.

[75] Ko, F. K., et al., "Engineering Properties of Spider Silk," in *Handbook of Properties of Textile and Technical Fibres* (second ed.), Woodhead Publishing, 2018, pp. 185–220, https://doi.org/10.1016/B978-0-08-101272-7.00006-7.

[76] Schmuck, B., et al., "Strategies for Making High-Performance Artificial Spider Silk Fibers," *Advanced Functional Materials*, Vol. 34, No. 35, 2023, https://doi.org/10.1002/adfm.202305040.

[77] Kiseleva, A. P., et al., "Recent Advances in Development of Functional Spider Silk-Based Hybrid Materials," *Frontiers in Chemistry*, Vol. 8, 2020, https://doi.org/10.3389/fchem.2020. 00554.

[78] Parker, A. R., et al., "Water Capture by a Desert Beetle," *Nature*, Vol. 414, 2001, pp. 33–34, https://doi.org/10.1038/ 35102108.

[79] Naidu, S. G., et al., "Water Balance and Osmoregulation in *Stenocara gracilipes*, a Wax-Blooming Tenebrionid Beetle from the Namib Desert," *Journal of Insect Physiology*, Vol. 47, No. 12, 2001, pp. 1429–1440, https://doi.org/10.1016/S0022-1910(01)00133-0.

[80] Nørgaard, T., et al., "Fog-Basking Behaviour and Water Collection Efficiency in Namib Desert Darkling Beetles," *Frontiers in Zoology*, Vol. 7, 2010, p. 23, https://doi.org/10.1186/1742-9994-7-23.

[81] Lei, J., et al., "A Fog-Collecting Surface Mimicking the Namib Beetle: Its Water Collection Efficiency and Influencing Factors," *Nanoscale*, Vol. 12, 2020, pp. 6921–6936, https://doi.org/10.1039/C9NR10808D.

[82] Toledano, J. C. F., et al., "Optimizing Fog Harvesting by Biomimicry," *Physical Review Fluids*, Vol. 7, 2022, p. 033604, https://doi.org/10.1103/PhysRevFluids.7.033604.

[83] van Wassenbergh, S., et al., "Boxfish Swimming Paradox Resolved: Forces by the Flow of Water Around the Body Promote Manoeuvrability," *Journal of the Royal Society Interface*, Vol. 12, No. 103, 2015, p. 20141146, https://doi.org/10.1098/rsif.2014.1146.

[84] Chowdhury, H., et al., "Biomimetics of Boxfish: Designing an Aerodynamically Efficient Passenger Car," in *Biomimicry for Aerospace: Technologies and Applications*, Elsevier, 2022, pp. 211–235, https://doi.org/10.1016/B978-0-12-821074-1.00007-4.

[85] Yang, C.- M., et al., "Analysis of Mercedes-Benz Concept Car Using Biomimicry Design Spiral and Template Analysis–An Exploratory Study," *International Journal of Innovation in Management*, Vol. 7, No. 2, 2019, pp. 49– 56.

[86] Cecchini, R., et al., "Alessandro Volta and His Battery," *IEEE Antennas & Propagation Magazine*, Vol. 34, No. 2, April 1992, pp. 30–37, https://doi.org/10.1109/74.134307.

[87] Fabbrizzi, L., et al., "The Invention of Volta's Pile and its Diffusion in Europe at War in the Year 1800," *Pure and Applied Chemistry*, Vol. 95, No. 11, 2023, pp. 1141–1157, https://doi.org/10.1515/pac-2023-0504.

[88] Traeger, L. L., et al., "A Tail of Two Voltages: Proteomic Comparison of the Three Electric Organs of the Electric Eel," *Sci. Adv.*, Vol. 3, No. 7, 2017, p. e1700523, https://doi.org/10.1126/sciadv.1700523.

[89] Xu, J., et al., "The Third Form Electric Organ Discharge of Electric Eels," *Sci. Rep.*, Vol. 11, No. 1, 2021, p. 6134, https://doi.org/10.1038/s41598-021-85715-3.

[90] Mate, S. E., et al., "Integrated Genomics and Proteomics of the *Torpedo californica* Electric Organ: Concordance with the Mammalian Neuromuscular Junction," *Skeletal Muscle*, Vol. 1, No. 20, 2011, https://doi.org/10.1186/2044-5040-1-20.

[91] Gotter, A. L., M. A. Kaetzel, and J. R. Dedman, "Electrocytes of Electric Fish," in *Cell Physiology Source Book*, Elsevier, 2001, pp. 1025–1038.

[92] Nazarian, J., et al., "The Proteome Survey of an Electricity-Generating Organ (*Torpedo californica* Electric Organ)," *Proteomics*, Vol. 7, No. 4, 2007, pp. 617–627, https://doi.org/10.1002/pmic. 200600686.

[93] Shao, H., "A Review of Aerodynamic Optimization Design for Automobiles: Technical Pathways and Energy Efficiency Improvement," *ACE*, Vol. 148, 2025, https://doi.org/10.54254/ 2755-2721/2025.23353.

3

Optical Fibers: From Fluid Analogies to Waveguide Theory

After reading this chapter, you should be able to:

- Explain how fluid analogies inspired the development of optical fibers;
- Describe how waveguide theory and materials breakthroughs enabled fiber communications;
- Discuss how the mass deployment of optical fibers revolutionized global communications.

3.1 Introduction

Optical fiber technology evolved from curious nineteenth-century experiments into a cornerstone of modern communications. Early scientists drew analogies between light and flowing water, imagining "pipes" that could carry light just as plumbing carries fluids. These fluid dynamic analogies, coupled with emerging waveguide theory, guided researchers toward the concept of confining light within a medium. Over time, theoretical insights and materials science breakthroughs converged to make low-loss optical fibers a reality. This chapter traces the discovery and development of optical fibers, from early observations of light guidance by total internal reflection (TIR) to the sophisticated waveguide theory that underpins modern fiber design. Key breakthroughs, including the pioneering work of Charles Kao and George Hockham [1] on low-loss fiber, the invention of ultratransparent glass by Corning, novel fiber manufacturing methods like modified chemical vapor deposition (MCVD), and the eventual commercialization of fiber-optic systems, are highlighted. Each milestone is discussed in an academic, technical style suitable for engineers and researchers, with references to foundational papers and technologies.

As illustrated in Figure 3.1, the propagation of light within an optical fiber can be compared to the behavior of water flowing through a hose. In the water hose analogy, shown in part (a), water is constrained by the hose walls and directed along its length, even when the hose bends. Similarly, in part (b), the optical fiber guides light through its high-index core surrounded by a lower index cladding. The change in refractive indices between these two regions ensures that light rays striking the interface at sufficiently high angles undergo total internal reflection. This phenomenon keeps the light confined within the core, allowing it to follow the fiber's curvature without escaping, much like the water remains within the hose until it exits. Such confinement is the fundamental principle that enables optical fibers to transmit information over long distances with minimal loss and high reliability.

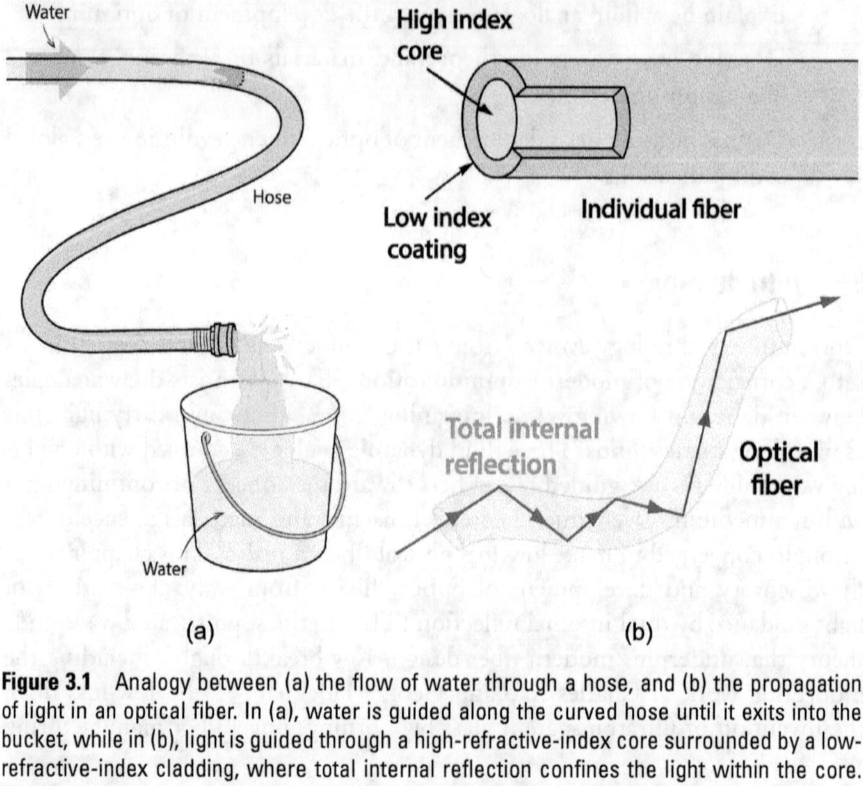

Figure 3.1 Analogy between (a) the flow of water through a hose, and (b) the propagation of light in an optical fiber. In (a), water is guided along the curved hose until it exits into the bucket, while in (b), light is guided through a high-refractive-index core surrounded by a low-refractive-index cladding, where total internal reflection confines the light within the core. This illustrates how optical fibers transmit light efficiently over long distances by continuously reflecting it internally, similar to how a hose confines and directs water flow.

3.2 Early Observations of Light Guidance

Early hints of optical fiber principles came from experiments in the 1800s. In 1841, Jean-Daniel Colladon and Jacques Babinet in Paris, demonstrated that light could be guided along a jet of water, an effect due to refraction and TIR [2, 3]. By directing light into a flowing water stream, they created a light fountain, the light stayed inside the curved water stream much as water flows through a pipe, vividly illustrating how a medium can confine light. A dozen years later, in 1854, John Tyndall included this striking demonstration in his London lectures, further popularizing the concept [2–4]. Tyndall's work explained the underlying physics: when light inside a denser medium (like water or glass) hits the interface with a less dense medium at a shallow angle beyond a critical angle, it reflects entirely back inside: TIR. He quantified the critical angle (with air) for materials (e.g. 48° for water, 39° for glass). This phenomenon, TIR, is the fundamental principle that makes fiber-optic light guidance possible.

As shown in Figure 3.2, the concept of TIR can be understood by examining how a light ray behaves at the boundary between two media of differing refractive indices. In the first case, when the angle of incidence θ_i is smaller than the critical angle θ_c, part of the light is refracted into the second medium and part is reflected back. At the critical angle, the refracted ray travels exactly along the interface, indicating the threshold beyond which refraction ceases. When the incident angle exceeds θ_c, the refracted ray no longer exists, and all of the light is reflected internally. This complete internal reflection is the essence of TIR and is utilized in optical fibers and other light-guiding systems where minimal loss of energy is desired.

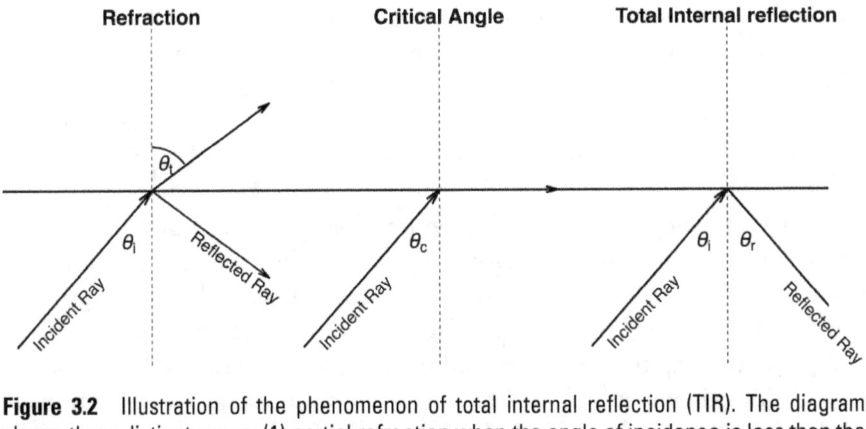

Figure 3.2 Illustration of the phenomenon of total internal reflection (TIR). The diagram shows three distinct cases: (1) partial refraction when the angle of incidence is less than the critical angle, (2) the condition at the critical angle where the refracted ray grazes along the interface, and (3) total internal reflection when the incident angle exceeds the critical angle, causing the ray to reflect entirely within the denser medium.

The water fountain analogy—light constrained in a stream as if flowing through a pipe—gave early investigators an intuitive model for guiding light. By the late nineteenth century, the idea of "light pipes" took hold. In Vienna, doctors experimented with bent glass rods to illuminate internal body cavities, essentially using glass to pipe light into otherwise inaccessible areas. Around 1880, an American engineer William Wheeler patented a system of hollow glass tubes with mirrored inner surfaces to distribute light in buildings, a design explicitly likened to plumbing, where light could be channeled through reflective pipes [5]. Wheeler's light pipes used metallic coatings (silvered glass) to reflect light down the tube (since optical TIR in air was not yet efficient enough), attempting to guide illumination from a central arc lamp to distant rooms [6]. Though Wheeler's approach did not achieve wide use (the pipes were inefficient and impractical), it underscored the appeal of the fluid analogy: if water or gas could be distributed by pipes, why not light? These nineteenth-century explorations established the basic concept of guided light and demonstrated TIR in dramatic fashion, setting the stage for further innovation.

3.3 The First Optical Fibers

In the early twentieth century, light-guiding techniques found niche applications and incremental improvements. By the 1920s, inventors were exploring image transmission using arrays of guided light. For example, Clarence Hansell (a radio experimenter) and John Logie Baird (a television pioneer) independently experimented with transmitting images through flexible tubes or rod arrays. In 1930, German medical student Heinrich Lamm became the first to successfully transmit an image through a bundle of optical fibers [3, 7]. Lamm bundled several small glass fibers and managed to relay a crude image of a light bulb filament through them, an achievement he hoped would aid medical endoscopy. However, Lamm's fibers were unclad (no protective lower-index coating), so light leaked between fibers and the image was dim and unclear. The potential was demonstrated, but the result was impractical and largely forgotten in subsequent decades.

As illustrated in Figure 3.3, an optical fiber consists of three concentric layers: the core, cladding, and protective jacket. The core, made of high-refractive-index glass, serves as the principal medium through which light signals propagate. Surrounding it, the cladding possesses a slightly lower refractive index, creating the necessary conditions for total internal reflection at the core–cladding boundary. This ensures that the optical energy remains confined within the core, significantly minimizing attenuation and preventing crosstalk between adjacent fibers. The outer jacket acts primarily as a mechanical shield, offering resistance to environmental stress, bending, and physical damage. The

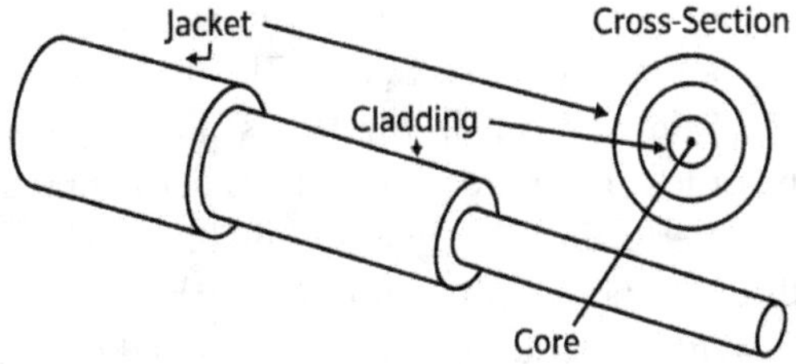

Figure 3.3 Schematic diagram of a single optical fiber showing its main structural layers. The core forms the central light-guiding region, surrounded by a cladding layer that prevents light leakage through total internal reflection. An outer jacket provides mechanical protection, while the cross-sectional view illustrates the concentric arrangement of these layers.

accompanying cross-sectional view provides a clear visualization of the fiber's cylindrical symmetry, emphasizing how the cladding effectively isolates the optical path from external perturbations.

A key improvement came in the early 1950s. In 1953, Dutch scientist Bram van Heel reported that adding a transparent cladding layer around a glass fiber dramatically reduced its light leakage and crosstalk [8, 9]. By surrounding the higher refractive-index core glass with a lower index material (initially using materials like acrylic or clear lacquer), van Heel created a true dielectric waveguide: light in the core reflected at the core-cladding boundary by TIR, rather than escaping. That same year, across the channel in London, physicist Harold Hopkins and his student Narinder S. Kapany at Imperial College built a coherent bundle of thousands of glass fibers, each fiber clad with a lower index coating [10, 11]. They succeeded in transmitting a clear optical image through a 75-cm-long fiber bundle composed of over 10,000 fibers, a revolutionary advancement at the time.

This invention of the fiberscope proved invaluable for medical endoscopy and industrial inspection, enabling one to see around bends and corners by piping an image through flexible glass fibers. The first practical medical fiber-optic endoscope soon followed. In 1956, Basil Hirschowitz and colleagues at the University of Michigan developed a flexible gastroscope using fiber optics, which they patented the same year [12–14]. During this project, team member Lawrence Curtiss drew some of the first glass-clad fibers, replacing earlier cladding materials (air gaps, oils, or wax) with a robust glass cladding fused to the fiber. This all-glass fiber (core and cladding) was more durable and had lower loss [15]. Thus, by the late 1950s, the essential elements of modern optical fibers were in place: fibers drawn from glass so pure and transparent that light could travel meters, clad by a lower index glass or polymer to ensure TIR guidance even when the fiber bent. Narinder Kapany, who had been instrumental in these developments, coined the term "fiber-optics" in 1960 and helped introduce the technology to a wider audience with a seminal *Scientific American* article and the first book on the subject [16, 17].

Despite these advances, fibers in the 1950s and 1960s still had very high attenuation (on the order of hundreds of decibels per kilometer), due to material impurities and imperfections. They were suitable for short imaging devices but not for long-distance data transmission. Nevertheless, the success of imaging fibers and the invention of the laser in 1960 sparked the notion that optical fibers might be used for communications if losses could be drastically reduced. Experiments confirmed that laser light could indeed be sent through a thin glass fiber, for instance, in 1961 Elias Snitzer and Will Hicks demonstrated a laser beam propagating through a small-core glass fiber [18–21]. These results, combined with the analogies to copper wires carrying electricity or pipes carrying fluids, tantalized researchers with the vision of a new communications medium: beams of light flowing through hair-thin glass waveguides.

3.4 Waveguide Theory and the Vision of Low-Loss Fiber Communications

By the early 1960s, the stage was set to seriously consider optical fibers for telecommunications. The challenge was fundamentally one of waveguide theory and materials science: Could an optical wave be guided over long distances with tolerable loss and distortion? Researchers approached this question through analogy, drawing from the well-established principles of microwave waveguides. This analogy was not a simple act of miniaturization within a shared electromagnetic framework, but a deeper structural comparison: although both systems obey Maxwell's equations, the physical realizations, boundary conditions, and dominant sources of loss differ significantly between the metallic and dielectric regimes. The analogy therefore resided in the transfer of conceptual structure rather than material identity—how electromagnetic energy might be guided, confined, and controlled by engineered boundaries, whether metallic or dielectric.

In microwave engineering, radio-frequency waves travel with low loss inside hollow metallic tubes, where reflection at conducting walls provides confinement [22, 23]. Optical fibers, by contrast, confine light by TIR at the interface between a high-index core and a lower index cladding. Recognizing this parallel allowed researchers to see that the same mathematical formalism used to describe guided modes in microwave systems could be extended to dielectric media at optical frequencies, provided that the differing material responses and scale-dependent losses were accounted for. The analogy thus provided a powerful conceptual bridge—suggesting that light, like microwaves, could be guided efficiently if one could design a structure and choose materials that minimized loss at the relevant frequency range.

Theoretical groundwork for optical waveguides had already been laid. In 1961, Elias Snitzer published a comprehensive analysis of modes in a cylindrical dielectric fiber, showing that for a small enough core diameter only a single electromagnetic mode would propagate (the concept of single-mode fiber) [18–20]. This work directly applied Maxwell's equations to the optical domain, confirming that such a fiber could guide light in discrete modes—analogous to how microwave waveguides support specific propagation modes—but governed by TIR rather than metallic boundaries. In this way, the analogy between the two systems illuminated both their unity under electromagnetic theory and their divergence in mechanism, helping researchers understand how a dielectric waveguide could serve as the optical counterpart of the microwave guide.

The crucial leap came from recognizing that high losses were not inevitable but could be overcome. In 1965, Charles K. Kao and George A. Hockham at Standard Telecommunication Laboratories (STL) in the United Kingdom, performed a landmark analysis of fiber attenuation [1]. They carefully measured and extrapolated loss data and argued that the excessive attenuation (then around 1,000 dB/km) was due to impurities in the glass, not a fundamental limitation of physics. Using the waveguide analogy as a reasoning scaffold, Kao and Hockham reframed the problem: if, as in microwave guides, wave confinement could be achieved by optimizing boundary conditions and material properties, then the key to progress lay in engineering an optical medium pure enough to suppress absorption and scattering. They showed that if glass could be purified—particularly by removing transition metal ions and hydroxyl (OH^-) impurities—losses could be brought below 20 dB/km, a threshold they identified as practical for telecom systems. They also emphasized that silica (SiO_2) glass would be ideal, given its chemical stability and potentially low intrinsic loss if purified.

This was a bold claim at the time: most experts believed that light at optical frequencies simply would be absorbed or scattered too strongly in any material over long distances. Kao, however, was convinced by the analogy's logic—just as radio-frequency energy could be guided efficiently through metal waveguides by managing boundary losses, so too could optical waves be guided through glass if material impurities could be eliminated. In their 1966 paper, *Dielectric-fibre surface waveguides for optical frequencies*, Kao and Hockham essentially outlined the roadmap to fiber-optic communications—specifying the need for ultratransparency and even discussing the optimal refractive index profile and fiber dimensions for minimal loss and dispersion. Their insight earned Charles Kao the Nobel Prize in Physics 2009, after it led to the global fiber-optic revolution.

It is noteworthy that multiple groups worldwide had similar ideas. Kao's work did not occur in isolation; researchers in France (Jean-Claude Simon and colleagues at Compagnie Generale de Telegraphie Sans Fil (CSF)) and in the United States, were also examining "optical analogs of microwave guides," exploring glass waveguides as a way to carry high-frequency signals [6]. In Germany, Manfred Borner at Telefunken actually built a prototype fiber-optic telephone link in 1965 and filed a patent in 1966 [24]. Borner's system demonstrated the concept of using a laser and optical fiber to transmit data, but it was only operable over short distances due to the fiber's high attenuation. These concurrent efforts underscore how powerful the waveguide analogy was: scientists recognized that even though optical and microwave systems differ in scale, materials, and confinement mechanisms, the same guiding principle applied—if electrons in copper wires or microwaves in hollow tubes could be replaced by photons in glass fibers, one could potentially achieve enormous bandwidth and low-loss communications. The remaining hurdle was to invent a glass fiber pure enough to carry light for kilometers, an achievement that would soon follow.

3.5 Materials Science Breakthroughs: Low-Loss Glass Fibers

Reducing fiber attenuation from approximately 1,000 dB/km to the <20 dB/km range foreseen by Kao required a revolution in glass manufacturing. Throughout the late 1960s, groups around the world set out to create ultrapure optical glass. The breakthrough came in 1970 at Corning Glass Works (United States) when a team led by Robert Maurer, Donald Keck, Peter Schultz, and Frank Zimar succeeded in making a fiber with an attenuation of about 17 dB/km at 633 nm (red light). They achieved this by doping silica glass with titanium in the core and using high-purity silica cladding. Crucially, they employed a new chemical vapor deposition technique to fabricate the fiber preform: reacting silicon tetrachloride $SiCl_4$ in a flame to deposit pure silica soot, which was then consolidated into a solid preform and drawn into fiber. This approach minimized impurities far beyond what traditional glass melting could do. In 1971, the Corning team improved their results by switching to germanium dioxide (GeO_2) as the core dopant, achieving losses around 4 dB/km in the near-infrared. Within a few years, optical fiber attenuation was further reduced to below 1 dB/km in the optimal infrared wavelength windows. This was a stunning four-order-of-magnitude reduction in loss compared to fibers of the early 1960s, essentially turning Kao's predictions into reality. Several enabling technologies and discoveries underpinned this materials revolution.

Figure 3.4 illustrates the wavelength-dependent refractive index, or material dispersion, for both the core and cladding of a standard single-mode optical fiber. The core is typically made of fused silica doped with germanium dioxide

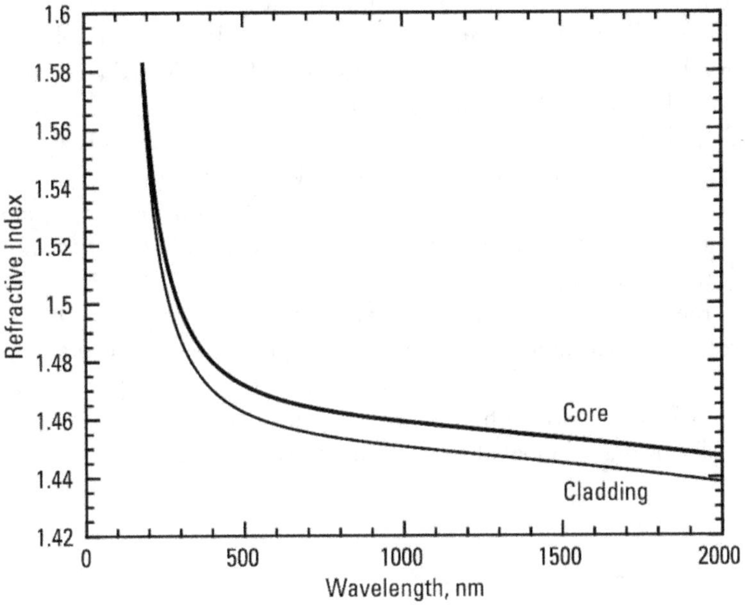

Figure 3.4 Dispersion of refractive index as a function of wavelength for the core and cladding materials of a typical silica optical fiber. The core, composed of germanium dioxide-doped fused silica, exhibits a slightly higher refractive index than the pure fused-silica cladding across the visible and near-infrared spectrum. Both curves show normal material dispersion, where the refractive index decreases with increasing wavelength. The small but precise difference in refractive index (Δn) between core and cladding enables total internal reflection and effective light confinement within the fiber core [25].

(GeO_2), which slightly increases its refractive index relative to the undoped silica cladding. This controlled difference in refractive index, on the order of 0.3% to 0.5%, creates the optical boundary necessary for total internal reflection, allowing light to remain confined within the core even when the fiber is bent or coiled.

As the plot shows, both materials exhibit normal dispersion: the refractive index decreases gradually as the wavelength increases from the visible range (below 500 nm) toward the near-infrared region (around 1,550 nm). The higher refractive index at shorter wavelengths arises from stronger electronic polarization effects in the glass. The dispersion characteristics shown here are crucial for determining the fiber's propagation constant, modal cutoff, and chromatic dispersion parameters. For example, around 1.3 μm and 1.55 μm—the standard telecommunication windows—the refractive indices of both core and cladding are nearly flat, minimizing pulse broadening and optical loss. Understanding and precisely engineering this refractive index profile is fundamental to achieving low-loss, single-mode operation and optimizing fiber performance across different wavelength regimes [25].

Figure 3.5 illustrates the wavelength dependence of the attenuation coefficient in optical fibers, comparing experimental results with a simplified theoretical model. The logarithmic vertical axis (in dB/km) highlights the dramatic variation in loss across the visible and infrared spectrum. At short wavelengths, particularly below 800 nm, the attenuation rises steeply due to Rayleigh scattering, an effect caused by microscopic density and compositional fluctuations within the glass. The scattering loss scales inversely with the fourth power of wavelength (λ^{-4}), making shorter wavelengths inherently less efficient for long-distance transmission.

As the wavelength increases toward the near-infrared region, attenuation decreases sharply, reaching a minimum in the range of approximately 1,300 to 1,550 nm. This region corresponds to the intrinsic transparency window of high-purity silica, where both scattering and material absorption are minimized. Within this window, the optical fiber exhibits attenuation values below

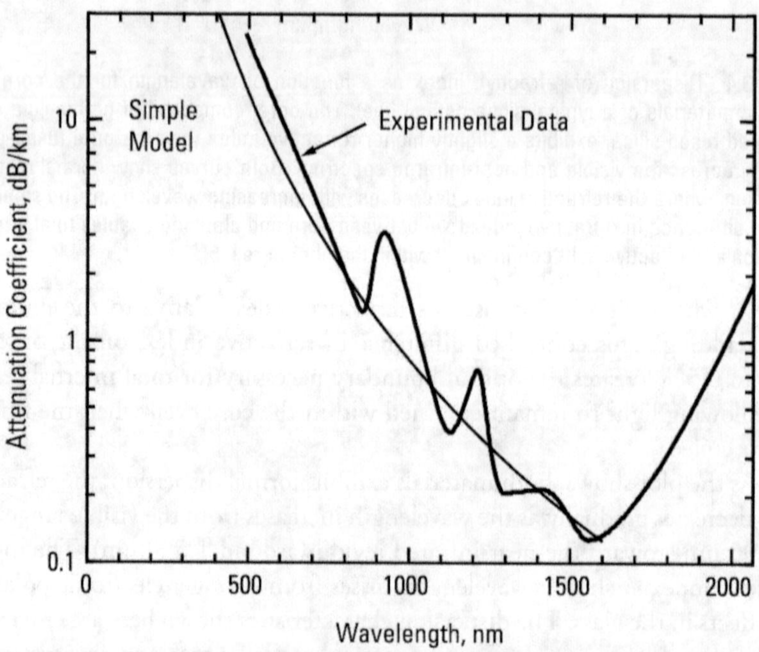

Figure 3.5 Spectral dependence of optical fiber attenuation. The plot compares experimental data with a simple theoretical model showing the variation of attenuation coefficient (in dB/km) as a function of wavelength. Attenuation is high at shorter wavelengths due to Rayleigh scattering, decreases toward the infrared region where material absorption is minimal, and rises again beyond 1,600 nm because of intrinsic infrared absorption. The characteristic low-loss window near 1,550 nm represents the optimal transmission region for modern fiber-optic communication systems [26].

0.2 dB/km, enabling signals to propagate over hundreds of kilometers with minimal amplification.

Beyond 1,600 nm, however, the attenuation rises again, primarily due to infrared absorption from vibrational modes associated with Si–O bonds in the glass network. Secondary peaks visible in the experimental curve result from trace hydroxyl (OH^-) impurities, which introduce absorption bands near 950, 1240, and 1380 nm. These residual absorption features were progressively reduced through advances in glass purification techniques such as vapor-phase deposition and dehydration processes.

The correspondence between the Simple Model and Experimental Data traces in the figure underscores how theoretical understanding of scattering and absorption mechanisms guided the design of low-loss fibers. The identification of the low-attenuation window around 1,550 nm ultimately defined the operating wavelengths for modern fiber-optic communication systems, laser sources, and amplifiers [26].

3.5.1 Ultrapure Fused Silica

One of the most essential achievements in the development of low-loss optical fibers was the advancement of ultrahigh-purity silica glass [27]. The early decades of fiber research made it abundantly clear that even trace impurities within the glass could catastrophically degrade performance. Scientists understood that light transmission in fiber would only be feasible if attenuation due to absorption and scattering could be brought down to levels acceptable for long-distance communication. To achieve this, they focused intensively on identifying and eliminating the causes of optical loss, most notably transition metal ions and hydroxyl (OH^-) groups. These impurities—often present at parts-per-million levels in conventional glass—posed formidable barriers. Transition metals such as iron, copper, and chromium introduced broad absorption bands across the visible and near-infrared spectrum, while OH groups created sharp absorption peaks, especially around 1.39 μm, right in the heart of the telecommunications window.

A breakthrough came with the realization that vapor-phase deposition methods could provide a route to building glass not from bulk melting of raw materials, but molecule by molecule, in a controlled gas-phase reaction. This approach offered dramatically higher purity. Instead of using silica sand, which often contained significant impurities, chemists could react highly purified silicon tetrachloride ($SiCl_4$) in a flame or furnace to deposit pure silica. Using this technique, they could drive down impurity concentrations to mere parts per billion, nearly a thousandfold reduction over earlier methods. The level of chemical control was unprecedented and provided the means to engineer glass with optical losses well below 1 dB/km.

By 1972, a team at Corning had perfected this method to reproducibly produce high-purity silica preforms using outside vapor deposition (OVD), a process in which soot generated from chlorosilane reactions was deposited onto a rotating rod and later consolidated into clear glass [28, 29]. The resulting fibers, once drawn from these preforms, achieved attenuation levels near 20 dB/km—a dramatic improvement over previous results and, crucially, within the range predicted by Charles Kao as the threshold for commercial viability. This advance validated the vision that light could be guided over long distances through glass and laid the foundation for every successive innovation in optical fiber communications. The ability to produce such chemically pure silica not only unlocked the potential of fiber optics but also marked a new era in glass science, where control at the atomic level translated directly into technological revolution.

3.5.2 Modified Chemical Vapor Deposition

The development of MCVD in 1974 marked a turning point in the evolution of fiber-optic manufacturing [30, 31]. Led by John MacChesney and his team at Bell Laboratories, this technique provided a powerful, controllable method for fabricating glass preforms with exceptionally low optical loss, an essential requirement for the practical deployment of long-haul optical fibers [32]. Prior to this innovation, producing high-purity glass suitable for telecommunications was fraught with difficulties. Traditional glass melting methods introduced unwanted impurities and lacked the fine control necessary to engineer complex refractive index profiles. MCVD addressed these limitations head-on by using gas-phase chemistry inside a precisely controlled environment.

In the MCVD process, precursor chemicals such as silicon tetrachloride ($SiCl_4$), germanium tetrachloride ($GeCl_4$), and other halides were introduced into a rotating silica tube, which acted as both the deposition chamber and the eventual outer cladding of the preform. An external traversing oxyhydrogen torch heated a moving hot zone within the tube, triggering chemical reactions that produced submicron particles of glass. These particles were deposited on the inner wall of the tube, forming soot layers that were subsequently sintered into transparent glass. By carefully adjusting the concentration and type of dopants, such as germanium or phosphorus, the refractive index of each layer could be precisely tailored. This enabled the creation of step-index profiles for single-mode fibers and smoothly varying graded-index profiles for multimode applications, giving fiber designers a high degree of flexibility [33].

What distinguished MCVD from earlier approaches was its cleanliness and reproducibility. Because the process occurred inside a sealed quartz tube using high-purity reactants, the resulting glass was free from many of the contaminants that plagued earlier melt-based techniques. This was critical for achieving attenuation levels below 1 dB/km, and eventually down to the 0.2 dB/km range that characterizes modern low-loss silica fibers. The technique's modularity and scalability also made it well-suited for industrialization. Within just a few years of its invention, MCVD had become a cornerstone of commercial fiber manufacturing, widely adopted by companies in the United States, Europe, and Japan. It supported the rapid expansion of both single-mode and multimode fiber deployment during the 1980s and 1990s, and remains a foundational process in optical fiber production to this day.

As illustrated in Figure 3.6, the MCVD technique involves a series of thermochemical steps performed inside a rotating silica tube to produce high-purity optical fiber preforms. In the first stage, gaseous precursors such as $SiCl_4$, $GeCl_4$, and $POCl_3$ are introduced into the tube while an external oxyhydrogen torch traverses along its length. The localized high temperature within the reaction zone promotes homogeneous oxidation reactions, forming submicron oxide soot particles that deposit uniformly on the inner tube surface. As the torch moves axially, these deposits are gradually sintered into a transparent glass layer, ensuring a dense, pore-free structure. Through multiple deposition and sintering cycles, precisely doped and undoped layers are produced, defining the desired refractive index profile of the fiber core and cladding. Finally, the tube is collapsed at elevated temperatures into a solid preform rod, which is subsequently drawn into optical fiber under controlled heating. This process yields fibers with exceptional compositional uniformity, low optical attenuation, and high mechanical integrity, establishing MCVD as a cornerstone technique in modern optical fiber manufacturing.

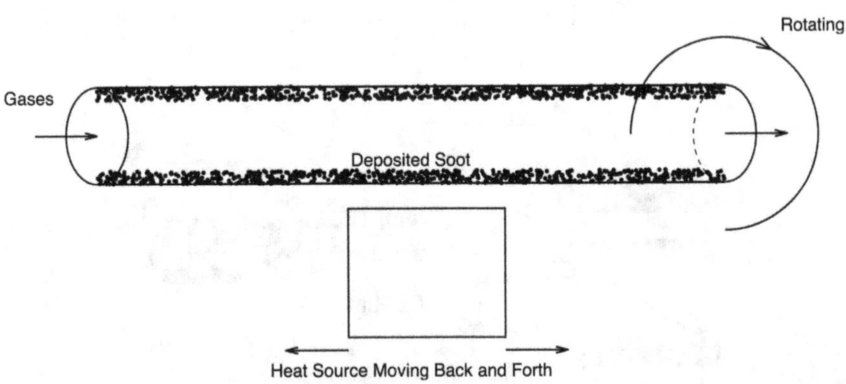

Figure 3.6 Schematic representation of the MCVD process for optical fiber preform fabrication. Introduction of gaseous precursors into a rotating silica tube, with an external traversing heat source driving oxidation and soot formation. Progressive sintering of the deposited soot into a transparent glass layer through successive torch passes, ensuring purity and uniformity. Final collapse of the tube into a solid preform rod, which is subsequently drawn into optical fiber.

As illustrated in Figure 3.7, the fundamental distinction between single-mode and multimode optical fibers lies in their core diameters and propagation characteristics. The 9/125-μm single-mode fiber possesses a narrow core that confines light to a single propagation mode, thereby reducing intermodal dispersion and enabling transmission over significantly longer distances with higher bandwidth. In contrast, the 50/125-μm multimode fiber has a much wider core that supports multiple light paths, or modes, which can interfere with one another as they travel. This multimodal behavior increases modal dispersion and limits transmission distance, but it allows for simpler and less expensive optical components, making multimode fibers advantageous for short-reach, high-density applications such as local area networks and data centers.

The figure clearly visualizes these differences while emphasizing that both fiber types maintain an identical 125-μm cladding diameter, ensuring compatibility in handling and connector interfaces.

3.5.3 Other Deposition Techniques

There are always two main steps in the manufacturing process of optical fibers: the preform production, and the drawing process. The preform is a solid glass rod that already has a core and a cladding, but their dimensions are far larger than in the final fiber. In the drawing process, the preform is heated and stretched by pulling with a tractor belt, forming the optical fiber this way.

The preform manufacturing process is usually based on the chemical reaction of gases $SiCl_4$ (silicon tetrachloride) and $GeCl_4$ (germanium tetrachloride) with O_2 (oxygen) to produce SiO_2 (silica) and GeO_2, also liberating Cl_2 (chlorine). Gas ratio can be modified depending on the desired final composition. Generally, the cladding is made of pure silica while the core has a higher

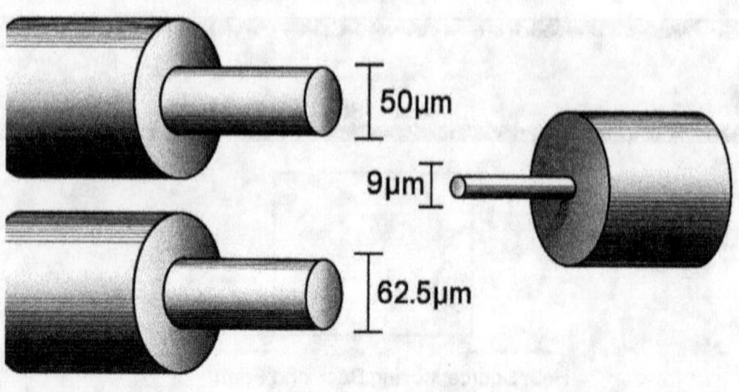

Figure 3.7 Comparison between 9/125-μm single-mode and 50/125-μm multimode optical fibers. The single-mode fiber (top) has a 9-μm core and supports a single light propagation path, ensuring minimal dispersion over long distances. The multimode fiber (bottom), with a 50-μm core, supports multiple propagation paths, which allows for higher light-gathering capacity but introduces modal dispersion over longer transmission lengths. Both fibers share the same 125-μm cladding diameter.

refractive index due to the presence of GeO_2. It must also be considered that multimode fibers typically have a higher content in GeO_2 than single-mode fibers.

Right now, there are three main techniques to produce the preform: (1) MCVD, (2) OVD, and (3) vapor axial deposition (VAD). Although there are differences among these three techniques, please consider that all of them include vapor deposition in their names. This means that they are processes that involve the deposition of materials in a vapor state on a substrate.

In the case of MCVD, gases ($SiCl_4$, $GeCl_4$) flow through a rotating tube of pure silica (which will be the cladding of the fiber), while heat is applied to the tube using a torch. Heat and oxygen cause the gases to react, leading to the formation of solid submicron particles, called soot, which deposit on the inner wall of the tube. The heat then transforms this soot into glass (SiO_2, GeO_2), starting to form the core region of the optical fiber. This process is repeated for hours to create the successive layers of the core. The parameters can be varied along the process to obtain fibers with different composition and properties. Once enough core layers have been manufactured, the temperature is increased so the tube collapses into the preform.

Regarding OVD, the manufacturing setup is very similar to that of the MCVD technique, but in this case, we start with a rotating inert rod instead of a tube of silica [34–36]. Again, gases are flowed along the rod and heat is applied to it with a torch. In this case, successive layers will be formed on the outside of the rod, first corresponding to the core, and then to the cladding. Once enough layers have been deposited, the rod is removed and heat is applied so the tube collapses into the preform.

Finally, in the case of VAD [33], a rotating inert rod is also employed, as it was done in OVD, but here it is vertically placed. The preform is manufactured of one of the tips of the rod, which is pulled upwards during the process. Two torches are employed, the one located in the lower part enables the deposition of the core glass, while the one placed above allows the deposition of the cladding glass.

As it has been previously mentioned, once the preform has been manufactured, the fiber is drawn. The speed of the fiber draw depends on the preform, fiber type, and available equipment, and it can vary between a few meters per minute to up to more than 1 kilometer per minute. During the drawing process the diameter of the fiber is controlled with high precision. Then, a protective jacket made of two layers (a soft inner coating and a hard outer one) is added to the fiber. Finally, different tests are performed to check everything is correct, including optical (attenuation, bandwidth, numerical aperture, etc.), mechanical (tensile strength), or environmental (temperature dependence, aging) properties.

All of these techniques shared a common objective: to fabricate high-purity glass preforms starting from gas-phase precursors. This approach not only minimized impurities but also allowed unprecedented control over core and cladding compositions. By the late 1970s, multiple manufacturers in the United States, Europe, and Asia had adopted one or more of these vapor deposition techniques, resulting in a global capability to produce low-loss single-mode and multimode fibers at industrial scales. This capacity formed the essential foundation for the fiber-optic communication revolution that would follow in the 1980s and beyond.

Figure 3.8 illustrates the three principal techniques used for producing optical fiber preforms: MCVD, OVD, and VAD. Although all three rely on vapor-phase deposition of silica and doped silica (typically GeO_2) particles, they differ in geometry and deposition location.

Table 3.1 summarizes the main differences among the most common vapor deposition techniques used in optical fiber preform fabrication, OVPO, MCVD, and VAD. While all three methods rely on the oxidation of gaseous precursors such as $SiCl_4$ and $GeCl_4$ to produce high-purity glass soot, their geometrical configurations and process dynamics vary significantly. In the OVPO process, material deposition occurs externally around a rotating mandrel through flame hydrolysis, building successive outer layers that later form the preform.

The MCVD technique, on the other hand, involves the chemical vapor reaction inside a rotating silica tube, where the soot forms and sinters along the internal wall, producing the preform core region. Finally, the VAD process distinguishes itself by its vertical arrangement, in which deposition proceeds

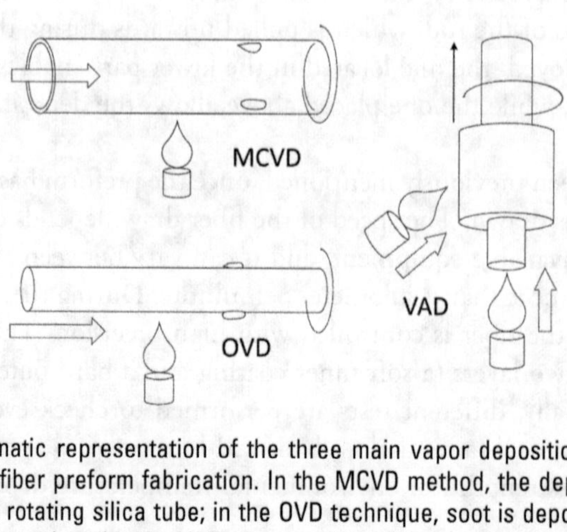

Figure 3.8 Schematic representation of the three main vapor deposition techniques employed for optical fiber preform fabrication. In the MCVD method, the deposition occurs on the inner wall of a rotating silica tube; in the OVD technique, soot is deposited on the outer surface of a rotating rod; and in the VAD process, deposition occurs vertically along the axis of a growing preform pulled upward.

Table 3.1
Comparison Optical Fiber Fabrication Techniques

—	OVPO	MCVD	VAD
Reaction Type	Flame Hydrolysis	Chemical Vapor Dimension	Flame Hydrolysis
Dimensional Direction	Outside Layer Deposition	Inside Layer Deposition	Axial Layer Deposition
RI Profile Formation	Layer Approximation	Layer Approximation	Simultaneous Formation
Process	Batch	Batch	Continuous

axially at the end of a rotating seed rod, allowing simultaneous formation of both core and cladding layers. Unlike the batch-based OVPO and MCVD methods, the VAD technique is inherently continuous, making it highly efficient for large-scale industrial production. Each method thus presents distinct advantages depending on the desired fiber type, refractive index control, and manufacturing throughput.

3.5.4 Refractive Index Profiling

Early fibers were step-index, a uniform high-index core and uniform lower index cladding. In 1972, researchers proposed and demonstrated graded-index (GI) fiber, where the core's refractive index gradually decreases from center to edge. This gradient causes light rays to follow curved paths and helps equalize the transit times of different rays, drastically reducing modal dispersion in multimode fibers [37–41]. The development of graded-index multimode fiber (using precise dopant concentration gradients) improved the bandwidth of early fiber links, an important factor for communication applications in the 1970s.

3.5.5 Intrinsic Loss Limits

As fiber losses dropped, researchers identified fundamental loss mechanisms. Rayleigh scattering from microscopic density fluctuations in glass sets a lower bound on loss, which decreases with wavelength (λ^{-4}). This pointed to longer wavelengths (1.3 μm and 1.55 μm) as better for even lower loss. Indeed, by 1979 fiber attenuation at 1.55 μm was measured below 0.2 dB/km, near the theoretical limit for silica. Silica's intrinsic infrared absorption rises beyond 2 μm, but the window around 1.5 μm is a sweet spot of minimal loss, which modern systems exploit.

Through these advances in materials and fabrication, optical fibers became an extraordinarily transparent medium–clearer than purified water. A visual comparison is often made: a typical pane of window glass, only a few millimeters thick, appears perfectly transparent to the eye. However, if that same glass were made a kilometer thick, it would appear completely opaque because of cumulative absorption and scattering within the material. By contrast, modern optical fiber, manufactured from ultrapure silica, can transmit light over distances exceeding 100 km with only a few percent power loss (thanks to occasional amplifiers). This level of transparency is what makes today's high-speed, long-distance fiber-optic communication possible.

3.6 Fiber Manufacturing and Waveguide Refinements

Producing optical fiber involves first making a glass preform with the desired core and cladding characteristics, then drawing it into a thin fiber. Early on, some fibers were made by a double-crucible method (melting two glasses and drawing fiber from a dual-nozzle), but the purity was limited. The 1970s vapor deposition processes (MCVD, OVD, VAD) solved this and dominated fiber fabrication. By 1980, Corning and others could mass-produce fiber with consistent quality. Initially, drawing speeds were modest, about 1 to 2 meters per second, due to the need for careful cooling and flaw prevention. Incremental improvements in machinery and materials led to higher speeds. In 1983, for instance, Corning's drawing towers, aided by work of engineers like Thomas Mensah, achieved draw speeds over 50 m/s [42, 43], dramatically lowering manufacturing costs such that optical fiber became cheaper per meter than copper cable. Meanwhile, fiber design evolved. Engineers refined the waveguide theory for fibers, calculating modal propagation constants, dispersion, and cutoff conditions for single-mode operation. They found that standard single-mode fiber has a core around 8 to 10 μm and a cutoff wavelength near 1.26 um, meaning it guides only the fundamental mode above that wavelength. This single-mode fiber offers the highest bandwidth since it avoids modal dispersion entirely. By contrast, multimode fibers (with core > 50 μm) were easier to couple to light sources but suffered dispersion; the graded-index profile was the remedy for those. The interplay between core size, numerical aperture, and index profile was well understood by the late 1970s through waveguide modeling (using Maxwell's equations and digital simulations, tools not available to the nineteenth-century pioneers who relied on analogies alone). The result was a portfolio of fiber types optimized for different applications, from long-haul single-mode fibers to short-run plastic fibers.

3.7 Commercialization of Optical Fiber Systems

With low-loss fibers and semiconductor lasers in hand, the late 1970s saw the translation of fiber optics from lab to field. In April 1977, the first live telephone traffic through fiber optics was carried in field trials, one by AT&T in Chicago (using fibers run underground in city conduits), and another by GTE in Long Beach, California [44–46]. Each system operated at 850-nm wavelength using GaAs semiconductor lasers and could carry the equivalent of a bundle of voice channels (e.g., 672 voice channels per fiber was reported). That same year in Europe, the first metropolitan fiber cable was deployed in Turin, Italy, by the research center Centro Studi e Laboratori Telecomunicazioni (CSELT) in collaboration with Corning [47, 48]. These early systems were modest in size—distances of a few kilometers between electronics, data rates on the order of tens of megabits per second—but they proved definitively that fiber-optic communication was practical outside the laboratory. Fiber-optic cables were soon installed to connect central telephone offices, forming the new telecom backbone in cities. Throughout the 1980s, fiber-optics technology advanced rapidly, and large-scale commercial deployment took off. Key developments in this era include the following areas.

3.7.1 Higher Bandwidth Systems

By the beginning of the 1980s, optical fiber technology had progressed from its experimental roots into early commercial deployment. At that stage, fiber-optic links were already demonstrating impressive capabilities compared to legacy copper systems, delivering data rates of 45 megabits per second over several kilometers using multimode fibers paired with light-emitting diodes (LEDs) or short-wavelength lasers operating near 850 nm. These systems were deployed in metropolitan and campus-scale networks, where repeater spacing could be kept relatively short and alignment tolerances remained manageable. However, multimode fibers, while easier to couple to and align, suffered from modal dispersion, a phenomenon in which different light paths through the fiber arrived at the receiver at slightly different times, limiting the overall bandwidth-distance product of the link.

As the limitations of multimode systems became increasingly apparent, attention shifted toward single-mode fiber, which allows only one propagation mode and thus eliminates modal dispersion altogether. This transition represented a significant leap in fiber-optic performance and was enabled by improved theoretical understanding, fabrication techniques, and optoelectronic components. Around 1983 and 1984, leading telecom operators, including British Telecom, made strategic decisions to deploy nationwide single-mode fiber networks. These systems operated primarily at 1,300 nm, a wavelength chosen because it coincides with the fiber's zero-dispersion point and offered relatively low attenuation of about 0.5 decibels per kilometer [33].

The shift to 1,300 nm brought dramatic improvements in repeater spacing and data rates. Links could now span tens of kilometers without the need for signal regeneration, greatly simplifying infrastructure and reducing operational costs. System capacities climbed to hundreds of megabits per second, enabling broader rollout of digital voice and data services across national and international backbones.

By the late 1980s, another pivotal development unfolded. Advances in laser sources and photodetectors enabled reliable operation at 1,550 nm—a spectral region where fiber attenuation drops to its absolute minimum, around 0.2 dB/km [49]. This further extended the reach of each fiber span and made gigabit-per-second data rates feasible on a practical scale. At the same time, dispersion-shifted and dispersion-compensated fiber designs were introduced to manage the chromatic dispersion that began to rise at 1,550 nm, keeping pulse spreading under control and preserving signal integrity over longer distances. Together, these innovations laid the groundwork for the high-speed optical networks that would come to define global communication in the decades to follow.

3.7.2 Optical Amplifiers

One of the most transformative innovations in the evolution of fiber-optic communication came with the advent of the erbium-doped fiber amplifier, or EDFA, in the mid-1980s [50–52]. Before its invention, optical signals traveling over long distances required periodic regeneration—conversion from optical to electronic form for amplification and reshaping, then conversion back into light to continue their journey through the fiber. This optical-electrical-optical (O-E-O) conversion process introduced latency, increased system complexity, consumed significant power, and raised the cost of maintaining long-haul networks. As the demand for higher capacity and longer distance fiber-optic links intensified, particularly with the emergence of undersea communication cables, the need for a purely optical solution became increasingly urgent.

The breakthrough arrived when researchers led by David Payne at the University of Southampton and Emmanuel Desurvire at Bell Laboratories independently demonstrated that a section of optical fiber doped with the rare-earth element erbium could be used to amplify light in the 1,550-nanometer range—the very window in which silica fiber exhibits its lowest attenuation [50, 53]. This insight was not accidental; erbium's energy levels are well-matched to produce stimulated emission at wavelengths around 1,550 nm when appropriately pumped by a secondary laser, typically operating at 980 or 1,480 nm. The result is the direct amplification of the signal light within the fiber itself, using the same guiding medium as the transmission path.

The significance of this development cannot be overstated. By eliminating the need for electronic regeneration, EDFAs allowed optical signals to be boosted midroute without disrupting their phase, coherence, or data encoding. This preserved the analog nature of the transmission and significantly reduced system overhead. Perhaps most importantly, EDFAs could amplify a broad range of wavelengths simultaneously, a property that later became essential with the rise of wavelength-division multiplexing (WDM). As WDM systems matured and began to carry dozens of channels within a single fiber, the EDFA's capability to amplify all these wavelengths uniformly, without demultiplexing, proved indispensable.

By the early 1990s, EDFAs had become standard components in both terrestrial and submarine optical networks. Their deployment marked a dramatic shift in how long-distance fiber-optic systems were engineered. Instead of needing an electronic repeater every 80 to 100 km, network designers could extend optical reach with amplifiers spaced hundreds of kilometers apart, dramatically reducing installation costs and increasing system reliability. Undersea cables, in particular, benefited from this technology, where the cost and difficulty of maintenance made reliability paramount. The laying of TAT-8, the first transatlantic fiber-optic cable, in 1988—just as EDFAs were being commercialized—heralded a new era of global communications. Subsequent cables rapidly adopted the new amplifier technology, enabling higher data rates and more robust links across vast distances of ocean floor [54].

The EDFA did more than improve fiber optics, it redefined the limits of optical communication. It allowed the vision of a seamless, global network to become reality by making long-haul transmission both technically and economically viable. Its simplicity, compatibility with existing fiber infrastructure, and ability to scale with the rise of multiplexed channels cemented its role as a foundational technology in modern telecommunications. As networks continue to grow in capacity and reach, the legacy of the erbium-doped fiber amplifier remains clear: a quiet revolution, buried in glass, that made the information age possible.

As illustrated in Figure 3.9, the EDFA operates by transferring energy from a pump laser, typically at 980 or 1,480 nm, into the erbium ions embedded within the fiber core. The schematic in Figure 3.9 demonstrates the integration of the EDFA within an optical communication link, where the input signal at 1,550 nm is first isolated to prevent back reflections and then coupled with the pump light through a WDM. The combined optical fields propagate through the erbium-doped fiber (EDF), where stimulated emission amplifies the signal. Long-period gratings embedded in the EDF are often used to control mode coupling and optimize gain uniformity across the signal bandwidth. The amplified output is then passed through an isolator to maintain unidirectional propagation.

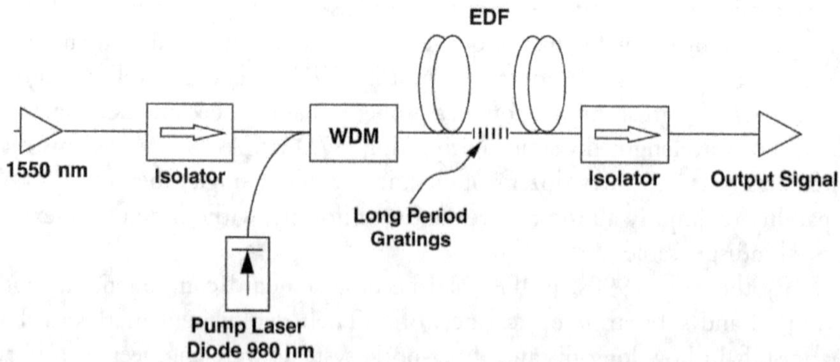

Figure 3.9 Schematic representation of an EDFA showing the propagation of the 1,550-nm signal through an isolator, WDM, EDF, and output isolator. The pump laser diode operating at 980 nm injects energy into the EDF via the WDM, while long-period gratings assist in optimizing the amplification.

3.7.3 Mass Deployment

The 1990s marked a pivotal decade in the evolution of global communications, driven by an unprecedented expansion in the deployment of optical fiber infrastructure. For decades prior, copper wires had served as the primary medium for telephone and data transmission. However, the limitations of copper—especially its bandwidth ceiling, susceptibility to electromagnetic interference, and signal attenuation over long distances—became increasingly problematic as demand for high-speed, high-volume data transport began to surge. The arrival of fiber optics, after years of maturation in laboratories and targeted deployments in the 1980s, offered a clear and decisive alternative. Capable of transmitting light over hundreds of kilometers with minimal signal degradation, optical fibers began to displace copper at a rapid pace, particularly in the long-haul and trunk segments of telecommunication networks.

Telecom companies, recognizing the economic and technical advantages of fiber, launched widespread replacement campaigns, upgrading their core infrastructure. Simultaneously, new services and markets began to emerge that were designed from inception around fiber capabilities. Cable television providers, for instance, began using hybrid fiber-coaxial networks to deliver clearer signals and more channels to suburban and urban households. The internet, transitioning from a niche academic network to a public utility, demanded backbone connections with capacity and reliability far beyond what existing copper lines could provide. This demand led to the laying of vast terrestrial fiber routes across continents, connecting data centers, internet exchange points, and metropolitan hubs.

The notion of an information superhighway, popularized in political and media discourse during the early 1990s, was not merely metaphorical.

Fiber-optic networks gave tangible form to this concept, offering the physical infrastructure necessary to carry not only voice and television but also the exponentially increasing volumes of internet data generated by web browsing, email, e-commerce, and emerging multimedia content. The deployment of dense wavelength division multiplexing (DWDM) during this period multiplied the capacity of individual fibers by allowing dozens or even hundreds of distinct optical channels to be carried simultaneously, each at gigabit or terabit speeds. This technological leap enabled fiber to scale in tandem with Moore's law, keeping pace with the accelerating demands of digital computing and networking.

By the late 1990s and early 2000s, global fiber-optic networks had reached a level of maturity and scale that was previously unimaginable. Submarine fiber-optic cables, such as the Fiber-optic Link Around the Globe (FLAG) [55, 56] series, connected continents, replacing older satellite and undersea coaxial systems with faster, more reliable links. These cables formed the backbone of a truly global internet, carrying the majority of international data traffic with latency and bandwidth suitable for real-time applications like video conferencing, voice over IP (VoIP), and streaming media. Terabit-scale throughput, once a theoretical goal, became routine in the operation of core network infrastructure.

This sweeping transformation was the culmination of over a century of research into waveguide theory, materials science, optical physics, and systems engineering. What made the 1990s unique was not just the maturation of the technology, but its mass adoption and integration into the fabric of everyday life. The transition from analog to digital communications, the commercialization of the internet, and the rise of a globally interconnected economy all rested, quite literally, on a lattice of glass—threaded through cities, buried beneath oceans, and linking people across vast distances in the blink of an eye.

It is worth noting that optical fiber technology continued to diversify well beyond the turn of the millennium. Earlier advances such as polarization-maintaining fiber (for preserving light polarization) [57], dispersion-shifted fiber (tailoring dispersion properties for 1,550 nm) [58], and plastic optical fiber (for short-distance and consumer applications) [59] laid the groundwork for a new generation of innovations. In the mid-1990s, researchers developed photonic crystal fiber [60, 61], which uses a microstructured arrangement of air holes running along the fiber length to guide light by photonic bandgap effects instead of TIR. The concept of photonic crystal fiber and bandgap will be presented in further detail in the following chapter. Photonic crystal fibers, first commercialized around 2000, enabled exotic waveguide properties such as extremely large mode areas, endlessly single-mode operation, and precise dispersion control across a wide wavelength range.

Since 2000, the field has advanced rapidly, driven by both scientific and industrial demands. Hollow-core photonic bandgap fibers and antiresonant fibers have emerged, allowing light to propagate predominantly through air, thus reducing nonlinear effects and latency. Multicore and few-mode fibers have been developed to increase capacity through space-division multiplexing, a key technology for next-generation high-capacity networks [62, 63]. In parallel, advances in fiber lasers and amplifiers have led to record power levels and new wavelength regimes for applications in manufacturing, medicine, and sensing. Specialized fibers, such as fiber Bragg grating sensors for structural monitoring [64, 65], doped fibers for ultrafast laser systems [66], and radiation-hardened fibers for aerospace [67, 68], have expanded the scope of fiber optics far beyond telecommunications. More recently, research into hollow-core and quantum-compatible fibers has opened pathways toward ultralow-latency data links and secure quantum communication networks [69–72].

These developments over the past 25 years demonstrate that the field of guided-wave optics remains vibrant and continually evolving. Modern innovations build upon the foundational analogies, theories, and materials breakthroughs of the twentieth century, while extending optical fiber technology into entirely new domains of science and engineering.

3.8 Optical Fibers Concepts and Applications

The field of optical fiber technology stands at the core of modern photonics and telecommunication systems, enabling the global exchange of information at unprecedented speeds and capacities. Since their invention, optical fibers have revolutionized how data, light, and energy are transmitted over vast distances with minimal loss, forming the backbone of the internet, high-speed data networks, sensing systems, and advanced imaging instruments. Their versatility arises from the ability to manipulate light propagation in a controlled manner—through variations in geometry, material composition, and refractive index design—allowing them to be tailored for specific applications that range from long-haul communication to biomedical diagnostics. This section explores the fundamental concepts and diverse applications of optical fibers, with particular focus on their role in data communication and imaging systems.

At the most fundamental level, optical fibers operate on the principle of total internal reflection, confining light within a core surrounded by a lower refractive-index cladding. However, the performance and functionality of a fiber depend heavily on its internal structure and the nature of light propagation it supports. For instance, multimode fibers (MMFs) can transmit multiple spatial modes of light simultaneously, making them ideal for short-range, cost-effective communication links such as those found in local area networks and data centers. Conversely, single-mode fibers (SMFs) support only the fundamental optical mode, ensuring minimal dispersion and signal degradation, thus serving as the preferred medium for long-distance, high-capacity data transmission.

As the demand for bandwidth and data throughput continues to grow exponentially, innovative multiplexing techniques have been developed to increase the effective capacity of optical communication systems without requiring proportional increases in physical infrastructure. Among these, WDM exploits the spectral domain by transmitting multiple data channels at distinct wavelengths, while space division multiplexing (SDM) leverages the spatial dimension of light propagation, using either multiple cores or spatial modes within a single fiber to carry independent data streams. Together, these technologies exemplify how optical engineering continues to push the limits of data transmission efficiency and scalability.

Beyond communication, the unique properties of optical fibers have been harnessed for imaging and sensing applications. Fiber bundles, in particular, enable flexible image transmission in environments where traditional optical systems are impractical. Their adaptability makes them indispensable in medical devices such as endoscopes and in precision instruments for spectroscopy and industrial inspection. Furthermore, advanced fiber assembly configurations, including groove-mounted arrays, lensed coupling systems, and V-groove alignments, demonstrate the intricate mechanical and optical design strategies used to ensure alignment precision, optical integrity, and mechanical robustness.

The following subsections delve into these topics in greater detail, beginning with an examination of multimode fibers for data communication, followed by discussions on space division multiplexing, wavelength division multiplexing in single-mode fibers, and finally, fiber bundles for imaging and assembly configurations. Together, these explorations provide a comprehensive understanding of the diverse and evolving landscape of optical fiber technology.

3.8.1 Multimode Fiber for Data Communication

In modern optical communication systems, the choice between single-mode and multimode fibers is fundamentally dictated by the balance between transmission distance, bandwidth requirements, and system cost. MMFs, characterized by their relatively large core diameters typically in the range of 50 to 62.5 μm are designed to support the simultaneous propagation of multiple optical modes. Each mode represents a distinct light path through the fiber core, determined by the launch angle and the refractive index profile. This ability to carry multiple spatial modes allows MMFs to transmit large volumes of data using inexpensive light sources such as vertical-cavity surface-emitting lasers (VCSELs) or LEDs, making them particularly attractive for short-reach communication applications in local area networks, data centers, and interconnects.

However, the multipath nature of MMFs introduces a key limitation: modal dispersion. Because each propagating mode experiences a slightly different optical path length and effective refractive index, light pulses injected into the fiber tend to spread out in time as they propagate. This temporal spreading, also known as pulse broadening, reduces the maximum achievable data rate and limits the effective transmission distance. The severity of this effect depends on the number of supported modes, the refractive index profile of the core (step-index or graded-index), and the optical wavelength. Graded-index multimode fibers, for instance, are engineered with a parabolic refractive index profile that mitigates modal dispersion by allowing higher order modes to travel faster than lower order modes, thereby partially compensating for differential delay.

Despite these challenges, multimode fibers continue to play a critical role in high-speed data communication where link distances typically range from a few meters to several hundred meters. The simplicity of their alignment, coupled with lower component costs and ease of installation, makes them an efficient and scalable solution for short-haul networks. As data rates continue to increase, techniques such as modal conditioning, advanced equalization algorithms, and improved fiber designs have been employed to extend the performance limits of MMFs.

Figures 3.10 and 3.11 provide a visual understanding of the underlying propagation mechanisms and dispersion effects in multimode fibers. Figure 3.10 contrasts the mode propagation characteristics of single-mode and multimode fibers, highlighting how multiple light paths give rise to modal dispersion. Building on this concept, Figure 3.11 illustrates how pulse broadening

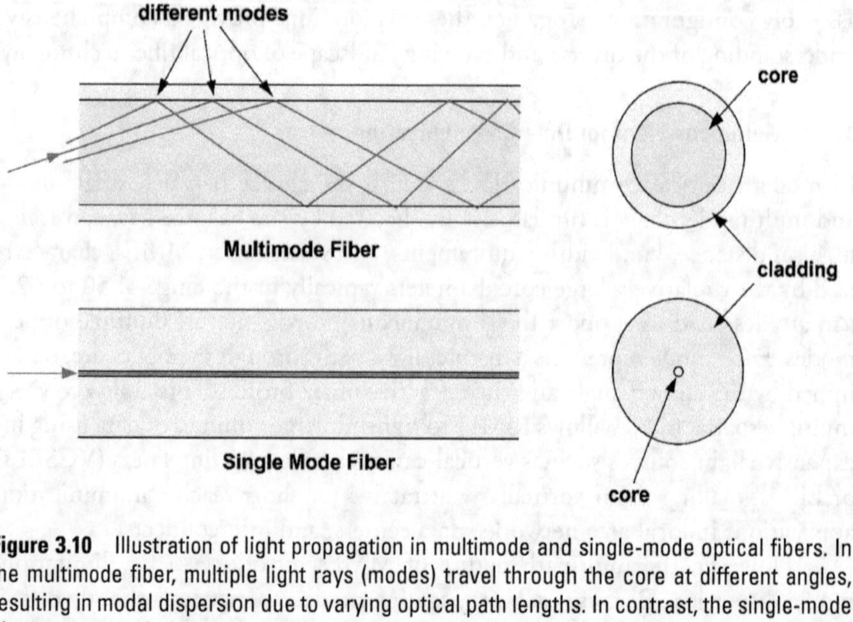

Figure 3.10 Illustration of light propagation in multimode and single-mode optical fibers. In the multimode fiber, multiple light rays (modes) travel through the core at different angles, resulting in modal dispersion due to varying optical path lengths. In contrast, the single-mode fiber supports only one propagation path, minimizing dispersion and maintaining signal integrity over long distances.

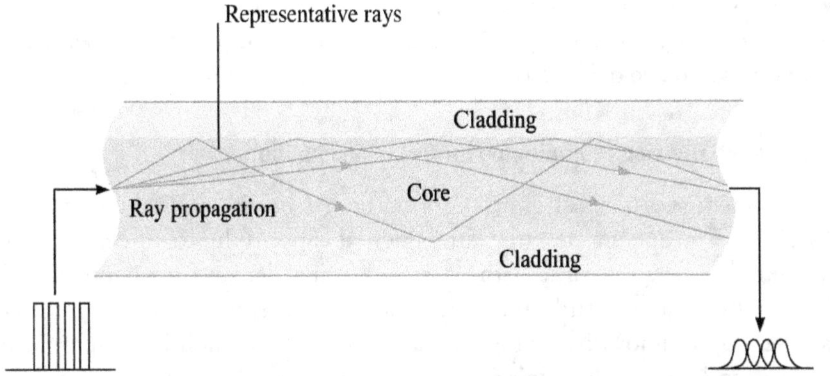

Figure 3.11 Illustration of modal dispersion in a multimode optical fiber. Multiple rays of light propagate through the fiber core at different angles and path lengths, leading to varying propagation times. This causes temporal broadening of optical pulses as they exit the fiber, resulting in signal distortion and potential ISI that increases the bit error rate (BER) in digital communication systems.

occurs in multimode fibers due to differential propagation times, ultimately influencing the signal integrity and bandwidth performance of optical communication links.

As shown in Figure 3.10, multimode and single-mode fibers differ significantly in their mode propagation characteristics. In a multimode fiber, the relatively large core diameter allows multiple light rays, or modes, to propagate simultaneously along different paths within the core. Each mode experiences a distinct optical path length, leading to a temporal spread of the transmitted signal at the output, known as modal dispersion. This phenomenon limits the bandwidth and transmission distance of multimode fibers. Conversely, single-mode fibers feature a much smaller core diameter that supports only the fundamental propagation mode, effectively eliminating modal dispersion. This results in higher transmission efficiency and signal fidelity over extended distances, making single-mode fibers preferable for long-haul and high-speed communication links.

Figure 3.11 demonstrates the effect of modal dispersion in a multimode optical fiber, where multiple propagation paths lead to pulse broadening. When short optical pulses are launched into the fiber, each mode travels a different optical path within the core and thus experiences a distinct propagation time. As a result, the initially sharp pulses at the input become temporally spread at the output. This pulse broadening reduces the temporal separation between successive bits, which increases the likelihood of overlapping signal levels, a phenomenon known as intersymbol interference (ISI). The presence of ISI degrades the signal integrity and increases the bit error rate (BER), thereby limiting both the transmission bandwidth and the effective communication distance of multimode fibers. Understanding and mitigating modal dispersion is essential for maintaining reliable data transmission in optical communication systems.

3.8.2 Space Division Multiplexing for Data Communication

The exponential growth of global data traffic has pushed optical communication systems to operate near the fundamental capacity limits of conventional single-mode fibers. To sustain the ever-increasing demand for bandwidth in telecommunications, cloud computing, and data center interconnects, researchers have sought innovative strategies that go beyond the traditional approaches of wavelength, time, and polarization multiplexing. Among these emerging techniques, SDM has gained significant attention as a transformative solution for scaling transmission capacity by exploiting the spatial dimension of light propagation [73]. Instead of relying on a single optical path per fiber, SDM enables multiple, spatially distinct channels to coexist within the same physical medium, thereby multiplying the total data throughput without requiring a proportional increase in fiber count.

The concept of SDM can be realized through two principal architectures: multicore fibers (MCFs) and multimode fibers (MMFs). In an MCF, several independent cores are embedded within a shared cladding, with each core functioning as an isolated transmission channel. This design minimizes intercore crosstalk through careful control of core spacing and refractive index distribution, making it highly suitable for long-haul and high-capacity networks. In contrast, MMF-based SDM utilizes distinct spatial modes within a single large core to transmit parallel data streams. Although this configuration offers fabrication simplicity and higher spatial density, it introduces complex challenges such as mode coupling and differential modal delay (DMD), often necessitating advanced digital signal processing techniques such as multiple-input multiple-output (MIMO) equalization. Both approaches, however, represent vital pathways toward achieving petabit-per-second transmission capacities in next-generation communication infrastructures.

The ongoing development of SDM technologies has also prompted the evolution of fiber designs, progressing from traditional single-core structures to few-core and multicore configurations. These designs balance manufacturability, mechanical robustness, and spatial channel density to optimize performance for various deployment scenarios. Moreover, SDM-compatible fibers are engineered to maintain compatibility with existing optical components and connector standards, ensuring practical scalability for network operators.

Figure 3.12 introduces the two primary implementations of SDM, multicore and multimode fibers, highlighting their distinct transmission mechanisms [73]. Subsequent figures expand on these concepts by showing the evolution of

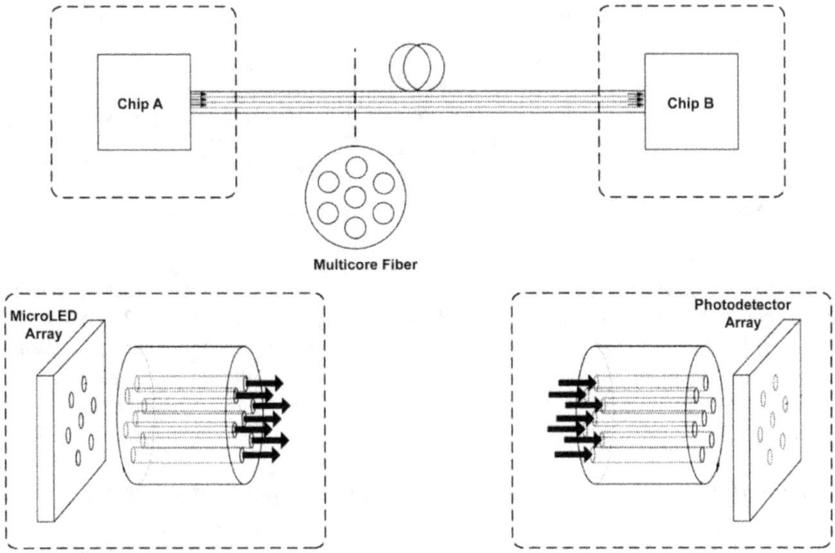

Figure 3.12 Schematic representation of SDM using a multicore fiber where multiple independent transmitters (Tx) and receivers (Rx) are linked through distinct cores within the same cladding, enabling parallel data transmission with low crosstalk.

fiber geometries, key dimensional parameters, and comparative architectures, providing a comprehensive overview of how spatial multiplexing enables a new era of high-capacity optical communication.

Figure 3.12 illustrates two principal implementations of SDM in optical fiber systems. In Figure 3.12(a), a MCF structure is shown, where several independent cores are embedded within a common cladding. Each core acts as a separate transmission path, enabling multiple data channels to be transmitted simultaneously with minimal intercore interference, provided adequate spacing and refractive index profiling are maintained. This architecture is particularly advantageous for high-capacity, long-haul transmission systems where low crosstalk and scalability are crucial. In contrast, Figure 3.12(b) depicts a MMF, where distinct spatial modes within a single large core are used to transmit multiple signals. Although MMFs offer simpler fabrication and high-density transmission, mode coupling and DMD introduce significant signal processing challenges, often requiring MIMO equalization at the receiver. Together, these configurations represent the two primary physical-layer strategies for realizing SDM in next-generation optical communication networks.

Figure 3.13 illustrates the evolution of optical fiber structures based on the number of cores, progressing from single-core to multicore configurations. A conventional single-core fiber guides light through a single optical path, which limits its capacity to a single communication channel. To overcome the capacity limitations imposed by single-core designs, researchers have developed

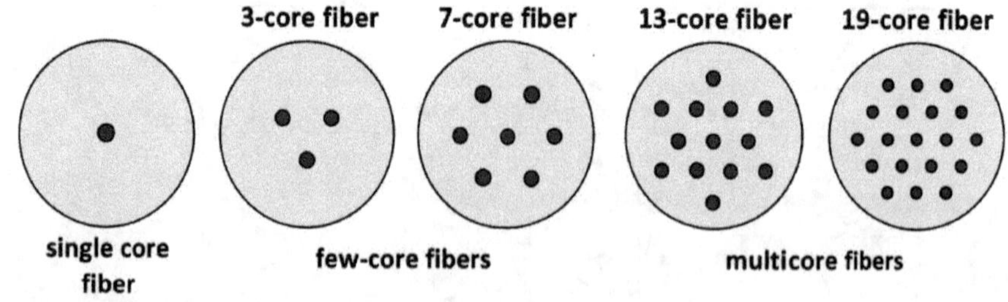

Figure 3.13 Illustration of optical fibers classified by the number of cores. The single-core fiber contains one light-guiding core, while few-core fibers (e.g., 3-core and 7-core) contain multiple spatially separated cores within a common cladding to support parallel transmission channels. Multicore fibers (e.g., 13-core and 19-core) further increase spatial density, enabling high-capacity SDM and improved data throughput in modern optical communication systems.

few-core and multicore fibers, where multiple independent cores are embedded within the same cladding. Each core operates as a distinct transmission path, enabling simultaneous propagation of multiple optical signals and thus supporting SDM. As shown, few-core fibers (such as 3-core or 7-core designs) strike a balance between fabrication complexity and increased data capacity, while multicore fibers with 13 or 19 cores offer significantly enhanced spatial channel density. These multicore architectures are crucial for meeting the growing bandwidth demands of next-generation telecommunication and datacenter networks by dramatically improving transmission efficiency and overall spectral utilization.

Figure 3.14 illustrates the fundamental geometrical distinctions between a single-core fiber (SCF) and a MCF. In the SCF configuration, light is guided through a single core surrounded by a cladding layer that ensures total internal reflection. This traditional design limits data transmission to a single spatial channel. In contrast, the MCF structure incorporates multiple light-guiding cores within the same cladding region, enabling parallel data transmission channels through SDM. Key dimensional parameters govern the optical and mechanical performance of MCFs, such as the core pitch (Λ), which determines the distance between neighboring cores and influences crosstalk behavior, and the outer cladding thickness (OCT), which controls mechanical strength and bending tolerance. Additionally, the overall cladding diameter must balance compactness with sufficient optical isolation. Understanding these parameters is essential for optimizing MCF performance in high-capacity communication systems, where minimizing intercore crosstalk and maximizing spatial channel density are critical design objectives.

As shown in Figure 3.15, commercially available optical fibers are designed with standardized dimensional parameters to balance performance, manufacturability, and mechanical robustness. The conventional SCF features

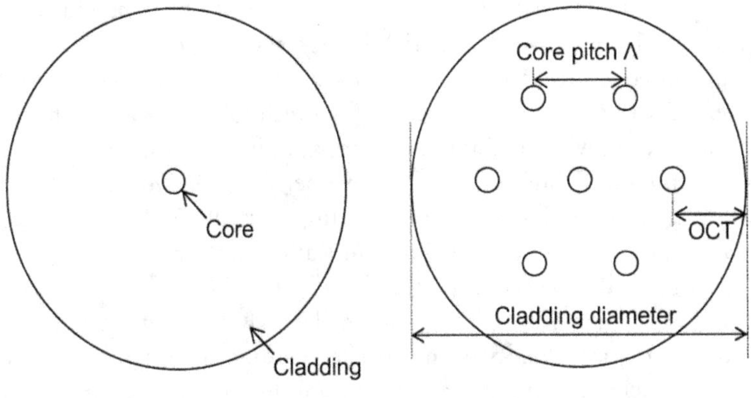

Figure 3.14 Comparison of the key structural dimensions of (a) a SCF, and (b) a MCF. The single-core fiber comprises a single light-guiding core surrounded by cladding, whereas the multicore fiber contains multiple cores embedded within a shared cladding region. The critical parameters for MCF design include the core pitch (Λ), which defines the center-to-center spacing between adjacent cores, and the OCT, which affects optical isolation between neighboring cores and mechanical robustness.

a 125-μm cladding diameter that encloses a single light-guiding core, surrounded by a protective coating that ensures mechanical durability and environmental resistance. In contrast, the MCF integrates several individual cores, each functioning as an independent optical channel, within a common cladding region of approximately 180 μm in diameter. The entire structure is further

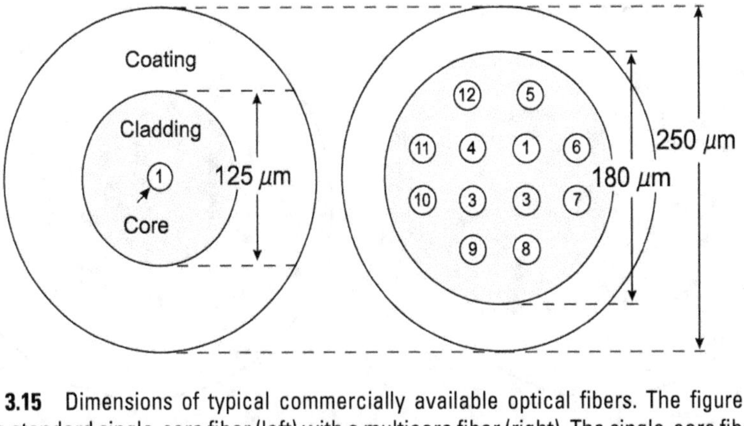

Figure 3.15 Dimensions of typical commercially available optical fibers. The figure compares a standard single-core fiber (left) with a multicore fiber (right). The single-core fiber has a cladding diameter of approximately 125 μm, while the multicore fiber incorporates multiple light-guiding cores within a larger cladding region of about 180 μm, and an overall coating diameter of 250 μm. These dimensions illustrate the compact yet scalable design approach used to accommodate multiple transmission channels within a single fiber structure.

enclosed by an outer coating layer extending to about 250 μm. This compact yet efficient design enables high spatial density, allowing multiple data streams to be transmitted simultaneously through a single fiber strand via SDM. Such fibers exemplify the ongoing evolution of fiber technology toward higher capacity and integration, while maintaining compatibility with existing cabling and connector standards. Understanding these dimensional relationships is essential for optimizing fiber performance, minimizing crosstalk, and ensuring reliable large-scale deployment in modern communication networks.

Figure 3.16 illustrates the structural distinction between a multicore fiber in (MCF) and a conventional fiber bundle. The MCF shown on the left contains seven cores embedded within a single cladding, maintaining a uniform core-to-core spacing denoted by *d*. This monolithic structure enables coherent light transmission and controlled coupling between adjacent cores, making it suitable for applications in spatial-division multiplexing and high-resolution imaging. In contrast, the fiber bundle on the right consists of multiple fine fibers arranged in a densely packed configuration, where each fiber acts as an independent transmission channel. The outer diameter of each fine fiber, represented as *d*′, determines the packing density and overall bundle diameter. Unlike the MCF, where the cores share a common cladding, the individual fibers in the bundle are mechanically independent, allowing greater flexibility but less intercore optical control. This comparison highlights the trade-off between integration and flexibility when selecting fiber structures for imaging, sensing, or data-transmission applications.

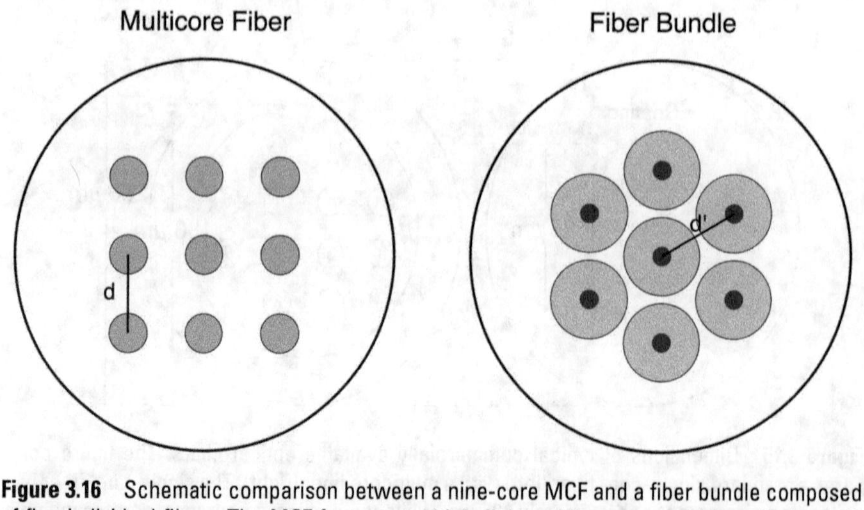

Figure 3.16 Schematic comparison between a nine-core MCF and a fiber bundle composed of fine individual fibers. The MCF features multiple light-guiding cores within a single cladding, while the fiber bundle consists of separate fibers closely packed together. The parameters *d* and *d*′ represent the core center-to-center spacing in the MCF and the outer diameter of individual fine fibers in the bundle, respectively.

As illustrated in Figure 3.20, the optical fiber assembly is designed to reformat the spatial distribution of light from a circular bundle at the input to a linear array at the output. This configuration is often used in spectroscopic applications, where the circular input efficiently collects light from a source or imaging system, while the linear output aligns with the entrance slit of a spectrometer. The callout on the left highlights the densely packed circular arrangement of individual fibers, optimizing collection efficiency, whereas the right callout shows the same fibers rearranged in a precise linear sequence to ensure uniform illumination across the spectrometer slit. The flexible, metal-sheathed construction provides both mechanical protection and routing flexibility, making it suitable for laboratory and industrial environments.

As shown in Figure 3.18, the fiber assembly demonstrates a bifurcated configuration where a single, densely packed input bundle divides into multiple smaller output bundles. The left-hand side of the figure depicts the primary composite bundle, consisting of numerous closely arranged fibers within a single sheath, designed to collect or deliver light uniformly from a source or sample. On the right-hand side, the bundle splits into several branches, each containing a subset of fibers drawn from the main input group. The magnified insets illustrate how the internal structure of each branch preserves the local fiber arrangement, ensuring consistent optical alignment and minimal crosstalk during signal transmission. This type of bifurcated design is commonly employed in applications such as illumination systems, multichannel spectroscopy, and

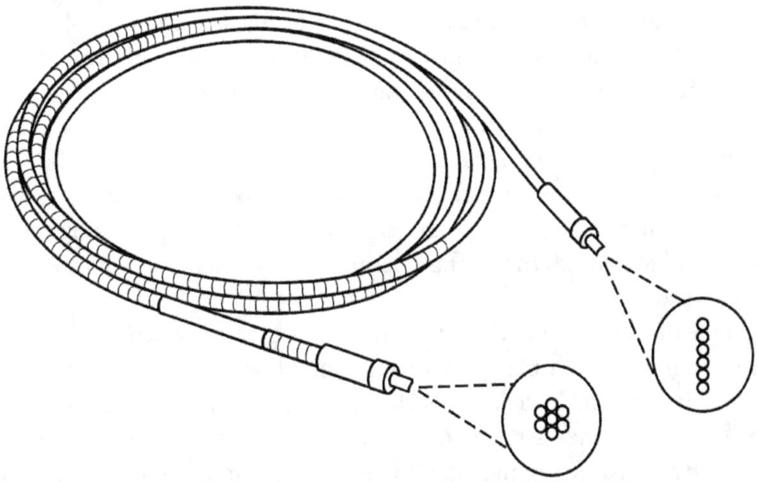

Figure 3.17 Custom-made optical fiber bundle featuring a circular fiber arrangement at one end and a linear array configuration at the opposite end. The dashed callouts illustrate the transformation of the fiber geometry from a compact circular bundle to an aligned linear output.

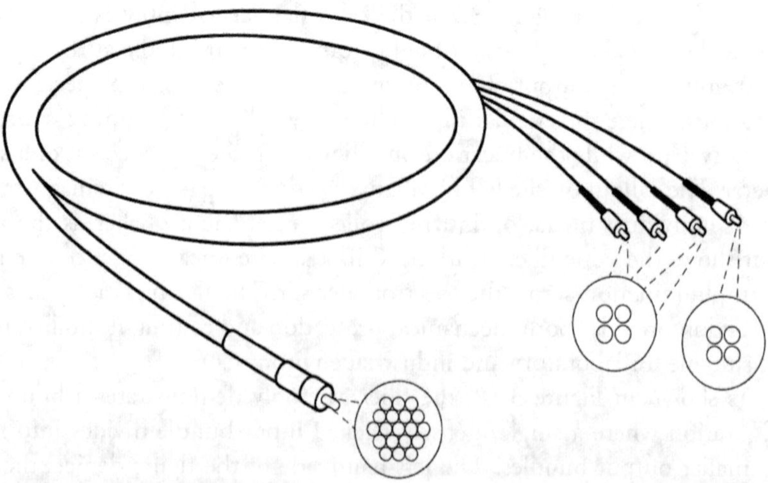

Figure 3.18 Schematic of a bifurcated fiber bundle showing a single composite input bundle on one end and multiple smaller output bundles on the other. The magnified insets illustrate how individual fibers from the main bundle are distributed among the smaller branches, each maintaining a consistent core geometry for efficient light splitting and transmission.

distributed sensing, where light needs to be shared or collected from multiple points while maintaining high transmission efficiency and spatial coherence.

3.8.3 Single-Mode Fibers for Wavelength Division Multiplexing

Single-mode optical fibers form the backbone of modern high-speed communication networks, enabling the transmission of massive volumes of data with exceptional fidelity and minimal signal degradation. Unlike multimode fibers, which support multiple propagation paths, single-mode fibers confine light to a single transverse mode, effectively eliminating modal dispersion and ensuring that optical signals maintain their integrity even over thousands of kilometers (e.g., [74]). This intrinsic property makes single-mode fibers ideally suited for long-distance and high-capacity data transmission, forming the physical layer foundation of global internet infrastructure, metropolitan networks, and high-performance data centers.

As the demand for bandwidth continues to grow exponentially, merely increasing the number of fibers or boosting individual signal power proves insufficient and inefficient. Instead, optical engineers have turned to WDM, a technique that exploits the vast spectral bandwidth available in optical fibers (e.g., [75, 76]). By transmitting multiple independent data channels, each encoded on a distinct wavelength of light, WDM allows simultaneous data streams to coexist within a single fiber without mutual interference. This effectively multiplies the information-carrying capacity of the optical link, achieving higher aggregate data rates while minimizing infrastructure expansion.

The implementation of WDM requires precise wavelength control and stable optical components, such as multiplexers (MUX) and demultiplexers (DEMUX), which combine and separate the different wavelength channels, respectively. Advances in laser stability, optical filtering, and fiber manufacturing have enabled both coarse WDM (CWDM) and dense WDM (DWDM) systems, tailored to various communication scales, from short-reach interconnects to transcontinental links. These innovations have transformed single-mode fiber networks into highly scalable and efficient communication backbones capable of supporting modern digital ecosystems.

Figure 3.19 illustrates the fundamental concept of WDM, a key technology for enhancing data throughput in optical communication systems. In this scheme, multiple data channels are transmitted simultaneously over a single optical fiber, with each channel assigned a unique wavelength ($\lambda_1, \lambda_2, \lambda_3, \lambda_4$). At the transmitter side, a MUX combines these wavelength channels into a composite optical signal, which propagates through the fiber as a single beam. Upon reaching the receiver end, a DEMUX separates the combined signal back into its individual wavelength components, allowing independent detection and data recovery for each channel. This approach significantly increases the fiber's capacity without the need for additional physical infrastructure. WDM can be implemented in two primary forms, CWDM and DWDM, depending on the wavelength spacing and channel count. The technology plays a crucial role in long-haul, metro, and data center networks, supporting terabit-scale transmission and enabling scalable optical interconnects for modern high-performance communication systems.

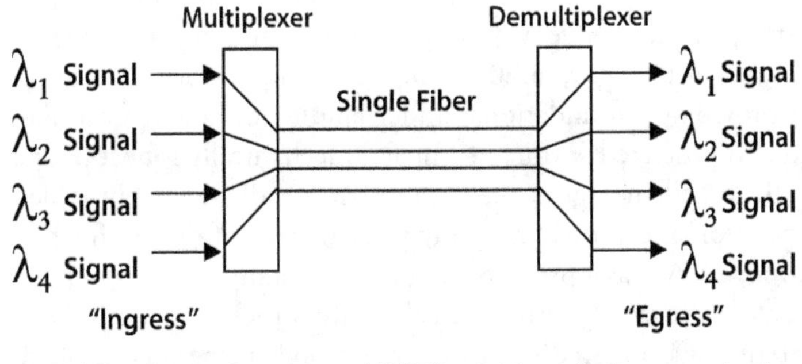

Figure 3.19 Schematic representation of WDM used in optical communication systems. Multiple optical signals at distinct wavelengths ($\lambda_1, \lambda_2, \lambda_3, \lambda_4$) are combined by a multiplexer into a single optical fiber for transmission. At the receiver end, a demultiplexer separates the composite signal back into its individual wavelength channels, thereby enabling simultaneous high-capacity data transmission through a single optical medium.

3.8.4 Fiber Bundle Assembly Configurations

The precise assembly and alignment of optical fiber bundles play a crucial role in determining the overall efficiency and performance of photonic systems [77–79]. As modern optical communication and sensing applications continue to demand higher integration density, lower insertion losses, and improved mechanical robustness, the structural design of fiber assemblies becomes increasingly significant [80, 81]. A well-engineered fiber bundle not only serves as a conduit for guided light transmission but also functions as a mechanically stable and environmentally protected interface between optical components. Achieving this balance requires careful consideration of both optical and mechanical constraints—where even minute deviations in alignment or spacing can result in significant coupling inefficiencies.

To meet these stringent requirements, various configuration strategies have been developed for assembling and packaging optical fibers into precise geometric arrangements. Central to these methods is the need to ensure accurate positioning, protection from external stress, and consistent optical performance across all fibers in the bundle. Techniques such as groove-based alignment [82, 83], encapsulation with transparent materials, and the use of optical coupling elements like lenses are often employed to maintain structural integrity while optimizing light transmission characteristics. The mechanical design, choice of materials, and assembly process must all work in concert to minimize optical losses due to scattering, misalignment, or microbending effects.

In the following sections, we present and illustrate several representative configurations that exemplify these design principles. Each configuration demonstrates a specific approach to achieving high-precision alignment and stable optical coupling between fibers and associated components. The first configuration focuses on the fiber bundle assembly, where a flat-bottom groove and glass lid are used to secure and encapsulate an array of optical fibers. This design emphasizes both protection and alignment stability, ensuring that the geometric uniformity of the bundle is preserved during handling and integration. The schematic shown in Figure 3.20 highlights the key structural features of this assembly, providing a foundational understanding of how optical fibers can be organized and protected within a compact, mechanically robust package.

As illustrated in Figure 3.20, the fiber bundle assembly is designed to securely position and protect a tightly packed array of optical fibers. The flat-bottom groove serves as a precision alignment channel, ensuring that the fibers remain parallel and evenly distributed along the structure's length. Once placed, a transparent glass lid is adhered over the groove using a controlled amount of glue, providing both mechanical stability and environmental sealing. This encapsulation method not only preserves the geometric integrity of the fiber bundle but also minimizes optical misalignment and mechanical stress during subsequent handling or integration into optical systems.

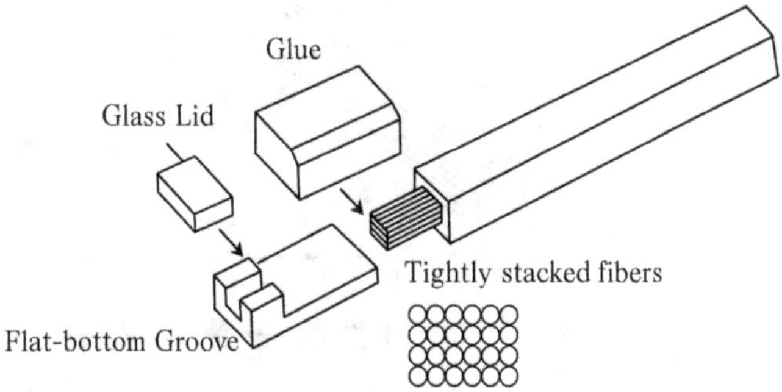

Figure 3.20 Schematic illustration of the fiber bundle assembly. A flat-bottom groove holds tightly stacked optical fibers, which are sealed in place with a glass lid using a thin adhesive layer of glue. This configuration ensures alignment stability and protection of the fiber array within the encapsulating structure.

As illustrated in Figure 3.21, a pair of ball lenses is strategically positioned between two opposing optical fibers to facilitate efficient light coupling. The first ball lens collects the divergent light emerging from the transmitting fiber and collimates it into a nearly parallel beam. This beam then propagates across the free-space gap before being refocused by the second ball lens into the core of the receiving fiber. Such an arrangement effectively reduces insertion losses caused by beam divergence and misalignment, making ball lenses highly suitable for compact fiber-to-fiber and fiber-to-detector coupling applications in integrated photonic systems.

As illustrated in Figure 3.22, the linear fiber bundle is carefully seated within a V-groove alignment substrate. This configuration allows each optical fiber to be accurately positioned by the geometrical constraint of the V-shaped channel, ensuring consistent pitch and parallelism across the array. The magnified inset highlights the cross-sectional alignment mechanism, where the groove profile securely nests each fiber to maintain uniform contact and minimize lateral displacement. Such a structure is commonly used in photonic packaging

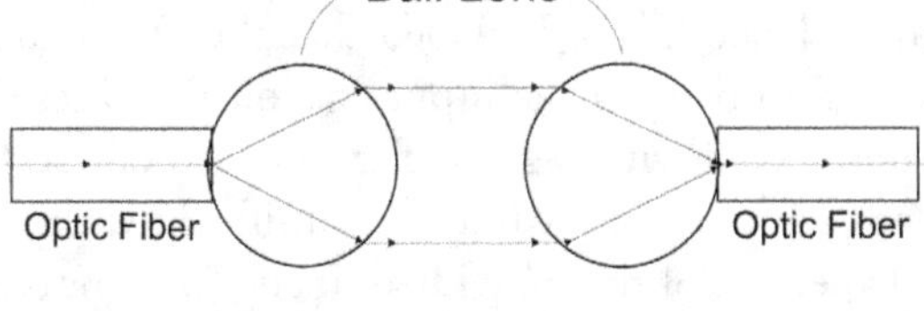

Figure 3.21 Schematic representation of optical coupling between two optic fibers using a pair of ball lenses [84]. The ball lenses act as collimating and focusing elements, allowing efficient light transfer between the fiber cores while minimizing divergence and alignment losses.

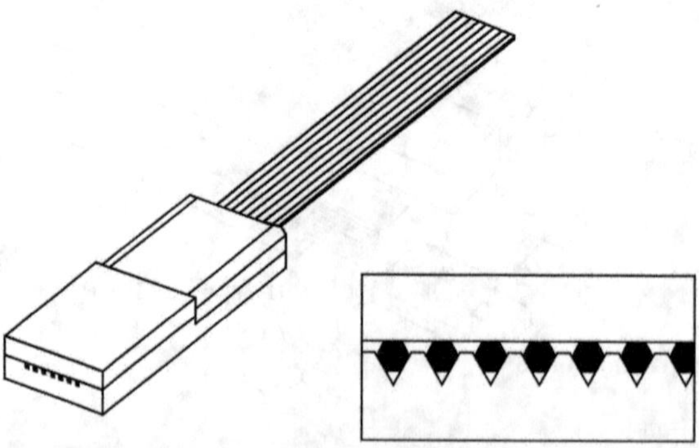

Figure 3.22 Schematic representation of a linear fiber bundle positioned within a V-groove substrate. The inset illustrates the cross-sectional view showing the V-shaped alignment structure that ensures precise positioning and uniform spacing of individual fibers.

and fiber array assemblies to achieve high coupling precision and mechanical stability.

3.8.5 Fiber Bundle for Imaging Applications

The use of optical fiber bundles in imaging systems represents a powerful intersection of photonics and precision engineering [85–89]. Unlike conventional optical paths that rely solely on lenses and mirrors for image relay, fiber bundles enable flexible and compact image transmission channels capable of preserving spatial information across significant distances or through constrained geometries. This capability has made them indispensable in applications where conventional imaging setups are impractical or physically inaccessible, such as in medical diagnostics, industrial inspection, and remote sensing. Each fiber within the bundle functions as a discrete light conduit, collectively forming a coherent array that reproduces the spatial intensity distribution of the object under observation.

At the heart of such systems lies the principle of pixelated optical transmission; the image formed at the input face of the bundle is decomposed into thousands of individual optical channels—each corresponding to a fiber core—that transmit light independently. Provided that the fiber bundle maintains a fixed spatial relationship between its input and output faces, the transmitted image can be reconstructed with high fidelity. The overall performance of the system thus depends critically on the fiber core uniformity, alignment precision, and the numerical aperture of the individual fibers. Any imperfections, such as crosstalk or core misalignment, can degrade resolution and contrast, underscoring the importance of meticulous bundle fabrication and assembly.

Beyond the fiber bundle itself, the integration of imaging and relay optics further refines the system's performance. The imaging optics capture the object and project its image onto the proximal face of the bundle, while the relay optics on the distal side refocus the transmitted light onto a detector or eyepiece. This modular configuration allows for scalable system designs adaptable to various imaging conditions and wavelength ranges. Moreover, the flexibility and compactness of fiber bundles enable imaging in confined or curved environments, extending the reach of optical systems to domains previously inaccessible.

The following figures illustrate representative configurations of such imaging systems. Figure 3.23 depicts the fundamental concept of fiber-bundle-based image relay, highlighting how optical information is transmitted from the object plane to a detector via the bundle. Building upon this foundation, Figure 3.24 presents a practical implementation in the form of a fiber-optic endoscope, demonstrating how these principles are employed in minimally invasive medical imaging to visualize internal structures with high precision and flexibility.

As illustrated in Figure 3.23, the imaging system employs a fiber bundle to relay optical information from the object plane to the detector. The imaging optics first form an intermediate image of the object at the input face of the fiber bundle, where each fiber core acts as an individual light channel. The image is then transmitted through the bundle to the output face, maintaining spatial correspondence across fibers. Relay optics capture and refocus the transmitted light onto the camera sensor, enabling digital acquisition of the optical signal. This configuration is particularly useful in applications requiring flexible or remote imaging, such as endoscopy, where direct line-of-sight imaging is not feasible.

As illustrated in Figure 3.24, the fiber-optic endoscope comprises a compact light source connected to a flexible insertion tube that transmits both illumination and image data. The distal end contains a light source bundle for delivering light into the cavity, an objective lens system for focusing the reflected image onto the image bundle, and an aspiration channel for fluid or debris removal during the procedure. The proximal end features an eyepiece that enables direct observation of the illuminated field. This arrangement allows

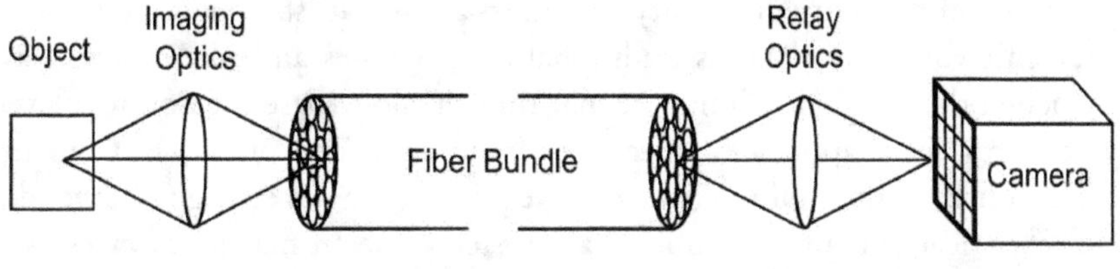

Figure 3.23 Schematic representation of an optical fiber bundle imaging system. The object is imaged through the imaging optics onto the proximal face of the fiber bundle, which transmits the image to its distal end. The relay optics then project the transmitted image onto a camera sensor for acquisition and analysis.

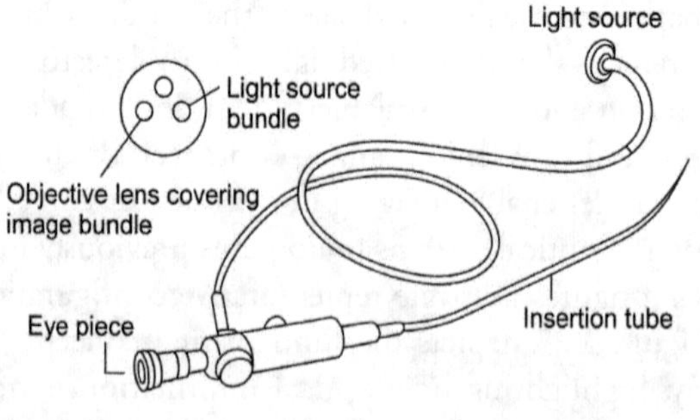

Figure 3.24 Schematic representation of a fiber-optic endoscope showing its main components. The device consists of a light source unit connected via a flexible insertion tube that houses the light source bundle, objective lens covering image bundle, and an aspiration channel. The distal end of the insertion tube provides illumination and image transmission, while the proximal end includes an eyepiece for direct visualization by the operator.

clinicians to visualize internal anatomical structures in real time with minimal invasiveness, demonstrating the integration of optical, mechanical, and ergonomic design principles in modern medical instrumentation.

3.9 Conclusion

The journey that led to the birth and global proliferation of optical fiber technology is a remarkable testament to the power of interdisciplinary thinking, scientific perseverance, and inspired engineering. What began with a simple observation of light traveling in a stream of water evolved into one of the most transformative technologies of the modern world. In 1842, Jean-Daniel Colladon's demonstration of the "light fountain" in Geneva—where a beam of light followed the curve of a falling water stream—captured the imagination of many and planted an early seed. Though it was a demonstration of TIR, the implications of this phenomenon would remain largely dormant for over a century. It took the right convergence of physics, mathematics, and materials science to translate that spectacle into a practical, scalable means of transmitting information.

In the mid-twentieth century, engineers and scientists sought better ways to transmit voice and data, as traditional copper wires and microwave relays approached their capacity limits. At that time, analogy-based reasoning played a pivotal role in shaping new ideas: researchers mapped principles from the familiar source domain of microwave waveguides onto the emerging target domain of optical guidance. In microwave systems, electromagnetic waves were confined within hollow metallic tubes, but this physical implementation could not be directly transferred to the optical regime because metals are highly lossy at visible and near-infrared frequencies. The creative leap, therefore, was to reinterpret the concept of guided waves—previously realized through metallic boundaries—as a guiding mechanism achieved by TIR in dielectric materials. This analogy allowed scientists to envision glass not as a passive window but as an engineered medium capable of directing light, just as a waveguide channels radio waves, though governed by entirely different physical constraints. It was within this framework of analogical mapping between domains that Charles Kao and George Hockham proposed a radical idea: that if impurities could be sufficiently reduced in glass, it might be possible to transmit light over kilometers rather than just a few meters. In their groundbreaking 1966 paper, they argued that attenuation levels of 20 dB/km or less would be enough to make optical fiber communication viable—an idea initially met with skepticism but later proven prophetic [1].

Turning theory into practice required not just vision but innovation in materials and manufacturing. Glass, long prized for its optical properties, had to be reengineered at a molecular level to achieve the purity and uniformity needed to meet Kao's benchmark. In 1970, researchers at Corning, led by Donald Keck, Peter Schultz, Robert Maurer, and Frank Zimar—succeeded in producing the first fiber with attenuation below 20 dB/km, using doped silica prepared through vapor-phase deposition. This single breakthrough turned skepticism into momentum. What followed was a wave of innovation: improvements in fiber geometry to reduce modal dispersion, developments in splicing and connector technologies to minimize insertion loss, and the creation of semiconductor lasers that could efficiently couple into fiber cores. Each of these advancements was necessary, and each built upon the foundational insight that light, like water or electrons, could be guided with precision and stability.

As the capabilities of optical fibers grew, so too did the ambition of their applications. Long-distance telephony was among the first beneficiaries, followed quickly by undersea cables and broadband internet infrastructure. The 1990s saw an explosion in fiber deployment, propelled by the rise of the internet and the realization that optical channels could be multiplexed to carry enormous amounts of data. DWDM allowed dozens, even hundreds, of data streams to coexist within a single fiber, vastly increasing bandwidth without requiring new physical infrastructure. Optical amplification, especially via erbium-doped fiber amplifiers, extended reach without repeated optical-electrical-optical conversions, making transcontinental and transoceanic communication truly seamless.

Today, fiber-optic cables span the globe, forming the backbone of the internet, powering cloud computing, and supporting global commerce, education, and entertainment. More than a million kilometers of fiber are installed each year, silently and reliably carrying light-encoded information across continents. Yet the story does not end here. Research continues into fibers capable of handling high-power laser applications, new geometries like hollow-core fibers for ultralow latency, and quantum-compatible fibers for secure communications using entangled photons. Optical fibers are also finding renewed roles in sensing applications, from structural health monitoring in bridges and aircraft to precision temperature and pressure sensing in oil wells and medical diagnostics.

What began as a curious dance of light in water has become a cornerstone of the digital age. The success of optical fibers is not merely a feat of materials or mechanics but a triumph of analogy refined by scientific clarity and driven by technological necessity. It reminds us that new paradigms often begin with familiar ideas viewed through an unfamiliar lens—and that, with persistence, even something as intangible as a beam of light can be captured, shaped, and made to serve the needs of an increasingly interconnected world.

References

[1] Kao, K., and G. Hockham, "Dielectric-Fibre Surface Waveguides for Optical Frequencies," *Proceedings of the Institution of Electrical Engineers*, Vol. 113, No. 7, 1966, pp. 1151–1158, https://doi.org/10.1049/piee.1966.0189.

[2] Hecht, J. et al., "Illuminating the Origin of Light Guiding," *Optics & Photonics News*, Vol. 10, No. 10, 1999, p. 26, https://doi.org/10.1364/OPN.10.10.000026.

[3] Ghatak, Ajoy, et al., "The Optical Fiber: A Historical Perspective," *Journal of Optics*, Vol. 33, No. 3, 2004, pp. 131–143, https://doi.org/10.1007/BF03354759.

[4] Gaebelein, J., et al., "In Colladon's Footsteps: Water Jet Laser Evolution MMXX–Applying Modern IR Fiber Lasers to Cut Challenging Applications," *Physik in unserer Zeit*, 2020, https://doi.org/10.1002/phvs.202000033.

[5] Wheeler, W., "Apparatus for Lighting Dwellings or Other Structures," 1881, United States Patent.

[6] Hecht, Jens, *City of Light: The Story of Fiber Optics*, Oxford, UK: Oxford University Press, 2004.

[7] Berci, G., et al., "History of Endoscopy: What Lessons Have We Learned from the Past?" *Surgical Endoscopy*, Vol. 14, No. 1, 2000, pp. 5–15, https://doi.org/10.1007/s004649900002.

[8] van Heel, A. C. S., et al., "A New Method of Transporting Optical Images without Aberrations," *Nature*, Vol. 173, No. 4392, 1954, p. 39, https://doi.org/10.1038/173039a0.

[9] Maciolek, K. A., et al., "A Race to Imaging Revolution: Pioneers in Fiber Optics," *Urology*, Vol. 112, 2018, pp. 52–55, https://doi.org/10.1016/j.urology.2017.10.027.

[10] Hopkins, H. H., et al., "A Flexible Fibrescope, Using Static Scanning," *Nature*, Vol. 173, No. 4392, 1954, pp. 39–41, https://doi.org/10.1038/173039b0.

[11] Hopkins, H. H., et al., "Transparent Fibres for the Transmission of Optical Images," *Optica Acta*, Vol. 1, No. 4, 1955, pp. 164–170, https://doi.org/10.1080/713818685.

[12] Hirschowitz, B. I., et al., "Demonstration of a New Gastroscope, 'the Fiberscope'," *Gastroenterology*, Vol. 35, No. 1, 1958, pp. 50–53, https://doi.org/10.1016/S0016-5085(19)35579-9.

[13] Hirschowitz, B. I., et al., "Endoscopic Examination of the Stomach and Duodenal Cap with the Fiberscope," *The Lancet*, Vol. 277, No. 7188, 1961, pp. 1074–1078, https://doi.org/10.1016/S0140-6736(61)92308-X.

[14] Hirschowitz, B. I., et al., "A Personal History of the Fiberscope," *Gastroenterology*, Vol. 76, No. 4, 1979, pp. 864–869, https://doi.org/10.1016/S0016-5085(79)80190-0.

[15] Hirschowitz, B., and L. E. Curtiss, 1961, *Flexible Light Transmitting Tube*, US Patent 3010357.

[16] Kapany, N. "Fiber Optics," *Scientific American*, Vol. 203, No. 5, 1960, pp. 72–81.

[17] Nair, R. V., et al., "Transmitting Photons for Humanity: Scientific Contributions of N. S. Kapany," *Resonance*, Vol. 26, No. 11, 2021, pp. 1489–1498, https://doi.org/10.1007/s12045-021-1257-2.

[18] Snitzer, E., et al., "Cylindrical Dielectric Waveguide Modes," *Journal of the Optical Society of America*, Vol. 51, No. 5, 1961, pp. 491–498, https://doi.org/10.1364/JOSA.51.000491.

[19] Snitzer, E., et al., "Observed Dielectric Waveguide Modes in the Visible Spectrum," *Journal of the Optical Society of America*, Vol. 51, No. 5, 1961, pp. 499–505, https://doi.org/10.1364/JOSA.51.000499.

[20] Snitzer, E., et al., "Proposed Fiber Cavities for Optical Masers," *Journal of Applied Physics*, Vol. 32, No. 1, 1961, pp. 36–39, https://doi.org/10.1063/1.1735955.

[21] Koester, C. J., et al., "Amplification in a Fiber Laser," *Applied Optics*, Vol. 3, No. 10, 1964, pp. 1182–1186, https://doi.org/10.1364/AO.3.001182.

[22] Southworth, G. C., "Hyper-Frequency Wave Guides—General Considerations and Experimental Results," *Bell System Technical Journal*, Vol. 15, No. 2, 1936, pp. 284–309, https://doi.org/10.1002/j.1538-7305.1936.tb00733.x.

[23] Page, L., et al., "Electromagnetic Waves in Conducting Tubes," *Physical Review*, Vol. 52, No. 6, pp. 647–651, https://doi.org/10.1103/PhysRev.52.647.

[24] Borner, M., 1966, *Electro-Optical Transmission System Utilizing Lasers*, US Patent 3845293.

[25] Okamoto, K., *Fundamentals of Optical Waveguides, 2nd ed.*, Amsterdam: Elsevier, 2006.

[26] Keiser, G., et al., "Review of Diverse Optical Fibers used in Biomedical Research and Clinical Practice," *Journal of Biomedical Optics*, Vol. 19, No. 8, August 2014, p. 080902, https://doi.org/10.1117/1.JBO.19.8.080902.

[27] Kao, C. K., et al., "Spectrophotometric Studies of Ultra Low Loss Optical Glasses I: Single Beam Method," *Journal of Physics E: Scientific Instruments*, Vol. 1, No. 11, 1968, pp. 1063–1068, https://doi.org/10.1088/0022-3735/1/11/303.

[28] Keck, D. B., et al., "On the Ultimate Lower Limit of Attenuation in Glass Optical Waveguides," *Applied Physics Letters*, Vol. 22, No. 7, 1973, pp. 307–309, https://doi.org/10.1063/1.1654649.

[29] Schultz, P. C., et al., "Preparation and Properties of Low-Loss Glass Optical Waveguides," *Journal of Solid State Chemistry*, Vol. 12, No. 3, 1975, p. 176, https://doi.org/10.1016/0022-4596(75) 90302-3.

[30] MacChesney, J. B., et al., "A New Technique for the Preparation of Low-Loss and Graded-Index Optical Fibers," *Proceedings of the IEEE*, Vol. 62, No. 9, 1974, pp. 1278–1279, https://doi.org/10.1109/PROC.1974.9608.

[31] Nagel, S. R., et al., "An Overview of the Modified Chemical Vapor Deposition (MCVD) Process and Performance," *IEEE Journal of Quantum Electronics*, Vol. 18, No. 4, 1982, pp. 459–476, https://doi.org/10.1109/JQE.1982.1071596.

[32] MacChesney, J. D., "MCVD: Its Origin and Subsequent Development," *IEEE Journal of Selected Topics in Quantum Electronics*, Vol. 6, No. 6, 2000, pp. 1305–1306, https://doi.org/10.1109/2944. 902182.

[33] Senior, J. M., *Optical Fiber Communications: Principles and Practice (2nd ed.)*, Prentice Hall, 1992.

[34] Blankenship, M. G., "Current Status of Outside Vapor Deposition: Process and Performance," *Optical Fiber Communication (OSA Technical Digest)*, Vol. 1982, p. TuCC1, https://doi.org/10.1364/OFC.1982.TuCC1.

[35] Graham, G. M., et al., "Experimental Study of the Outside Vapor Deposition Process," *Aerosol Science and Technology*, Vol. 14, No. 1, 1991, pp. 69–76, https://doi.org/10.1080/02786829108959514.

[36] He, J., et al., "Numerical Study on Inlet Operational Conditions in the Outside Vapor Deposition Process," *Powder Technology*, Vol. 411, 2022, p. 117915, https://doi.org/10.1016/j.powtec.2022.117915.

[37] Kapron, F. P., et al., "Radiation Losses in Glass Optical Waveguides," *Applied Physics Letters*, Vol. 17, No. 10, 1970, pp. 423–425, https://doi.org/10.1063/1.1653255.

[38] Nishizawa, J., et al., "Focusing Diffused Waveguides," *Opto-Electronics*, Vol. 5, 1973, pp. 309–321, https://doi.org/10.1007/BF02057130.

[39] Gloge, D., et al., "Multimode Theory of Graded-Core Fibers," *The Bell System Technical Journal*, Vol. 52, No. 9, 1973, pp. 1563–1578, https://doi.org/10.1002/j.1538-7305.1973.tb02033.x.

[40] Gloge, D., "Bending Loss in Multimode Fibers with Graded and Ungraded Core Index," *Applied Optics*, Vol. 11, No. 11, 1972, pp. 2506–2513, https://doi.org/10.1364/AO.11.002506.

[41] Kawakami, S., et al., "An Optical Waveguide with the Optimum Distribution of the Refractive Index with Reference to Waveform Distortion," *IEEE Transactions on Microwave Theory and Techniques*, Vol. 16, No. 10, 1968, pp. 814–818, https://doi.org/10.1109/TMTT.1968.1126797.

[42] Blyler, L. L., et al., "Fiber Drawing and Coating: An Overview," *OSA Technical Digest (Optical Fiber Communication Conference 1983)*, p. TuF1, https://doi.org/10.1364/OFC.1983.TuF1.

[43] Jaluria, Y., et al., *Manufacture of Optical Fibers: Drawing and Coating Processes*, Springer, 2018, https://doi.org/10.1007/978-3-319-76983-78.

[44] Basch, E. E. B., et al., "Optical Transmission for Interoffice Trunks," *IEEE Transactions on Communications*, Vol. 26, No. 7, 1978, pp. 1007–1014, https://doi.org/10.1109/TCOM.1978.1094181.

[45] Maione, T. L., et al., "Experimental Fiber Optic Transmission System for Interoffice Trunks," *IEEE Transactions on Communications*, Vol. COM-25, No. 5, 1977, pp. 517–523, https://doi.org/10.1109/TCOM.1977.1093853.

[46] Schwartz, M. I., et al., "Atlanta Fiber System Experiment: The Chicago Lightwave Communications Project," *Bell System Technical Journal*, Vol. 57, No. 6, 1978, pp. 1881–1888, https://doi.org/10.1002/j.1538-7305.1978.tb00132.x.

[47] Cocito, G., et al., "COS 2 Experiment in Turin: Field Test on an Optical Cable in Ducts," *IEEE Transactions on Communications*, Vol. 26, No. 7, 1978, pp. 1028–1036, https://doi.org/10.1109/TCOM.1978.1094185.

[48] Cocito, G., "A National Review on Fiber and Integrated Optics Activity in Italy," in *Fibre and Integrated Optics Activity in Italy*, Springer, 1978, https://doi.org/10.1007/978-1-4613-2937-4 21.

[49] Smith, R. G., "Optical Power Handling Capacity of Low Loss Optical Fibers as Determined by Stimulated Raman and Brillouin Scattering," *Applied Optics*, Vol. 11, No. 11, 1980, pp. 2489–2494, https://doi.org/10.1364/AO.11.002489.

[50] Mears, R. J., et al., "Low-Noise Erbium-Doped Fibre Amplifier Operating at 1.54 μm," *Electronics Letters*, Vol. 23, No. 19, 1987, pp. 1026–1028, https://doi.org/10.1049/el:19870719.

[51] Desurvire, E., et al., "Amplification of Spontaneous Emission in Erbium-Doped Single-Mode Fibers," *Journal of Lightwave Technology*, Vol. 7, No. 5, 1989, pp. 835–845, https://doi.org/10.1109/50.19124.

[52] Morkel, P. R., et al., "Theoretical Modeling of Erbium-Doped Fiber Amplifiers with Excited-State Absorption," *Optics Letters*, Vol. 14, No. 19,1989, pp. 1062–1064, https://doi.org/10.1364/OL.14.001062.

[53] Desurvire, E., et al., "High-Gain Erbium-Doped Traveling-Wave Fibre Amplifier," *Optics Letters*, Vol. 12, No. 11, 1987, pp. 888–890, https://doi.org/10.1364/OL.12.000888.

[54] Desurvire, E., *Erbium-Doped Fiber Amplifiers: Principles and Applications*, Hoboken, NJ: Wiley-Interscience, 1994.

[55] Ma, X., et al., "Global Submarine Cable Network and Digital Divide," *Journal of Geographical Sciences*, Vol. 35, No. 6, 2025, pp. 1204–1232, https://doi.org/10.1007/s11442-025-2364-x.

[56] Perin, J. K., et al., "Importance of Amplifier Physics in Maximizing the Capacity of Submarine Links," *Journal of Lightwave Technology*, Vol. 37, No. 9, 2019, pp. 2076–2086, https://doi.org/10.1109/JLT.2019.2897831.

[57] Liu, T.- G., et al., "Complete Characterization of Polarization-Maintaining Fibers using Distributed Polarization Analysis," *Journal of Lightwave Technology*, Vol. 32, No. 11, 2014, pp. 1997–2004, https://doi.org/10.1109/JLT.2014.2377091.

[58] Bhagavatula, V. A., et al., "Dispersion-Shifted Segmented-Core Single-Mode Fibers," *Optics Letters*, Vol. 9, No. 5, 1984, pp. 186–188, https://doi.org/10.1364/OL.9.000186.

[59] Prabu, R. T., et al., "Various Graded Index Plastic Optical Fiber Performance Signature Capability with the Optimum Dispersion Control for Indoor Coverage Applications,"

Journal of Optical Communications, Vol. 46, No. 2, 2025, pp. 299–306, https://doi.org/10.1515/joc-2024-0069.

[60] Knight, J. C., et al., "All-Silica Single-Mode Optical Fiber with Photonic Crystal Cladding," *Optics Letters*, Vol. 21, No. 19, 1996, pp. 1547–1549, https://doi.org/10.1364/OL.21.001547.

[61] Birks, T. A., et al., "Endlessly Single-Mode Photonic Crystal Fiber," *Optics Letters*, Vol. 22, No. 13, 1997, pp. 961–963, https://doi.org/10.1364/OL.22.000961.

[62] Richardson, D. J., et al., "Space-Division Multiplexing in Optical Fibres," *Nature Photonics*, Vol. 7, No. 5, 2013, pp. 354–362, https://doi.org/10.1038/nphoton.2013.94.

[63] Li, G., et al., "Space-Division Multiplexing: The Next Frontier in Optical Communication," *Advances in Optics and Photonics*, Vol. 6, No. 4, 2014, pp. 413–487, https://doi.org/10.1364/AOP.6.000413.

[64] Kinet, D., et al., "Fiber Bragg Grating Sensors Toward Structural Health Monitoring in Composite Materials: Challenges and Solutions," *Sensors*, Vol. 14, No. 4, 2014, pp. 7394–7419, https://doi.org/10.3390/s140407394.

[65] Wu, Q., et al., "Ultrasonic Structural Health Monitoring using Fiber Bragg Grating," *Sensors*, Vol. 18, No. 10, p. 3395, https://doi.org/10.3390/s18103395.

[66] Chang, G., et al., "Ultrafast Fiber Lasers: An Expanding Versatile Toolbox," *iScience*, Vol. 23, No. 5, 2020, p. 101101, https://doi.org/10.1016/j.isci.2020.101101.

[67] Rovera, A., et al., "Fiber Optic Sensors for Harsh and High Radiation Environments in Aerospace Applications," *Sensors*, Vol. 23, No. 5, 2023, p. 2512, https://doi.org/10.3390/s23052512.

[68] Girard, S., et al., "Radiation Hardening Techniques for Er/Yb Doped Optical Fibers and Amplifiers for Space Application," *Optics Express*, Vol. 20, No. 8, 2012, pp. 8457–8465, https://doi.org/10.1364/OE.20.008457.

[69] Borzycki, K., et al., "Hollow-Core Optical Fibers for Telecommunications and Data Transmission," *Applied Science*, Vol. 13, No. 19, 2023, p. 10699, https://doi.org/10.3390/app131910699.

[70] McGarry, C., et al., "Microstructured Optical Fibers for Quantum Applications: Perspective," *APL Quantum*, Vol. 1, No. 3, 2024, p. 030901, https://doi.org/10.1063/5.0211055.

[71] Kong, W., et al., "Enhanced Coexistence of Quantum Key Distribution and Classical Communication Over Hollow-Core and Multi-Core Fibers," *Entropy*, Vol. 26, No. 7, 2024, p. 601, https://doi.org/10.3390/e26070601.

[72] Chen, X.- Y., et al., "High-Fidelity, Low-Latency Polarization Quantum State Transmissions Over a Hollow-Core Conjoined-Tube Fibre at Around 800 nm," *Photonic Research*, Vol. 9, No. 5, 2021, pp. 1036–1044, https://doi.org/10.1364/PRJ.409521.

[73] Ahmed, K., *MicroLED Devices and Systems*, Norwood, MA: Artech House, 2024.

[74] Nakazawa, M., et al., "Rayleigh Backscattering Theory for Single-Mode Optical Fibers," *Journal Opt. Soc. Am.*, Vol. 73, No. 9, p. 1175–1180, https://doi.org/10.1364/JOSA.73.001175.

[75] Guan, X., et al., "Ultra-Dense Wavelength-Division Multiplexing with Microring Modulator," *Journal of Lightwave Technology*, Vol. 39, No. 13, 2021, pp. 4300–4306, https://doi.org/10.1109/JLT. 2021.3070515.

[76] Venkatesan, K., et al., "Design Analysis for Wavelength Division Multiplexing Technique in Optical Fibre Networks," *Multimed. Tools Appl.*, Vol. 83, No. 11, 2024, pp. 8279–8296, https://doi.org/10.1007/s12596-024-02319-6.

[77] Feng, H., et al., "Assembly Coupling Technology of Microlens Array and Optical Fiber Bundle," *Journal of Flow Visualization and Image Processing*, Vol. 32, No. 1, 2025, pp. 61–79, https://doi.org/10.1615/JFlowVisImageProc.2024053799.

[78] Yan, X., et al., "Design of Microlens Arrays Coupling with Imaging Fiber Bundle," *Optical Engineering*, Vol. 52, No. 7, 2013, p. 071301, https://doi.org/10.1007/s11801-013-3016-4.

[79] Sawyer, T. W., et al., "Tolerancing the Alignment of Large-Core Optical Fibers, Fiber Bundles and Light Guides Using a Fourier Approach," *Applied Optics*, Vol. 56, No. 12, 2017, pp. 3303–3310, https://doi.org/10.1364/AO.56.003303.

[80] Sanders, T. M., et al., "Side-Polished Coherent Fiber Bundle Assemblies: A Pathway to Large Imaging Arrays," *Journal of Lightwave Technology*, Vol. 42, No. 19, 2024, pp. 6940–6947, https://opg.optica.org/jlt/ abstract.cfm?URI=jlt-42-19-6940.

[81] Patel, K., et al., "Fiber Bundle-Based Beam Tracking Approach for Free-Space Optical Communications Link Optimization," *Optical Engineering*, Vol. 63, No. 4, 2024, p. 041212, https://doi.org/10.1117/1.OE.63.4.041212.

[82] Pigram, P. J., et al., "Keyed Optical V-Fiber to Silicon V-Groove Interconnects," *Optical Engineering*, Vol. 33, No. 8, 1994, pp. 2594–2599, https://doi.org/10.1117/12.177113.

[83] Lo, J. C. C., et al., "Experimental Assessment of Passive Alignment of Optical Fibers with V-Grooves on Silicon Optical Bench," *Electronics Packaging Technology*, Vol. 6, No. 4, 2004, pp. 375–380, https://doi.org/10.1115/IMECE2003-43902.

[84] Gauthier, R. C., et al., "Self-Centering of a Ball Lens by Laser Trapping: Fiber-Ball-Fiber Coupling Analysis," *Applied Optics*, Vol. 42, No. 9, 2003, pp. 1610–1619, https://doi.org/10.1364/AO.42. 001610.

[85] Chang, Y., et al., "Compact High-Resolution Endomicroscopy Based on Fiber Bundles and Image Stitching," *Optics Letters*, Vol. 43, No. 17, 2018, pp. 4168–4171, https://doi.org/10.1364/OL.43. 004168.

[86] Orth, A., et al., "Optical Fiber Bundles: Ultra-Slim Light Field Imaging Probes," *Science Advances*, Vol. 5, No. 10, 2019, p. eaav1555, https://doi.org/10.1126/sciadv.aav1555.

[87] Shao, J., et al., "Fiber Bundle Image Restoration Using Deep Learning," *Optics Letters*, Vol. 44, No. 5, 2019, pp. 1080–1083, https://doi.org/10.1364/OL.44.001080.

[88] Liu, S., et al., "Fiber Bundle Optical System: Imaging Modeling and Analysis," *Optical Engineering*, Vol. 61, No. 11, 2022, p. 114106, https://doi.org/10.1117/1.OE.61.11.114106.

[89] Jiang, Z., et al., "Optical Fiber Bundle Differential Compressive Imaging," *Optics Letters*, Vol. 49, No. 9, 2024, pp. 2297–2300, https://doi.org/10.1364/OL.519161.

4

Photonic Crystals: From Biological Inspiration to Photonic Engineering

After reading this chapter, you should be able to:

- Explain how the analogy between electronic band gaps in semiconductors and optical band gaps in periodic dielectrics led to the concept of photonic crystals;
- Describe how advances in micro- and nanofabrication techniques enabled the realization of practical photonic crystal devices across different dimensions and applications;
- Discuss how natural photonic structures and emerging concepts in topology inspire ongoing innovations in photonic crystal research and technology.

4.1 Introduction

Following the framework of Chapter 1, the natural world serves here as a rich source domain (structural color in feathers, beetle shells, butterfly wings). The target domain is engineered control of light in devices. The crucial abstraction is that geometry at subwavelength scales, rather than chemistry, governs spectral response. This abstraction unlocks later mapping to periodic dielectric lattices and engineered photonic band structures.

Long before the clean room and the lithographic mask, long before lasers and silicon wafers, the study of how light interacts with structured materials had already begun—not in the realm of human invention, but in the silent elegance of the natural world. The iridescent flashes on a peacock's tail or the shifting, metallic blues of a Morpho butterfly's wing were, for centuries, phenomena of wonder rather than inquiry. In the seventeenth century, Robert Hooke, with his crude but revolutionary microscope, took one of the first analytical glimpses into this mystery. Peering into the fine structures of feathers and insect scales, he noted that their dazzling colors did not arise primarily from chemical dyes or traditional colorants, but from minute architectures that interact with light [1, 2]. Modern understanding refines this picture: in Morpho butterflies the color originates from nanoscale ridges and lamellae on the wing scales that act as multilayer interferometric structures; by contrast, in peafowl the iridescence arises from quasi-two-dimensional photonic architectures—periodic arrays of melanin rods (melanosomes) embedded in a keratin matrix—whose lattice spacing and refractive-index contrast set the reflected hue. In both cases, absorptive melanin can suppress incoherent scatter and enhance saturation, but it is the geometry of the photonic structure that governs the interference responsible for the vivid, angle-dependent colors.

As illustrated in Figure 4.1, the remarkable iridescent blue coloration of the Morpho butterfly does not arise from pigmentation but from intricate nanoscale architecture on its wing scales. Each scale comprises a series of parallel ridges, within which stacked lamellae form a periodic multilayer structure

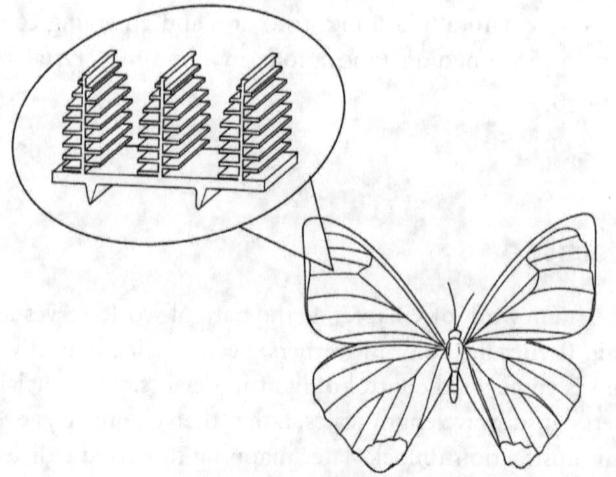

Figure 4.1 Schematic illustration of the Morpho butterfly and the nanostructured ridges observed in cross-sectional scanning electron microscope (SEM) images of its wing scales. The periodic multilayered structures act as natural photonic crystals, producing vivid structural colors through constructive interference of reflected light.

resembling a natural photonic crystal. When light impinges on these nanostructures, selective wavelengths undergo constructive interference due to multiple reflections within the layered system, while other wavelengths destructively interfere. The resulting optical effect manifests as an intense and angle-dependent blue hue. Minor variations in the layer spacing, refractive-index contrast, or viewing angle can modulate the reflected spectrum, giving rise to the shimmering color shifts characteristic of Morpho wings. Such naturally engineered nanostructures have inspired biomimetic designs in photonic materials and anticounterfeiting technologies.

As scientific tools improved and the theoretical framework of optics developed, the nineteenth century brought new clarity. Thomas Young's famous double-slit experiment [3], while best known for confirming the wave nature of light, also demonstrated the core principle that interference from periodic structures could generate colors without absorption-based colorants. When light waves of different wavelengths encounter surfaces with repeating structures—typically on the order of hundreds of nanometers—they can interfere constructively or destructively depending on several factors: the angle of incidence, the refractive indices of the incident medium and substrate, the refractive-index contrast among the structural elements themselves, and the specific geometry and periodicity of the structure. This interplay of geometry and refractive-index contrast determines which wavelengths are preferentially reflected or suppressed, giving rise to vivid iridescent colors. In systems such as the wings of Morpho butterflies, these interference effects produce an intense blue coloration that varies in brightness and spectral composition with viewing angle—sometimes shifting toward violet or showing desaturated tones rather than a change in hue [4–6]. What Young observed in principle, the natural world had already perfected through millions of years of evolution.

These naturally occurring optical structures are now recognized as some of the earliest and most refined examples of what we call structural color. Their properties stem not solely from the material's geometry, but from a combination of geometric arrangement and the intrinsic optical properties—specifically, the refractive indices—of their constituent materials. In some bird feathers, some beetle cuticles, and even opals, light is not primarily absorbed, as it is with dyes or colorants, but instead scattered, redirected, and filtered through a lattice of nanostructures whose refractive-index contrast determines the strength and spectral position of reflection. Within these architectures, successive interfaces and periodic variations in refractive index cause partial reflections and interference, selectively amplifying certain wavelengths while suppressing others. Together, these factors, geometry and refractive index, govern the interference patterns responsible for the brilliant, angle-dependent colors observed in nature.

It is, however, important to note that in many biological systems, absorbing materials and structural coloration act together. For instance, the striking blue-violet hues of Morpho butterflies originate from multilayered Christmas-tree-like ridges that produce interference effects, while underlying melanin layers absorb incoherent scattered light, thereby enhancing color saturation. Similarly, the iridescent feathers of peafowl result from periodic arrays of melanin rods embedded in keratin, with both structural and absorptive contributions combining to yield vivid hues [7]. Thus, while geometry governs the optical interference responsible for the hue and angular dependence, melanin and other absorbing materials modulate brightness and contrast, producing the saturated and stable colors we perceive.

This realization—that color can be controlled by structure, often in interplay with absorption by melanin and other absorbing materials—was a milestone in the understanding of natural optics; however, it did not directly lay the groundwork for modern photonics. The conceptual and technological foundations of photonic crystals emerged later from solid-state physics and semiconductor science in the late 1980s (see Section 4.2). The photonic community revisited biological interference phenomena predominantly in the late 1990s and early 2000s, when researchers such as Helen Ghiradella, Pete Vukusic, and Joanna Aizenberg leveraged these natural systems to inspire bioinspired photonic designs. In this sense, nature provided a profound analogy and enduring source of inspiration—not the direct origin—of photonic engineering.

4.2 The Analogy: Electrons and Photons

Here we perform explicit mapping between the periodic potential of electrons in solids and the periodic refractive index for photons. The abstraction is periodicity ⇒ band gaps, independent of the carrier (electron versus photon). The immediate insight is the possibility of a photonic bandgap and semiconductor-like functionality for light.

By the mid-twentieth century, the emergence of semiconductors had transformed the landscape of technology. At the heart of this revolution was a deceptively simple concept: the electronic bandgap. In a crystal lattice, electrons cannot occupy certain energy levels—these forbidden zones, known as bandgaps, form the foundation of devices like diodes, transistors, and all of modern electronics [8]. With this discovery came the ability to manipulate electron flow with precision, to allow or block current based on external signals, and ultimately to build the logical architecture behind computing. However, in the world of optics, no such parallel yet existed. Light, despite being a wave, could not be corralled in quite the same way. Traditional materials could bend it, reflect it, or absorb it, but they could not define a region in space where certain wavelengths of light were fundamentally prohibited from existing. That level of control, the kind semiconductors offered for electrons, remained elusive.

Figure 4.2 illustrates the conceptual analogy between electronic band structures in solids and photonic band structures in photonic crystals (PCs) [9]. Panel (a) represents the conventional energy dispersion relation for electrons in a periodic atomic potential, where allowed and forbidden energy regions form due to the interaction between the electron wave functions and the lattice periodicity. The resulting forbidden zone, or energy gap, separates the valence and conduction bands, governing the electrical and optical properties of the material.

Panel (b), in contrast, depicts the analogous behavior in a one-dimensional photonic crystal. Here, instead of electrons, photons experience periodic modulation in the refractive index, leading to photonic band formation and the emergence of a photonic bandgap. Within this frequency range, light propagation is prohibited, much like electrons are forbidden from occupying states within the electronic bandgap. The slope of the dispersion curves corresponds to the group velocity of light, and the boundary conditions at $k = \pm\pi/a$ (the edges of the Brillouin zone) determine the allowed photonic modes. This analogy forms the foundational framework for understanding how photonic crystals can control light propagation and confine electromagnetic energy through engineered periodic structures.

Then, in 1987, a transformative question was asked by Eli Yablonovitch, a physicist working at Bell Communications Research [10]. Could one create a semiconductor for light? Not metaphorically, but literally—a material in which light waves, like electrons in a solid, encounter a forbidden range of frequencies they cannot occupy. Yablonovitch envisioned a periodic dielectric structure, something crafted with a repeating pattern in its refractive index, much like the

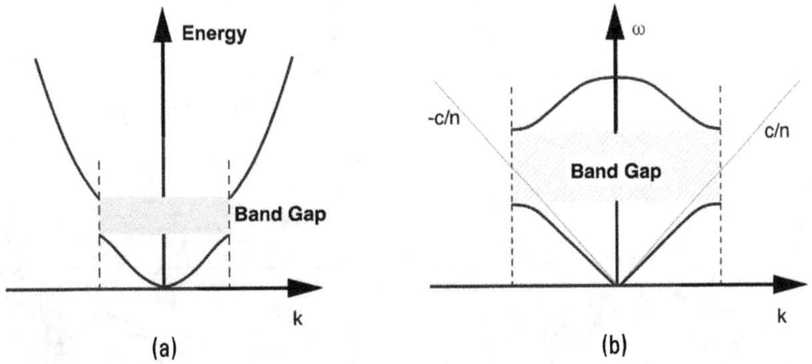

Figure 4.2 Comparison between (a) the electronic band structure of a crystalline solid, and (b) the photonic band structure of a one-dimensional photonic crystal. In both cases, the periodic potential or dielectric modulation introduces forbidden energy (or frequency) ranges, known respectively as the electronic energy gap and the photonic bandgap. This analogy highlights the correspondence between electron wave functions in solids and electromagnetic waves propagating through periodic dielectric media.

periodic potential that defines electronic behavior in atomic lattices. In such a structure, he reasoned, it should be possible to create a photonic bandgap: a spectral region where light, irrespective of direction or polarization, is simply not allowed to propagate. The implications were staggering. If successful, this would mean total control over the flow of light—not just bending it or reflecting it, but routing, trapping, or even suppressing it with engineered precision.

At nearly the same time, and working independently, physicist Sajeev John introduced a complementary idea [11]. Drawing inspiration from Anderson localization in disordered electronic systems [12], he proposed that structural disorder in a material could localize photons, preventing them from diffusing through space by forcing them into regions of constructive interference. While Yablonovitch sought control through periodicity, John sought confinement through randomness, but the underlying aim was the same: to manipulate photons with a fidelity previously reserved for electrons.

Together, their work opened the door to a new class of materials—photonic crystals—engineered structures that manipulate the flow of light through spatial periodicity or tailored disorder. By drawing a direct analogy between solid-state physics and optical engineering, they reframed our understanding of how light could be governed at the nanoscale. The term "photonic crystal" was coined to reflect this synthesis, encapsulating the idea that light, too, could be bound by a lattice, trapped in bandgaps, and harnessed with design principles once limited to electronics. This insight did more than bridge two fields—it launched an entirely new one, setting in motion decades of research that would ultimately reshape the future of communication, computation, and light-based technologies.

As shown in Figure 4.3, photonic crystals can be conceptualized as periodic arrangements of dielectric materials whose refractive indices vary spatially in 1D, 2D, or 3D. The 1D crystal, depicted on the left, typically represents multilayer thin-film stacks where the refractive index alternates along a single axis. The 2D lattice in the center extends this periodicity into a plane, forming

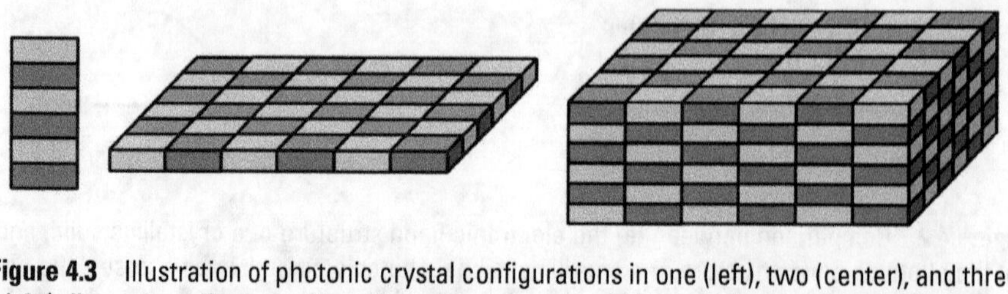

Figure 4.3 Illustration of photonic crystal configurations in one (left), two (center), and three (right) dimensions. Each configuration represents a periodic modulation of the refractive index, with light and dark shades indicating regions of higher and lower refractive index, respectively. The structures exemplify how dimensionality affects the periodic dielectric arrangement that governs the photonic bandgap properties.

structures such as arrays of dielectric rods or holes that confine and guide light in specific directions. The 3D crystal on the right exhibits full spatial periodicity, enabling complete photonic bandgaps that can inhibit light propagation in all directions. The contrast between light and dark regions visually emphasizes the modulation of refractive index, which is fundamental to controlling light–matter interactions in photonic systems.

4.3 From Concept to Reality

Scaling the idea to microwave Yablonovite exemplifies transfer from theory to prototype. Measurements provide validation. The discovery of a directional pseudogap, followed by redesign to a diamond-like lattice, demonstrates constraint matching and iterative refinement when geometry, symmetry, and index contrast constrain feasibility.

In the years immediately following the theoretical proposals for photonic bandgaps, excitement ran high, but so did the realization that the road to practical implementation would not be straightforward. The elegance of the idea, that light could be sculpted and constrained with the same mathematical tools that govern semiconductors, was undeniable. Yet translating this concept into a real, physical structure proved deeply challenging. The difficulty lay not in theory, but in fabrication. To create a three-dimensional photonic crystal with a bandgap in the visible or even near-infrared spectrum required manufacturing structures with submicron precision—something that, in the late 1980s, was simply beyond the reach of contemporary lithographic tools. The solution, as often happens in physics, came through a clever workaround.

To demonstrate the photonic bandgap in a tangible way, Eli Yablonovitch chose to scale up the concept. Rather than working at optical frequencies, he shifted the experiment into the microwave domain, where the corresponding wavelengths are orders of magnitude larger—on the scale of centimeters rather than nanometers. This meant that the required structural features, still governed by the same electromagnetic principles, could be fabricated using mechanical tools rather than clean-room nanofabrication. In 1989, along with his collaborator T. J. Gmitter, Yablonovitch built one of the earliest artificial photonic crystals by drilling a face-centered cubic lattice of air holes into a plastic block [13]. The lattice mimicked the regular arrangement of atoms in a crystal, with the air holes playing the role of low-dielectric regions. This structure, dubbed Yablonovite, became the first experimental demonstration of a material that exhibited a photonic bandgap—albeit at microwave frequencies.

Measurements confirmed that certain microwave frequencies were strongly attenuated by the structure, unable to propagate through it. This was a moment of profound validation. For the first time, it was possible to see the idea of a bandgap—so central to electronics—playing out in the behavior of light itself. However, as more rigorous analyses were conducted, it became clear that the initial design had limitations. Researchers such as K. M. Leung [14] examined the electromagnetic modes of the Yablonovite lattice in detail and found that the bandgap was directional. It existed only along certain crystallographic axes and allowed light to propagate along others. What had been observed was a pseudogap—a partial bandgap rather than a complete, omnidirectional one.

Rather than being a setback, this prompted further innovation. Yablonovitch and his team revisited the geometry of the lattice. They sought a configuration that would block all modes of light within a certain frequency range, regardless of direction. Inspired by the more complex topology of a diamond lattice, which has superior symmetry and uniformity in three dimensions, they designed a new structure. Intersecting air holes were drilled at tetrahedral angles, forming a lattice that more effectively prohibited wave propagation in all directions. By 1991, this refined design yielded what had eluded earlier efforts: a true, complete three-dimensional photonic bandgap. It was the clearest proof yet that the analogy to electronic semiconductors was more than metaphor; it was a framework that could be realized in matter and light alike.

4.4 Engineering the Crystal for Light

Moving from opals and inverse opals to high-index semiconductors (Si, GaAs) reflects constraint matching (index contrast, fabrication tolerance). The 1998 woodpile demonstration [15] is a decisive transfer into the infrared with experimental validation via spectroscopy.

As the theoretical foundation for photonic bandgaps grew more robust, attention quickly turned toward a far more daunting challenge: scaling these ideas down to wavelengths relevant to optical technologies. Demonstrating a photonic bandgap at microwave frequencies was an essential first step, but the ultimate goal was to control light in the visible and near-infrared spectrum, where photonic devices could interface with lasers, fiber optics, and next-generation optical circuits. Achieving this required more than clever geometry—it demanded submicron fabrication precision and materials capable of producing strong optical contrast at nanometer scales. These requirements pushed the limits of both materials science and nanotechnology.

Initial experimental efforts leaned on naturally occurring periodic structures, such as opals, which are formed by the self-assembly of silica spheres into close-packed lattices. These materials exhibit Bragg scattering and can produce vivid iridescence, a hint of underlying photonic behavior. Researchers attempted to exploit this natural periodicity by infiltrating opals with high-index materials or creating inverse opals—structures in which the voids between spheres were filled and the original template removed. While these techniques offered valuable insights and early visual demonstrations of structural color and light localization, the refractive-index contrast between silica and air, typically less than 1.5:1, was simply too low to open a complete photonic bandgap. Partial gaps were observed, but the full suppression of light propagation in all directions remained out of reach.

Progress required both materials with higher refractive indices and methods to pattern them with unprecedented precision. Silicon and gallium arsenide emerged as the leading candidates. These semiconductors, with refractive indices around 3.5 in the near-infrared, offered the high contrast necessary to reflect and trap light effectively within a periodic dielectric landscape. But their use introduced a new layer of complexity: fabricating intricate three-dimensional lattices in such materials required the adaptation of techniques like deep reactive ion etching, wafer bonding, and focused ion beam milling. The interplay between optical design and fabrication feasibility became a core theme of the field.

The field of photonic crystals took a monumental step forward in 1998 with the work of Shawn-Yu Lin and his colleagues at Sandia National Laboratories [16]. For decades, the concept of a photonic bandgap—a spectral range where light cannot propagate through a periodic dielectric structure—had remained mostly within the realm of theory and microwave-scale demonstrations. While the parallels to electronic bandgaps in semiconductors had sparked considerable interest, a persistent challenge remained: how to translate this elegant idea into a functional, scalable system that could operate at technologically relevant wavelengths. Lin's group was the first to bridge that gap convincingly. By fabricating a silicon-based three-dimensional photonic crystal with lattice constants on the order of a few microns, they moved photonic bandgap structures firmly into the infrared domain, precisely where modern optical communication technologies reside.

The structure they created was based on a woodpile architecture, in which layers of rods are stacked orthogonally in successive layers, mimicking the regularity of crystalline arrangements while introducing the vertical and lateral periodicity required for a full three-dimensional bandgap. Constructing such a geometry with micron-scale precision demanded a level of fabrication rigor previously reserved for advanced semiconductor devices. Using wafer bonding, planarization, and deep reactive ion etching, Lin's team was able to achieve the delicate balance of structural integrity and symmetry necessary for light manipulation. The experimental confirmation of a photonic bandgap using infrared spectroscopy provided more than just validation—it signaled a shift in the field from speculative physics to implementable photonic engineering [15].

This breakthrough was not merely a demonstration of theoretical principles; it opened the door to a new class of optical devices that leverage photonic bandgap effects for functionality. Suddenly, the possibility of guiding light through sharp bends without loss, trapping photons in subwavelength cavities, and switching optical signals without mechanical parts became attainable goals. The precision of semiconductor fabrication techniques, when combined with photonic crystal design principles, promised devices with unprecedented performance: zero-scattering waveguides, ultrahigh-Q resonators, and compact filters with engineered dispersion. The success of Lin's woodpile crystal thus did more than establish a milestone—it launched an era where the manipulation of photons, once governed by bulk optics and material interfaces, could be conducted with atomic-scale precision and architectural elegance.

4.5 2D Photonic Crystals: Compact and Practical

2D slab crystals map the 3D bandgap idea onto a planar platform, transfer functionality via engineered defects (line/point), and respect constraints of CMOS-compatible fabrication and integration. The first photonic crystal laser [17] exemplifies insight generation: new device architectures (ultrasmall cavities, low thresholds) that were not obvious from bulk optics.

While the early demonstrations of three-dimensional photonic crystals were celebrated for proving that complete photonic bandgaps could be engineered, their complexity made them less practical for widespread application. Fabricating intricate 3D geometries with submicron precision remained a significant challenge, even as nanofabrication techniques advanced. The sheer difficulty of aligning multiple layers and maintaining structural fidelity over a volume made mass production and integration into devices a formidable undertaking. As a result, much of the field's momentum shifted toward two-dimensional photonic crystals, which offered a more accessible path to functionality. These structures, formed by etching periodic patterns, typically holes, into thin slabs of high-index materials like silicon or indium phosphide, produced in-plane photonic bandgaps that were ideal for guiding and confining light within a single layer.

The simplicity of the two-dimensional geometry, when combined with the power of modern semiconductor processing, enabled a level of design flexibility and precision that had been hard to achieve in three dimensions. Engineers discovered that by introducing controlled imperfections into these lattices, such as omitting a single hole or creating a row of missing holes, they could form highly localized modes or tightly confined pathways for light. These intentional defects turned the periodic crystal into a functional photonic device. Line defects acted as waveguides, allowing light to travel around sharp bends with minimal loss, while point defects created optical microcavities with extraordinary confinement properties.

A defining moment came in 1999, when Oskar Painter and his team at Caltech demonstrated the first photonic crystal laser [17]. Using an InGaAsP semiconductor slab, they fabricated a two-dimensional photonic crystal with a carefully engineered defect that acted as a resonant cavity. Within this defect region, the photonic bandgap prohibited the escape of certain wavelengths of light, causing the light to bounce back and forth, amplifying with each pass. When combined with a gain medium embedded in the slab, this confinement led to the generation of coherent laser emission. What made this achievement so striking was not just the novelty of the laser's architecture, but its scale. The device was extraordinarily compact, a fraction of the size of conventional lasers, and it demonstrated that photonic crystals could be used to both guide and generate light at the microscale.

As fabrication techniques continued to improve, researchers pushed these designs to their physical limits. Photonic crystal cavities with quality factors exceeding a million and mode volumes approaching the theoretical minimum were soon realized. These cavities trapped light in spaces only a few optical wavelengths wide, dramatically enhancing the interaction between photons and matter. Such enhancement made them ideal for exploring quantum-optics phenomena, enabling strong coupling between single photons and quantum emitters like quantum dots or color centers [18]. At the same time, the small mode volumes reduced the threshold energy required for lasing, opening the door to ultralow power, on-chip light sources. What began as a pragmatic simplification of the 3D photonic crystal ideal had, in many ways, surpassed it—ushering in a generation of scalable, integrated, and tunable photonic devices tailored for the most demanding applications in classical and quantum information technologies [19].

Figure 4.4 presents two fundamental configurations of two-dimensional slab photonic crystals, which serve as building blocks for planar photonic devices. The left schematic depicts a square lattice composed of high-refractive-index dielectric pillars embedded in a low-index background such as air. This design primarily supports guided modes confined by total internal reflection, with the periodic index modulation leading to photonic bandgap effects for specific in-plane propagation directions. Conversely, the right structure represents a triangular lattice of air holes perforated into a high-index dielectric slab, a geometry widely used in integrated photonic circuits due to its superior bandgap symmetry and fabrication compatibility with semiconductor materials like silicon or gallium arsenide. In both cases, the spatial periodicity in two dimensions governs the control of light propagation, reflection, and localization, forming the basis for waveguiding, cavity formation, and slow-light phenomena in photonic crystal-based platforms.

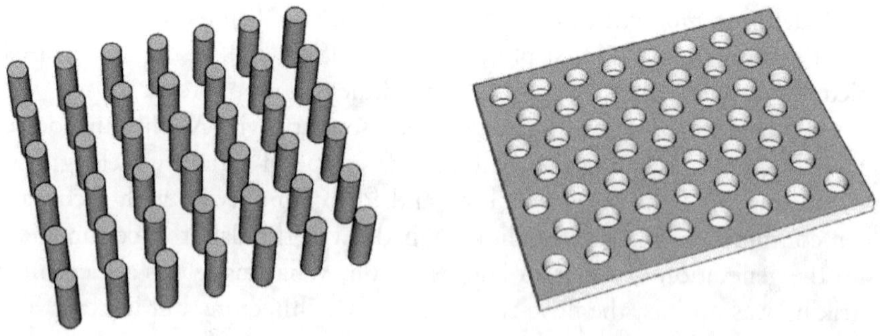

Figure 4.4 Schematic illustration of two-dimensional slab photonic crystal configurations: a square lattice of high-refractive-index (high-RI) dielectric pillars in air (left), and a triangular lattice of low-refractive-index (low-RI) air holes etched into a high-RI dielectric membrane (right). Both geometries exhibit in-plane periodic modulation of the refractive index, enabling photonic bandgap formation and light confinement within the slab plane.

4.6 The Fiber Revolution

Photonic crystal fibers embody transfer of bandgap guidance into cylindrical geometries. Constraint matching (low loss at mid-IR, bend flexibility, high-power delivery) motivates hollow-core designs. Clinical and industrial uses illustrate insight generation: new capabilities (flexible CO_2-laser delivery) unavailable with conventional TIR fibers [20]. This was discussed in Chapter 3 in the context of optical fibers.

In the mid-1990s, the world of fiber optics experienced a pivotal transformation when Philip Russell's group proposed an entirely new class of optical fibers [21–24]. Departing from the conventional model of light guidance through total internal reflection in a solid glass core, Russell introduced the concept of photonic crystal fibers—structures that relied on microstructuring the cladding of the fiber itself to control and confine light. The key innovation was the use of a two-dimensional photonic crystal arrangement, typically a regular lattice of air holes running along the length of the fiber, to manipulate the flow of light not merely through material interfaces, but through engineered band structures. This shift marked a fundamental rethinking of how light could be guided over long distances with precision, flexibility, and minimal loss.

Among the most striking variations of Russell's design was the hollow-core photonic crystal fiber. Instead of guiding light through a solid glass core, these fibers confined it within an air-filled central channel. Surrounding this hollow core was a cladding made from a periodic array of high-contrast dielectric layers, designed to exhibit a photonic bandgap. Much like the photonic crystals in planar devices, this cladding prevented specific wavelengths of light from escaping, forcing them to remain tightly bound within the core—even though that core contained no solid material. This approach was not only a conceptual breakthrough but also a practical one. Guiding light in air dramatically reduced scattering and absorption losses, particularly in spectral regions like the mid-infrared, where traditional glass materials become increasingly lossy. For wavelengths in the 10 to 11 micron range, including those emitted by CO_2 lasers, the performance gains were dramatic.

Figure 4.5 depicts the structural design of a photonic crystal fiber (PCF), a microstructured optical fiber that employs a periodic array of air holes running along its length to manipulate light propagation. As shown, the fiber features a solid silica core at the center, encircled by a cladding region patterned with a hexagonal lattice of air holes. This periodic microstructure gives rise to unique guiding mechanisms: in solid-core PCFs, light is confined by a modified form of total internal reflection due to the effective refractive index contrast between the core and the air-hole lattice; in contrast, hollow-core variants utilize photonic bandgap effects to guide light within the central air region. The magnified inset emphasizes the high regularity and precision of the air-hole arrangement, which critically influences modal dispersion, confinement loss, and single-mode behavior. Such control over light–matter interaction makes PCFs valuable for nonlinear optics, supercontinuum generation, and high-power or ultrafast laser delivery systems.

As illustrated in Figure 4.6, the structural design of PCFs represents a fundamental evolution from conventional optical fibers. Traditional fibers rely on a homogeneous glass core surrounded by a cladding with a slightly lower refractive index, enabling light confinement by total internal reflection. This simple two-layer geometry limits the tunability of dispersion, confinement loss, and mode area.

In contrast, photonic crystal fibers introduce a periodic arrangement of microscopic air holes running along the fiber length, forming a photonic lattice that drastically alters light propagation characteristics. The effective refractive

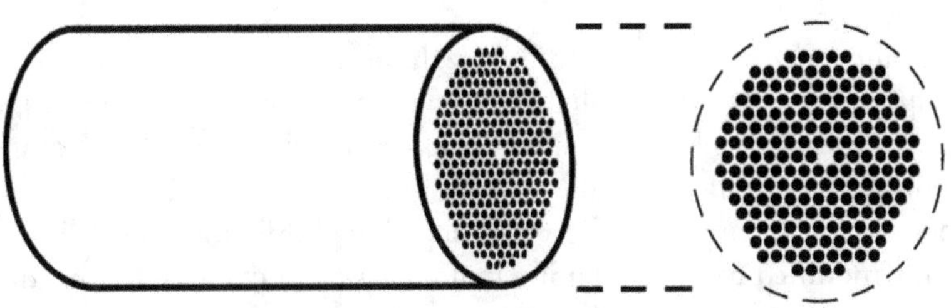

Figure 4.5 Schematic representation of a PCF structure. The figure illustrates the fiber in perspective (left), and cross-sectional (center) views, showing a solid core surrounded by a periodic array of air holes in the cladding region. The magnified inset (right) highlights the hexagonal arrangement of air holes that form the photonic crystal lattice, responsible for controlling light confinement and propagation through photonic bandgap or modified total internal reflection mechanisms.

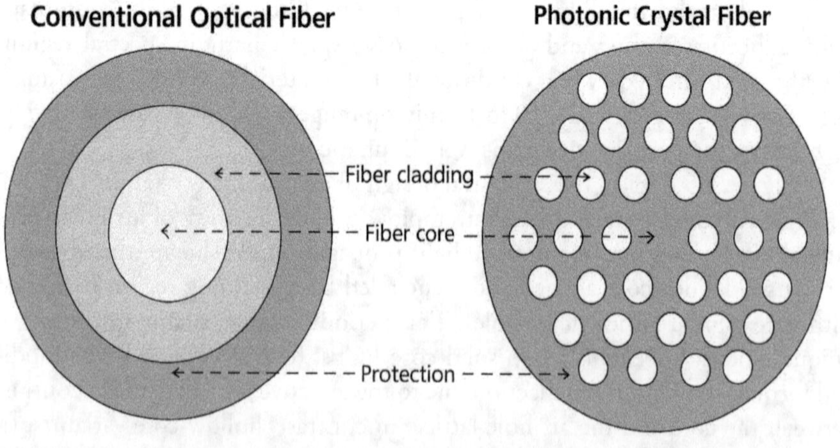

Figure 4.6 Comparison between a conventional optical fiber and a PCF. The conventional fiber consists of a solid glass core surrounded by cladding of slightly lower refractive index, guiding light through total internal reflection. In contrast, the photonic crystal fiber incorporates an array of air holes in the cladding region, creating a periodic microstructured pattern that enables light guidance through photonic bandgap or modified total internal reflection mechanisms.

index distribution can be engineered by adjusting the size, spacing, and arrangement of these air holes, offering unprecedented control over modal confinement, birefringence, and nonlinear effects. Depending on the design, light can be guided either by a modified form of total internal reflection (index-guiding PCFs), or through a photonic bandgap mechanism (bandgap guiding PCFs), where light is confined in a low-index core surrounded by a periodic microstructure that forbids certain optical modes from propagating in the cladding. This versatility enables PCFs to achieve single-mode operation over wide wavelength ranges, enhanced supercontinuum generation, and high-power transmission, extending their utility well beyond the limits of conventional fiber technologies.

As illustrated in Figure 4.7, PCFs exhibit a variety of architectures that enable fundamentally different light-guiding mechanisms compared with conventional optical fibers [25]. Panel (a) depicts an index-guiding PCF composed of a solid silica core surrounded by a regular lattice of air holes, which lowers the effective refractive index of the cladding. This configuration confines light through a modified form of total internal reflection and allows flexible dispersion engineering and large single-mode operation regions.

Panel (b) illustrates a hollow-core PCF, where guidance arises not from total internal reflection but from a two-dimensional photonic bandgap created by the periodic air-glass structure. In this case, the air core supports propagation within a spectral range where light is forbidden to leak into the cladding due to bandgap constraints. Such designs enable low-loss transmission and high-power

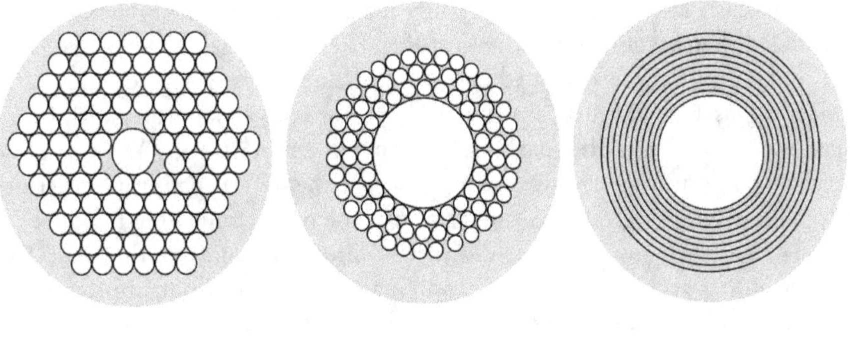

(a) (b) (c)

Figure 4.7 Representative geometries of PCFs: (a) an index-guiding PCF with a solid silica core surrounded by a periodic array of air holes acting as a reduced-index cladding; (b) a hollow-core PCF where light is confined within an air core through the photonic bandgap mechanism provided by the surrounding air-silica lattice; and (c) a concentric multilayer or Bragg-type PCF with alternating high- and low-index layers forming a radial photonic bandgap. Each configuration offers unique light-guiding characteristics, enabling control over dispersion, mode area, and confinement losses beyond the limitations of conventional optical fibers.

delivery in air- or gas-filled cores, which is particularly advantageous for nonlinear optics and spectroscopy applications.

Panel (c) shows a multilayer or Bragg-type PCF consisting of alternating concentric high- and low-index rings, effectively forming a radial photonic crystal. This geometry provides bandgap guidance through radial periodicity rather than a two-dimensional lattice, allowing for simpler fabrication and strong wavelength selectivity. Collectively, these PCF structures demonstrate the design flexibility of microstructured fibers, where geometric control of refractive index contrast at the subwavelength scale allows precise tailoring of modal, dispersion, and nonlinear properties, expanding optical fiber functionality far beyond traditional limits.

The promise of this technology was soon translated into real-world applications. One of the most transformative came through the efforts of OmniGuide, a company spun out of research at MIT. OmniGuide recognized the potential of hollow-core photonic crystal fibers for delivering infrared laser light with unprecedented precision and flexibility. By designing fibers specifically for CO_2 laser wavelengths, around 10.6 microns, they enabled the transmission of high-power, mid-infrared light through thin, flexible fibers, something that had long eluded conventional optics. This capability proved to be a game changer in the medical field. Traditional rigid optics for infrared surgery were limited in maneuverability and application, especially in minimally invasive procedures. With OmniGuide's fibers, surgeons could perform endoscopic operations using powerful CO_2 lasers, reaching previously inaccessible regions of the body with controlled thermal ablation and minimal collateral damage.

The implications extended beyond medicine. These fibers found roles in industrial laser machining, sensing, and spectroscopy, wherever mid-infrared light was needed in tight spaces or challenging environments. Yet it was the medical use—turning a physics concept from theoretical bandgap structures into a life-saving surgical tool—that stood as the most compelling narrative.

What began as an academic insight into photonic confinement had evolved into a technology that not only advanced science but improved and saved lives, illustrating with clarity how abstract theoretical physics could become a direct agent of healing and human benefit.

4.7 Returning to Nature: Biomimicry

Biological structures (e.g., Morpho) are the source domain. Optical function (mapping) is identified in ridge/lamellae geometry. Fabricated replicas via self-assembly and nanoimprint constitute transfer, with optical measurements as validation of structural color performance [7, 26].

In a poetic twist of scientific progress, photonic crystals eventually circled back to their origins—not in the realm of clean rooms or simulation software, but in the iridescent shimmer of a butterfly's wing. Among the most iconic examples is the Morpho butterfly, whose vibrant blue hues have captivated naturalists and scientists alike for generations. These colors arise primarily from a complex, hierarchical structure of nanoscale ridges and lamellae on its wing scales, which act as photonic multilayers producing interference and diffraction effects. Importantly, however, underlying melanin layers absorb stray light, enhancing saturation and contrast. The Morpho's wing, therefore, combines geometric and absorptive optical elements in a subtle synergy—a pattern that recurs widely across biological photonic systems, including peacock feathers and beetle shells [7, 26]. In effect, such biological materials are hybrid photonic-absorptive systems, evolved for both chromatic richness and optical robustness.

The idea that such naturally occurring structures could provide a roadmap for advanced optical materials had long been recognized. Yet it wasn't until the 2010s that researchers began to fully translate the intricate features of biological photonic systems into engineered analogs. Pioneering work by scientists such as Helen Ghiradella (State University of New York (SUNY Albany)), Joanna Aizenberg (Harvard University), and Pete Vukusic (University of Exeter) played a pivotal role. These researchers not only studied the morphological features of Morpho scales in extraordinary detail, using electron microscopy and spectroscopy to understand the interplay of geometry, refractive-index contrast, and absorption by melanin or other absorbing materials, but they also began replicating them using advanced fabrication techniques. The goal was not simply to mimic the aesthetics of the butterfly, but to harness the underlying photonic principles for practical applications.

Through techniques such as self-assembly, nanoimprinting, and layer-by-layer deposition, the teams of Ghiradella, Aizenberg, and Vukusic created or inspired synthetic surfaces that reproduced the intense, angle-dependent colors of the Morpho wing [27–30]. These artificial structures exhibited striking iridescence and durability, outperforming traditional colorants that tend to fade over time or degrade under UV exposure. Because the color arose primarily from geometry, with optional absorptive backing layers for tuning contrast, it remained vivid, stable, and environmentally resilient. This development opened new doors for creating colorfast materials in everything from textiles to automotive finishes, where long-lasting vibrancy is highly desirable.

Beyond aesthetics, the precision and uniqueness of these structural colors offered powerful new tools in the fight against counterfeiting. Inks and coatings based on bioinspired photonic crystals could be manufactured with complex, difficult-to-replicate nanoscale features that reflect light in distinct patterns—making them ideal for security printing, currency protection, and brand authentication. Since even slight deviations in structure alter the optical response, these coatings act like optical fingerprints—virtually impossible to forge without access to the same nanoscale fabrication capabilities.

The connection to e-readers and display technology was another promising frontier [26]. Photonic crystal coatings capable of reflecting specific wavelengths while remaining transparent to others could be tuned to create nonemissive displays with high contrast, readability in sunlight, and minimal energy use—attributes long sought after in electronic paper technologies.

In recent years, the principles of biomimicry in photonic structures have extended far beyond coloration and displays. For instance, bioinspired gas and vapor sensors exploit the same structural color mechanisms found in Morpho wings: minute changes in refractive index caused by gas adsorption alter the reflected spectrum, enabling sensitive and rapid detection of volatile compounds [31, 32]. Similarly, bioinspired strategies for enhancing LED emission have drawn from the nano-architectures of firefly lanterns and butterfly scales, where corrugated or layered geometries improve light extraction efficiency by reducing internal reflection losses [33, 34]. Other innovations include superhydrophobic and self-cleaning surfaces inspired by lotus leaves [35], antireflective coatings modeled after moth eyes [36], and structural adhesives and dry attachment systems inspired by gecko feet [37].

Each of these technologies embodies the same analogical reasoning central to this book: mapping functional strategies from the biological to the engineered domain. In each case, the inspiration from nature did more than validate the initial concept of photonic crystals; it became a wellspring of design strategies and optical mechanisms, reminding us that evolution, in its silent brilliance, had already explored much of the design space we now seek to engineer.

4.8 Technological Impact

Surface photonic textures on LEDs transfer band-structure ideas to light-extraction; integrated waveguides and slow-light sections yield insight generation in compact routing and nonlinear enhancement. Device yields and process compatibility reflect ongoing constraint matching with manufacturing realities.

Today, photonic crystals have matured from theoretical constructs and laboratory curiosities into indispensable tools for manipulating light in both commercial products and cutting-edge research. Their ability to control the flow of photons with high precision, derived from engineered band structures and spatial periodicity, has made them foundational to a variety of optical technologies. Perhaps one of the most visible applications lies in the domain of high-efficiency light-emitting diodes.

Traditional LEDs suffer from the fact that much of the light generated inside the semiconductor never escapes into free space [38–40]. This is due to total internal reflection and the presence of guided optical modes that remain trapped within the high-index material. By patterning the surface of LEDs with photonic crystal textures or nanostructured lattices, engineers can disrupt these guided modes, creating new escape pathways for photons [41–43]. The result is a dramatic increase in light extraction efficiency, making LEDs brighter without requiring additional electrical power. This improvement has not only enhanced the performance of display backlights and general illumination but has also contributed to the broader adoption of solid-state lighting as a sustainable alternative to incandescent and fluorescent technologies.

In the realm of optical communication and data processing, photonic crystals have enabled the miniaturization and enhancement of integrated photonic circuits [44–46]. Traditional waveguides, while effective, struggle with bending losses when light must change direction sharply on a chip. Photonic crystal waveguides, which rely on line defects within a photonic lattice to guide light, can manage tight bends with minimal radiation loss. This unique property makes them ideal for compact photonic chips, where dense routing of optical signals is essential. By harnessing the bandgap to confine light precisely and prevent leakage, these waveguides allow the construction of intricate, multilayered optical interconnects that support high-speed communication on chip-scale platforms. This is increasingly important as computing paradigms shift toward optical information processing and silicon photonics.

An even more striking capability of photonic crystals is their ability to slow down light [47–49]. By tailoring the dispersion relation of the photonic band structure, it is possible to engineer modes in which the group velocity of light approaches a tiny fraction of its speed in vacuum—sometimes less than 1% of *c*. This phenomenon, commonly referred to as slow light, has profound implications for nonlinear optics, where the strength of interaction between light and matter is enhanced by increasing the photon's dwell time within the medium [50–52]. Slow light in photonic crystal waveguides amplifies effects like four-wave mixing and second-harmonic generation, enabling all-optical signal processing, tunable delays, and enhanced sensitivity in sensing platforms. It also opens new avenues in buffering and synchronizing photonic data streams in time-sensitive communication networks.

Beyond conventional photonic and communication applications, the concept of slow light has also inspired developments in multidisciplinary domains such as photocatalysis [53–55]. In particular, slow-photon enhanced photocatalysis leverages the same principle of reduced group velocity to intensify light–matter interactions within semiconductor catalysts. By embedding or coating photocatalytic materials with photonic crystal structures designed to support slow-photon modes, one can significantly increase the local electromagnetic field and prolong photon residence times near reactive surfaces. This enhancement boosts the rate of photochemical reactions, offering new pathways for solar energy harvesting, environmental remediation, and artificial photosynthesis. The conceptual transfer between slow-light photonics and photocatalysis exemplifies how analogy thinking enables breakthroughs across traditionally separate scientific disciplines.

Quantum optics is another frontier where photonic crystals have demonstrated unique and transformative potential [56–59]. When a photonic crystal cavity—formed by introducing a carefully shaped point defect into a 2D or 3D lattice—is coupled to a single quantum emitter such as a quantum dot or nitrogen-vacancy center, the local density of optical states can be dramatically enhanced. This modification of the electromagnetic environment alters spontaneous emission rates through the Purcell effect, allowing efficient, directional, and deterministic coupling between the emitter and the cavity mode. Such configurations are critical for developing sources of single photons, quantum logic gates, and entanglement protocols essential to quantum information science. Photonic crystal platforms offer an unprecedented degree of control, enabling scalable quantum systems with integrated optical routing, switching, and feedback.

Finally, the precision with which photonic band structures can be engineered has made photonic crystals powerful tools for spectral control [60–62]. Optical filters, multiplexers, and demultiplexers based on photonic crystal slabs can separate wavelengths with sharp roll-offs and minimal insertion loss. These components are vital in dense wavelength division multiplexing systems that underpin modern fiber-optic communication networks. Moreover, the ability to create flat bands or highly dispersive regions enables the construction of optical buffers and delay lines—components that are notoriously difficult to realize with conventional optics. By precisely tuning geometry and refractive index contrast, photonic crystals allow photons to be slowed, trapped, or rerouted on demand, functioning as spectral architects with capabilities far beyond those of traditional thin-film filters or Bragg gratings.

In all these domains, the success of photonic crystals reflects their profound versatility. What began as a theoretical analogy to electron band structures in semiconductors has evolved into a real-world platform for manipulating the flow of light in ways once deemed impossible. From lighting to quantum computing, from laser surgery to optical networks, photonic crystals have established themselves not just as components, but as enablers of a new class of optical technologies—precise, scalable, and capable of reshaping the relationship between light and matter at the deepest levels.

4.9 Advanced Concepts: Topology and Beyond

Topological photonics applies abstractions from condensed-matter topology (Chern numbers, edge states) and maps them onto photonic lattices. Device demonstrations validate one-way, scattering-immune transport and enable new insights for robust photonic routing [63].

As photonic crystals matured into a foundational platform for controlling light, researchers began to look once more to the world of electronics for deeper inspiration—not just in the structure of semiconductors, but in the rich topological behaviors uncovered in condensed matter physics. In the 2000s, the analogy between photons and electrons evolved in a surprising and profoundly fertile direction, as topological phases—originally discovered in electronic systems—found new life in the realm of photonics. A pivotal breakthrough came in 2008, when researchers succeeded in demonstrating a photonic analog of the quantum Hall effect. In electronic systems, the quantum Hall effect is marked by the existence of edge states that conduct electricity in a single direction, immune to scattering from defects or impurities. Translating this concept into photonics meant building a lattice that could guide light along its edges in a one-way path, unperturbed by imperfections or disorder—an optical version of robust, topologically protected transport.

This was achieved by constructing a photonic crystal infused with magneto-optic materials, which break time-reversal symmetry, a key requirement for mimicking the conditions of the electronic quantum Hall regime. The result was a structure where light, instead of being reflected or scattered at corners or obstacles, flowed unidirectionally around the perimeter. These edge modes were inherently robust—immune to backscattering and capable of navigating imperfections without reflection. It was a stunning example of how ideas born in quantum condensed matter could reshape the design of photonic systems, not just in function but in conceptual architecture. Topological photonics became a new and rapidly expanding frontier, promising optical devices with unprecedented resilience and control.

At MIT, John Joannopoulos and collaborators deeply explored how analogies from semiconductor physics could inform photonic systems. In particular, they examined how localized defect states arise at interfaces between dissimilar photonic crystals (i.e., photonic heterostructures) and how such interfaces can trap light [64–68]. These heterojunction-based microcavities permit ultrahigh-Q resonances localized at the photonic band-structure discontinuity. Such structures introduced new levers in optical resonator and filter design, enabling extreme control over mode lifetimes, spatial confinement, and coupling to surrounding photonic crystal lattices.

4.10 Conclusion: The Power of Analogy

Across this chapter, progress followed the Chapter 1 loop: choose rich sources (nature; solid-state physics) → define targets (on-chip routing, colorfast coatings, mid-IR delivery) → perform mapping and abstraction (periodicity ⇒ bandgaps) → transfer via prototypes (microwave lattices, woodpiles, 2D slabs, fibers) with validation (spectroscopy, device metrics) → refine through constraint matching (index contrast, fabrication, loss) → generate insight (lasers, slow light, topological transport), which feeds back to richer sources and more ambitious targets.

The evolution of photonic crystals illustrates how analogy drives innovation while historical lines remain clear. The decisive conceptual origin of modern photonics and photonic crystals lies in late–twentieth century solid-state physics: translating the semiconductor bandgap framework to light-enabled materials with photonic bandgaps and device-like functionality for photons [11, 13]. Only after this foundation was established did the community return to biological structural color with renewed purpose. In the late 1990s and early 2000s, work spanning biology and physics—prominently by Helen Ghiradella, Pete Vukusic, Joann Aizenberg, and collaborators—showed how living systems exploit interference and hierarchical geometry, inspiring robust bioinspired coatings, colorants, and display concepts [7, 26].

Thus, natural iridescence did not found modern photonics; rather, it provided a rich source domain that became especially fruitful once the electronic-to-photonics analogy had been articulated. The dialogue is now bidirectional: semiconductor-inspired photonic crystals enable devices from on-chip lasers to hollow-core fibers, while insights from biology continue to suggest new architectures and fabrication strategies. In this reciprocity, the power of analogy thinking—sourcing, mapping, abstracting, transferring, validating, and refining—remains the throughline connecting history, mechanism, and technology.

References

[1] Jervis, M. A., "Robert Hooke's Micrographia: An Entomologist's Perspective," *Journal of Natural History*, Vol. 47, No. 39–40, 2013, pp. 2531–2573, https://doi.org/10.1080/00222933.2013.780270.

[2] Knellwolf, C., "Robert Hooke's Micrographia and the Aesthetics of Empiricism," *Seventeenth Century*, Vol. 16, No. 1, 2001, pp. 177–200, https://doi.org/10.1080/0268117X.2001.10555489.

[3] Young, T., "The Bakerian Lecture. On The Theory of Light and Colours," *Philosophical Transactions of the Royal Society of London*, Vol. 92, 1802, pp. 12–48, https://doi.org/10.1098/rstl.1802.0004.

[4] Niu, S., et al., "Angle-Dependent Discoloration Structures in Wing Scales of *Morpho menelaus* Butterfly," *Science China Technological Sciences*, Vol. 59, No. 5, 2016, pp. 749–755, https://doi.org/10.1007/s11431-016-6007-4.

[5] Giraldo, M. A., et al., "Colouration Mechanisms and Phylogeny of Morpho Butterflies," *Journal of Experimental Biology*, Vol. 219, No. 24, 2016, pp. 3936–3944, https://doi.org/10.1242/jeb.148726.

[6] Ding, Y., et al., "Structural Colors from *Morpho peleides* Butterfly Wing Scales," *Journal of Applied Physics*, Vol. 106, No. 7, 2009, p. 074702, https://doi.org/10.1063/1.3239513.

[7] Zhang, S., and Y. Chen, "Nanofabrication and Coloration Study of Artificial Morpho Butterfly Wings with Aligned Lamellae Layers," *Scientific Reports*, Vol. 5, 2015, p. 16637, https://www.nature.com/articles/srep16637.

[8] Bardeen, J., et al., "The Transistor, A Semi-Conductor Triode," *Physical Review*, Vol. 74, No. 2, 1948, pp. 230–231, https://doi.org/10.1103/PhysRev.74.230.

[9] Zheng, H., et al., "Bottom-Up Assembly and Applications of Photonic Materials," *Crystals*, Vol. 6, No. 5, 2016, p. 54, https://doi.org/10.3390/cryst6050054.

[10] Yablonovitch, E., "Inhibited Spontaneous Emission in Solid-State Physics and Electronics," *Physical Review Letters*, Vol. 58, 1987, p. 2059, https://doi.org/10.1103/PhysRevLett.58.2059.

[11] John, S., "Strong Localization of Photons in Certain Disordered Dielectric Superlattices," *Physical Review Letters*, Vol. 23, 1987, p. 2486, https://doi.org/10.1103/PhysRevLett.58.2486.

[12] Anderson, P. W., et al., "Absence of Diffusion in Certain Random Lattices," *Physical Review*, Vol. 109, No. 5, 1958, pp. 1492–1505, https://doi.org/10.1103/PhysRev.109.1492.

[13] Yablonovitch, E., and T. J. Gmitter, "Photonic Band Structure: The Face-Centered-Cubic Case," *Physical Review Letters*, Vol. 63, 1989, p. 1950, https://doi.org/10.1103/PhysRevLett.63.1950.

[14] Leung, K. M., et al., "Full Vector Wave Calculation of Photonic Band Structures in Face-Centered-Cubic Dielectric Media," *Physical Review Letters*, Vol. 65, No. 21, 1990, pp. 2646–2649, https://doi.org/10.1103/PhysRevLett.65.2646.

[15] Lin, S. Y., et al., "A Three-Dimensional Photonic Crystal Operating at Infrared Wavelengths," *Nature*, Vol. 394, 1998, pp. 251–253, https://doi.org/10.1038/28343.

[16] Lin, S.-Y., et al., "Experimental Demonstration of Guiding and Bending of Electromagnetic Waves in a Photonic Crystal," *Science*, Vol. 282, No. 5394, 1998, pp. 274–276, https://doi.org/10.1126/science.282.5387.274.

[17] Painter, O., et al., "Two-Dimensional Photonic Band-Gap Defect Mode Laser," *Science*, Vol. 284, 1999, pp. 1819–1821, https://doi.org/10.1126/science.284.5421.1819.

[18] Hennessy, K., et al., "Quantum Nature of a Strongly Coupled Single Quantum Dot–Cavity System," *Nature*, Vol. 445, No. 7130, 2007, pp. 896–899, https://doi.org/10.1038/nature05586.

[19] O'Brien, J. L., et al., "Photonic Quantum Technologies," *Nature Photonics*, Vol. 3, No. 12, 2009, pp. 687–695, https://doi.org/10.1038/nphoton.2009.229.

[20] Wong, G., et al., "Characterization of Chromatic Dispersion in Photonic Crystal Fibers Using Scalar Modulation Instability," *Optical Express*, Vol. 13, 2005, pp. 8662–8670, https://doi.org/10.1364/OPEX.13.008662.

[21] Birks, T. A., et al., "Full 2-D Photonic Band Gaps in Silica/Air Structures," *Electronics Letters*, Vol. 31, No. 22, 1995, pp. 1941–1943, https://doi.org/10.1049/el:19951306.

[22] Knight, J. C., et al., "All-Silica Single-Mode Optical Fiber with Photonic Crystal Cladding," *Optics Letters*, Vol. 21, No. 19, 1996, pp. 1547–1549, https://doi.org/10.1364/OL.21.001547.

[23] Birks, T. A., et al., "Endlessly Single-Mode Photonic Crystal Fiber," *Optics Letters*, Vol. 22, No. 13, 1997, pp. 961–963, https://doi.org/10.1364/OL.22.000961.

[24] Russell, P. S. J., "Photonic Crystal Fibres," *Science*, Vol. 299, No. 5605, 2003, pp. 358–362, https://doi.org/10.1126/science. 1079280.

[25] Knight, J. C., "Photonic Crystal Fibres," *Nature*, Vol. 424, 2003, pp. 847–851, https://doi.org/10.1038/nature01940.

[26] Ball, P., "Understanding How Animals Create Dazzling Colors Could Lead to Brilliant New Nanotechnologies," *Scientific American*, 2012, https://www.scientificamerican.com/article/natures-color-tricks.

[27] Ghiradella, H., et al., "Ultra-Violet Reflection of a Male Butterfly: Interference Color Caused by Thin Layer Elaboration of Wing Scales," *Science*, Vol. 178, No. 4066, 1972, pp. 1214–1217, https://doi.org/10.1126/science.178.4066.1214.

[28] Ghiradella, H., et al., "Light and Color on the Wing: Structural Colors in Butterflies and Moths," *Applied Optics*, Vol. 30, No. 24, 1991, pp. 3492–3500, https://doi.org/10.1364/AO.30.003492.

[29] Kolle, M., et al., "Bio-Inspired Band-Gap Tunable Elastic Optical Multilayer Fibers," *Advanced Materials*, Vol. 25, No. 15, 2013, pp. 2239–2245, https://doi.org/10.1002/adma.201203529.

[30] Pouya, C., et al., "Characterization of a Mechanically Tunable Gyroid Photonic Crystal Inspired by the Butterfly *Parides sesostris*," *Advanced Optical Materials*, Vol. 4, No. 1, 2016, pp. 99–105, 2016, https://doi.org/10.1002/adom.201500436.

[31] Potyrailo, R. A., H. Ghiradella, and A. Vertiatchikh, "Morpho Butterfly Wing Scales Demonstrate Highly Selective Vapour Response," *Nature Photonics*, Vol. 1, 2007, pp. 123–128, https://doi.org/10.1038/nphoton.2007.2.

[32] Pris, A. D., et al., "Towards High-Speed Imaging of Infrared Gases with Bio-Inspired Photonic Structures," *Nature Communications*, Vol. 3, 2012, p. 1216, https://doi.org/10.1038/nphoton.2011.355.

[33] Clapham, P. B., and M. C. Hutley, "Reduction of Lens Reflexion by the 'Moth Eye' Principle," *Nature*, Vol. 244, 1973, pp. 281–282, https://doi.org/10.1038/244281a0.

[34] Lakhtakia, A., and R. Messier, *Sculptured Thin Films: Nanoengineered Morphology and Optics*, SPIE press, 2005, Vol. 143.

[35] Barthlott, W., et al., "Purity of The Sacred Lotus, or Escape from Contamination in Biological Surfaces," *Planta*, Vol. 202, No. 1, 1997, pp. 1–8, https://doi.org/10.1007/s004250050096.

[36] Mokkapati, S., and K. R. Catchpole, "Moth-Eye Inspired Anti-Reflective Surfaces for Photovoltaics and Optics," *Journal of Applied Physics*, Vol. 112, 2012, p. 101101, https://doi.org/10.1088/ 1748-3190/aab738.

[37] Autumn, K., et al., "Evidence for van der Waals Adhesion in Gecko Setae," *Proceedings of the National Academy of Sciences*, Vol. 99, No. 19, 2002, pp. 12252–12256, https://doi.org/10.1073/pnas.192252799.

[38] Ee, Y.- K., et al., "Optimization of Light Extraction Efficiency of III-Nitride LEDs With Self-Assembled Colloidal-Based Microlenses," *IEEE Journal of Selected Topics in Quantum Electronics*, Vol. 15, No. 4, 2009, pp. 1218–1225, https://doi.org/10.1109/JSTQE.2009.2015580.

[39] Zhang, Z.- H., et al., "On the Light Extraction Efficiency for III-Nitride-Based Light-Emitting Diodes," in *Light-Emitting Diodes: Materials, Processes, Devices and Applications*, J. Li and G.-Q. Zhang (eds.), Springer, 2019, Vol. 4, https://doi.org/10.1007/978-3-319-99211-2 9.

[40] Schubert, E. F., *Light-Emitting Diodes*, Cambridge, UK: Cambridge University Press, 2006.

[41] Bergenek, K., et al., "Enhanced Light Extraction Efficiency from AlGaInP Thin-Film Light-Emitting Diodes With Photonic Crystals," *Applied Physics Letters*, Vol. 93, No. 4, 2008, p. 041105, https://doi.org/10.1063/1.2963030.

[42] Kwon, M.- K., et al., "Enhanced Emission Efficiency of GaN/InGaN Multiple Quantum Well Light-Emitting Diode With an Embedded Photonic Crystal," *Applied Physics Letters*, Vol. 92, No. 25, 2008, p. 251110, https://doi.org/10.1063/1.2948851.

[43] Wiesmann, C., et al., "Photonic Crystal LEDs–Designing Light Extraction," *Laser & Photonics Review*, Vol. 3, No. 3, 2009, pp. 262–286, https://doi.org/10.1002/lpor.200810053.

[44] Butt, M. A., et al., "Recent Advances in Photonic Crystal Optical Devices: A Review," *Optics and Laser Technology*, Vol. 142, p. 107265, https://doi.org/10.1016/j.optlastec.2021.107265.

[45] Marg, G., et al., "Recent Progress in Photonic Crystal Devices and their Uses," *Photonics*, Vol. 10, No. 11, 2023, p. 1199, https://doi.org/10.3390/photonics10111199.

[46] Caballero, L. E. P., et al., "A Review On Photonic Crystal Logic Gates," *Journal of Integrated Circuits and Systems*, Vol. 16, No. 1, 2021, pp. 1–12, https://doi.org/10.29292/jics.v16i1.478.

[47] Baba, T., "Slow Light in Photonic Crystals," *Nature Photonics*, Vol. 2, No. 8, 2008, pp. 465–473, https://doi.org/10.1038/nphoton.2008.146.

[48] Hirotani, K., et al., "Si Photonic Crystal Slow-Light Waveguides Optimized Through Informatics Technology," *Optics Letters*, Vol. 46, No. 17, 2021, pp. 4422–4425, https://doi.org/10.1364/OL.436118.

[49] Soljačič, M. et al., "Photonic-Crystal Slow-Light Enhancement Of Nonlinear Phase Sensitivity," *Journal of the Optical Society of America B*, Vol. 19, No. 9, 2002, pp. 2052–2059, https://doi.org/10.1364/JOSAB.19.002052.

[50] Monat, C., et al., "Slow Light Enhanced Nonlinear Optics in Silicon Photonic Crystal Waveguides," *IEEE Journal of Selected Topics in Quantum Electronics*, Vol. 16, No. 1, 2010, pp. 344–356, https://doi.org/10.1109/JSTQE.2009.2033019.

[51] Corcoran, W. P., et al., "Green Light Emission Through Slow-Light Enhanced Third-Harmonic Generation In Photonic-Crystal Waveguides," *Nature Photonics*, Vol. 3, No. 4, 2009, pp. 206–210, https://doi.org/10.1038/nphoton.2009.28.

[52] Inoue, K., et al., "Enhanced Third-Order Nonlinear Effects in Slow-Light Photonic-Crystal Slab Waveguides of Line-Defect," *Optics Express*, Vol. 17, No. 9, 2009, pp. 7206–7216, https://doi.org/10.1364/OE.17.007206.

[53] Toumazatou, A., et al., "Improving Visible Light Photocatalysis Using Optical Defects in CoOx-TiO2 Photonic Crystals," *Materials*, Vol. 17, No. 23, 2024, p. 5996, https://doi.org/10.3390/ ma17235996.

[54] Han, Y., et al., "Slow-Light-Driven Photocatalytic CO2 Reduction to CH4 Mediated by Photonic Crystal GDY-Cu/ZnO Z-Scheme System," *Chemical Engineering Journal*, Vol. 500, 2024, p. 157636, https://doi.org/10.1016/j.cej.2024.157636.

[55] Chen, Y. K., et al., "Advancing Photocatalysis Through Efficient Fabrication Of Morphology-Controlled Photonic Crystals with Slow-Light Effects," *Inorganic Chemistry Frontiers*, Vol. 12, 2025, pp. 5962–5988, https://doi.org/10.1039/d5qi01087j.

[56] Hauff, N. V., et al., "Chiral Quantum Optics in Broken-Symmetry and Topological Photonic Crystal Waveguides," *Physical Review Research*, Vol. 4, 2022, p. 023082, https://doi.org/10.1103/ PhysRevResearch.4.023082.

[57] Kuruma, K., et al., "Strong Coupling Between a Photonic Crystal Nanobeam Cavity and a Single Quantum Dot," *Applied Physics Letters*, Vol. 98, 2011, p. 173104, https://doi.org/10.1063/1. 3579535.

[58] Majumdar, A., et al., "Cavity Quantum Electrodynamics with a Single Quantum Dot Coupled to a Photonic Molecule Consisting of a Pair of Coupled Photonic Crystal

Cavities," *Physical Review B*, Vol. 86, 2012, p. 045315, https://doi.org/10.1103/PhysRevB.86.045315.

[59] Pernice, A., et al., "Hybrid Quantum Photonics Based on Artificial Atoms Placed Inside One-Dimensional Photonic Crystal Cavities," *ACS Photonics*, Vol. 8, No. 9, 2021, pp. 2635–2641, https://doi.org/10.1021/acsphotonics.1c00530.

[60] Kumar, G., et al., "Photonic Crystal Based Heterostructures in the Control of Emission and Diffraction Features," *ISSS Journal of Micro and Smart Systems*, Vol. 11, 2022, pp. 81–112, https://doi.org/10.1007/s41683-021-00086-1.

[61] Shafagh, S. G., et al., "Customized Bandgap Engineering in Plasmonic-Photonic Crystal Structures for Advanced Broadband and Narrowband Optical Filtering," *Optical and Quantum Electronics*, Vol. 57, 2025, https://doi.org/10.1007/s11082-025-08308-8.

[62] Jannesari, R., et al., "Design of a Narrow Band Filter Based on a Photonic Crystal Cavity for CO2 Sensing Application," *Sensors*, Vol. 23, No. 10, 2023, p. 4958, https://doi.org/10.3390/s23104958.

[63] Fan, S., et al., "High Extraction Efficiency of Spontaneous Emission from Slabs of Photonic Crystals," *Physical Review Letters*, Vol. 78, No. 17, 1997, p. 3294, https://doi.org/10.1103/PhysRevLett.78. 3294.

[64] Joannopoulos, J. D., et al., "Microcavities in Photonic Crystals: Mode Symmetry, Tunability, and Coupling Efficiency," *Physical Review B*, Vol. 54, 1996, pp. 7837–7842, https://doi.org/10.1103/PhysRevB.54.7837.

[65] Joannopoulos, J. D., et al., "Photonic Crystals: Molding the Flow of Light," in *Photonic Crystals*, J. D. Joannopoulos et al. (eds.), Princeton, New Jersey: Princeton University Press, 1995, https://doi.org/10.1515/9780691124568.

[66] Meade, Robert D., et al., "Photonic Bound States in Periodic Dielectric Materials," *Physical Review B*, Vol. 44, 1991, pp. 13772–13774, https://doi.org/10.1103/PhysRevB.44.13772.

[67] Fan, S., et al., "Guided and Defect Modes in Periodic Dielectric Waveguides," *Journal of Optical Society of America B*, Vol. 12, 1995, pp. 1267–1272, https://doi.org/10.1364/JOSAB.12.001267.

[68] Meade, Robert D., et al., "Novel Applications of Photonic Band Gap Materials: Low-Loss Bends and High Q Cavities," *Journal of Applied Physics*, Vol. 75, No. 9, 1994, pp. 4753–4755, https://doi.org/10.1063/1.355934.

5

Quantum Dot Lasers: From the Artificial Atom Analogy to Scalable Technology

After reading this chapter, you should be able to:

- Explain how analogy thinking—particularly the artificial atom or particle-in-a-box analogy—fueled the invention of quantum dot lasers;
- Describe how quantum dot lasers evolved into high-performance, scalable technology;
- Discuss how cross-domain analogies continue to drive future innovations in photonics.

5.1 Introduction

Quantum dot lasers stand as a remarkable success in photonic engineering, uniting concepts from quantum mechanics with practical device fabrication. A key conceptual foundation for these lasers emerged from analogies with quantum mechanical tunneling—the phenomenon where particles penetrate energy barriers that classical physics forbids [1]. By drawing parallels between electron behavior in nanoscale structures and tunneling through potential barriers, researchers developed new ways to confine charge carriers and manipulate light.

As illustrated in Figure 5.1, the concept of quantum tunneling can be intuitively understood through a classical analogy. In the classical world, a

Figure 5.1 Illustration of the quantum tunneling analogy. On the left, the lion is unable to cross a thick barrier, representing a classical particle blocked by an insurmountable potential wall. On the right, when the barrier becomes sufficiently thin, the lion appears on the other side, symbolizing a quantum particle that tunnels through a classically forbidden region. This simple analogy conveys how probability, rather than direct energy, governs motion at the nanoscale.

macroscopic object—like the lion shown on the left—cannot penetrate a solid barrier unless it possesses enough energy to overcome it. However, in the quantum regime, particles exhibit wavelike behavior, allowing a finite probability of existing on the other side of a potential barrier even when their energy is insufficient to surmount it. The right side of the figure humorously depicts this situation: when the barrier becomes sufficiently thin, the lion appears beyond it, much like an electron that tunnels through a narrow potential wall. This analogy captures the essence of tunneling, emphasizing how quantum mechanics replaces deterministic trajectories with probabilistic outcomes that defy classical intuition.

In terms of the analogy-thinking framework introduced in Chapter 1, quantum tunneling served as the source domain—a well-understood physical concept describing how particles behave under confinement—while semiconductor laser design represented the target domain seeking improved carrier control and temperature stability. Scientists performed systematic mapping between the two domains: electrons tunneling through potential barriers were reinterpreted as carriers confined in nanoscale potential wells. Through abstraction, they extracted the general principle of quantum confinement, independent of material specifics, and transferred it into the design of semiconductor nanostructures. As fabrication techniques evolved, iterative constraint matching—adapting to limits of materials, lattice mismatch, and heat management—transformed the analogy into viable devices. The resulting insight generation launched a new era of semiconductor lasers based on discrete quantum states rather than continuous bands.

This chapter explores how the tunneling and particle-in-a-box analogies inspired the invention of quantum dot (QD) lasers, and how analogy thinking continued to guide their technological evolution. Beginning with a historical overview, we trace how conceptual parallels became engineering practice, and how analogy-driven reasoning still fuels innovation across disciplines.

As illustrated in Figure 5.2, the dimensionality of carrier confinement fundamentally determines the electronic density of states in semiconductor nanostructures. A quantum well restricts carrier motion in only one dimension, allowing free propagation in the remaining two. This partial confinement leads to a step-like density of states $\rho(E)$, where each plateau corresponds to the onset of a new subband in the quantized direction.

By contrast, a quantum dot confines carriers in all three spatial dimensions, effectively trapping electrons and holes within a nanoscale box. As a result, the density of states becomes discrete, consisting of sharp energy levels that resemble atomic orbitals—hence the frequent analogy to artificial atoms. The visual contrast between the two structures in Figure 5.2 highlights how increasing the degree of confinement not only alters the electronic structure but also amplifies quantum effects such as energy quantization, carrier localization, and enhanced radiative efficiency. These properties form the physical basis for the superior performance of quantum dot lasers compared to their quantum well counterparts.

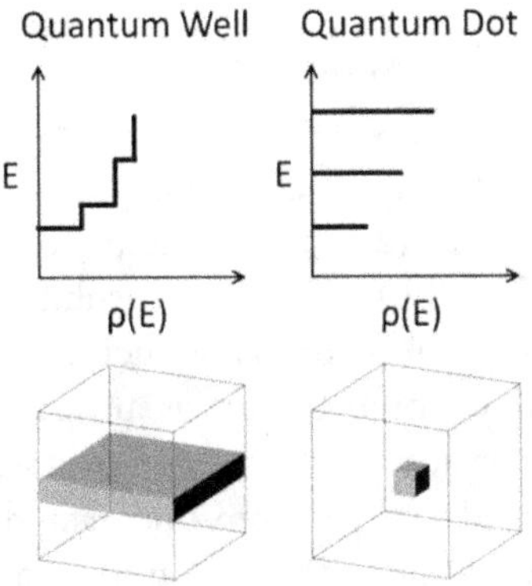

Figure 5.2 Comparison of quantum confinement and corresponding density of states for a quantum well (left), and a quantum dot (right). In a quantum well, carriers are confined in one dimension, resulting in a step-like density of states ρ(E) characteristic of 2D systems. In contrast, a quantum dot confines carriers in all three spatial dimensions, producing discrete energy levels and a delta-functionlike density of states similar to that of an atom. The lower panels visualize the dimensional confinement within each nanostructure.

5.2 From Quantum Tunneling to QD: Historical Foundations

Quantum tunneling provided the conceptual bridge that ultimately led to the idea of carrier confinement in quantum dot lasers. In quantum mechanics, tunneling arises because electrons possess wavelike properties, allowing their probability amplitude to extend through and even beyond potential barriers that are classically impenetrable. This phenomenon, first formalized in the 1920s, revealed that electrons are not localized particles but delocalized wave functions capable of coupling between regions separated by barriers. The realization that such coupling—and its converse, confinement—could be engineered marked a fundamental shift in how electronic and optical devices were conceived. By the mid-twentieth century, tunneling had evolved from a theoretical curiosity into a design principle embodied in devices such as the Esaki tunnel diode (1957) [1] and the resonant tunneling diode, both of which exploited controlled tunneling currents across nanoscale potential barriers.

This understanding of electron penetration and reflection at barriers inspired the notion that barriers could be used not only to enable tunneling but also to restrict it. In semiconductor heterostructures, when a low bandgap layer is sandwiched between higher bandgap materials, the resulting potential wells limit carrier motion, preventing tunneling leakage and concentrating electrons and holes where light emission occurs. This principle underpinned the double heterostructure laser of the 1960s, work that earned Z. Alferov and H. Kroemer the 2000 Nobel Prize, and represented the first deliberate application of tunneling-derived confinement concepts to laser design.

Once tunneling was understood as a manifestation of electron wave function overlap between potential wells, researchers began to ask how controlling that overlap could yield new quantum behaviors. Reducing the dimensions of the confinement region progressively weakened interwell coupling and thus suppressed tunneling, producing discrete energy levels. This logic drove the evolution from quantum wells (2D confinement) to quantum wires (1D) and ultimately to quantum dots (0D). In 1982, Yasuhiko Arakawa and Hiroyuki Sakaki proposed that a semiconductor nanostructure small enough to eliminate interwell tunneling entirely would behave as an artificial atom, with carriers confined in all three dimensions and exhibiting discrete quantized states [2]. The QD laser thus emerged directly from understanding tunneling as both a limitation (causing unwanted carrier escape) and an inspiration (suggesting the benefits of its complete suppression).

The eventual discovery of self-assembled InAs QDs on GaAs substrates in the mid-1980s [3] provided the material mechanism to realize this vision. These nanostructures naturally formed potential wells deep enough to localize carriers and suppress tunneling to adjacent sites, achieving the fully discrete energy spectrum predicted by theory. The first demonstration of QD laser emission by Kirstaedter et al. (1994) [4], confirmed that eliminating tunneling between dots and confining carriers within nanoscale boxes not only was feasible but also yielded superior performance, with lower threshold currents and improved temperature stability.

In summary, the QD laser's lineage traces directly to the physics of tunneling. The insight that electron wave functions could penetrate—or be prevented from penetrating—potential barriers inspired the very concept of quantum confinement. This intellectual progression follows the analogy-thinking cycle from Chapter 1: tunneling as the source domain, semiconductor lasers as the target domain, confinement as mapping and abstraction, epitaxial growth as transfer and constraint matching, and the QD laser as the resulting insight generation. Quantum tunneling, initially a paradox of barrier penetration, thus became the conceptual key to engineering barriers that confine.

Figure 5.3 juxtaposes two cornerstone concepts in quantum mechanics—tunneling and particle confinement—that together form the intellectual foundation for modern quantum nanostructures. On the left, the schematic of quantum tunneling depicts a particle wave function that penetrates a potential barrier despite possessing insufficient classical energy to overcome it. This nonclassical behavior, governed by the probabilistic nature of wave functions, enables phenomena such as electron tunneling in diodes, scanning tunneling microscopy, and tunneling injection in quantum dot lasers.

In contrast, the right-hand side shows the particle-in-a-box model, where the particle is confined by potential walls so high that its wave function vanishes at the boundaries. This perfect confinement produces discrete allowed energy states—an early model of quantum quantization. In real nanostructures such as quantum dots, this box analogy manifests physically as three-dimensional

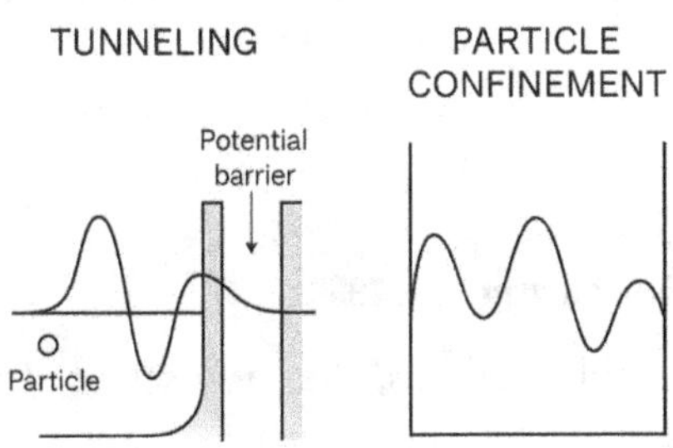

Figure 5.3 Conceptual comparison between quantum tunneling (left), and particle-in-a-box confinement (right). In quantum tunneling, a particle's wave function extends beyond the potential barrier, allowing a finite probability of transmission through classically forbidden regions. In contrast, in the particle-in-a-box model, the particle is fully confined within impenetrable boundaries, leading to discrete quantized energy levels. Together, these models illustrate the foundational principles of quantum confinement and carrier dynamics in nanostructures.

confinement of charge carriers, leading to energy level spacing that increases as the dot size decreases.

Together, the two panels of Figure 5.3 illustrate how modifying boundary conditions—whether by permitting partial penetration through barriers (tunneling) or enforcing total confinement (particle-in-a-box)—fundamentally alters the quantum behavior of carriers. These complementary principles provide the conceptual bridge between classical devices and the quantized operation of nanoscale lasers and emitters.

5.3 Evolution of QD Laser Technology

Having established the conceptual foundation, we now trace the technological evolution of QD lasers from the 1980s to the present. Each decade brought key breakthroughs that turned QD lasers from a lab curiosity into a mature technology.

5.3.1 Early Concepts and Theoretical Insights: 1980s

The theoretical proposal by Arakawa and Sakaki in 1982 [2, 3] marked the starting point, predicting that 0D carrier confinement would yield lasers with superior properties. Throughout the 1980s, other researchers (such as Asada, Miyamoto, and Suematsu in 1986 [5]) further analyzed QD laser dynamics, calculating that threshold current could be dramatically reduced if only the discrete ground state of QDs were utilized. However, experimentally creating QDs remained the chief hurdle. Researchers experimented with advanced epitaxial growth techniques, and late in the decade came the pivotal discovery that strained-layer epitaxy could spontaneously form nanometer islands (self-assembled QDs). By 1985 to 1986, the first optical characterizations of such QDs (for example, Ekimov's colloidal QDs [6], and later Merz and Petroff's epitaxial QDs [7]) showed discrete energy level behavior, setting the stage for device applications.

5.3.2 First QD Lasers Demonstrations: 1990s

After refinement of growth methods, the early 1990s saw rapid progress. In 1993 to 1994, separate groups in Europe, Japan, and the United States reported the first lasing from QD ensembles. The milestone 1994 paper by Kirstaedter et al. documented a QD laser diode with low threshold and high temperature stability, using InGaAs QDs on GaAs [4].

This was soon followed by improvements: by the late 90s, devices showed threshold current densities well below 100 A/cm^2 (a remarkable achievement at the time) and characteristic temperature T_0 values much higher than quantum well lasers, confirming the reduced temperature sensitivity. Spectral measurements in these years also demonstrated the zero-dimensional density of states: instead of a broad gain spectrum, QD lasers exhibited peaks corresponding to quantized energy levels, much like an atomic emission spectrum. However, early QD lasers did face issues such as relatively low output power and gain due to limited QD density and inhomogeneous dot sizes. Researchers mitigated these by stacking multiple QD layers and improving size uniformity.

In the 2000s, QD laser research moved from proof-of-concept toward practical devices [3]. Material growth advances (including better control of dot size distribution and the introduction of techniques like rapid thermal annealing to homogenize emission) led to significant performance gains. By mid-2000s, QD lasers achieved higher power and efficiency, in some cases exceeding the performance of commercial quantum well lasers. For instance, the first 10 Gb/s directly modulated QD lasers were demonstrated for telecom wavelengths around 1.3 μm, showing the feasibility of QDs in high-speed applications. Companies such as Innolume (in Germany) and QD Laser, Inc. (a spinoff in Japan) were founded to commercialize the technology. They produced prototype QD laser diodes for niche markets, like temperature-stable pump lasers and eye-safe lasers.

A notable breakthrough was the use of p-type modulation doping and tunneling injection structures to overcome the phonon bottleneck in QDs. Normally, carriers captured into a QD may linger in excited states or the surrounding "wetting layer" if they cannot quickly shed excess energy (due to limited phonon emission). This can limit modulation speed and raise thresholds. By adding a tunneling-coupled quantum well as a carrier injector, engineers found they could directly and rapidly feed carriers into QD ground states [8]. This tunneling injection QD laser design improved the high-speed modulation bandwidth and reduced sensitivity to hot carriers, as electrons no longer had to rely on slow phonon processes to populate the lasing state. Such innovations in the 2000s addressed many earlier shortcomings. By the end of the decade, quantum dot lasers had firmly established performance advantages: for example, devices showed lower threshold currents, higher T_0, and reduced feedback sensitivity compared to the best quantum well lasers [9–12].

As illustrated in Figure 5.4, the InAs/InP QD laser structure consists of a vertically stacked epitaxial configuration optimized for efficient carrier confinement and optical gain. The top contact layer is formed by a heavily doped (p^{++}) InGaAs layer, capped with a Ti/Au metallization to ensure low-resistance ohmic contact. Beneath it lies a thick p^{+}-doped InP layer (3 μm) and a thinner p-doped InP layer (0.5 μm), which together serve as the upper cladding and facilitate hole injection into the active region.

The central guiding region is composed of two 150-nm-thick quaternary alloy (Q1.18) layers that form the optical waveguide and provide carrier and optical confinement. Sandwiched between these layers is a single layer of

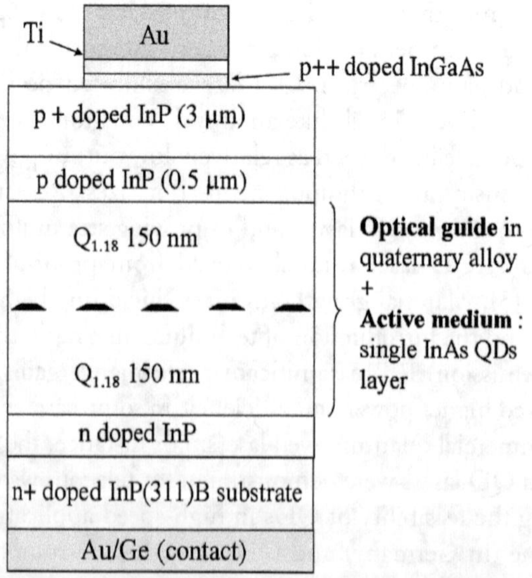

Figure 5.4 Schematic cross-sectional structure of an InAs/InP QD laser containing a single InAs QD layer. The device comprises a *p*++-doped InGaAs contact layer and successive *p*- and *n*-doped InP cladding layers surrounding a quaternary alloy optical guiding region. The active medium, consisting of a single InAs QD layer, is embedded between two 150-nm-thick quaternary alloy (Q1.18) barriers. Metal contacts of Au/Ti on the top and Au/Ge at the bottom provide electrical access to the device.

self-assembled InAs QDs, which constitutes the active medium responsible for light emission. Below this, an n-doped InP layer acts as the lower cladding, grown atop an n^+-doped InP(311)B substrate to ensure good electrical and crystalline quality. The bottom Au/Ge contact provides an efficient n-type ohmic interface, completing the device stack. This epitaxial design ensures strong optical confinement and efficient current injection, key for achieving low-threshold lasing operation in QD-based InP devices.

5.3.3 Towards Broad Adoption and Integration: 2010s

In the 2010s, the focus shifted to integrating QD lasers in new platforms and scaling up production. A significant driver was the need for on-chip light sources in silicon photonics. Silicon (widely used for electronic integrated circuits) is not a good light emitter, so adding lasers to silicon chips is challenging. Researchers found that QD lasers offered an enticing solution: their strong carrier confinement made them less sensitive to defects, meaning QD lasers could be grown on silicon substrates (which inherently have many crystal defects due to lattice mismatch) yet still perform well [13]. Pioneering work by groups at University of California Santa Barbara (UCSB) and elsewhere demonstrated InAs/GaAs QD lasers monolithically grown on silicon with excellent results. For instance, in 2016 a QD laser on Si achieved over 100,000 hours of continuous operation at 35°C, a reliability level approaching commercial needs [3]. The QD active layer's insensitivity to dislocations (defect tolerance) and reduced optical feedback sensitivity (less need for isolators) were crucial advantages in the silicon photonics context. By late 2010s, performance on silicon kept improving—devices showed output powers of tens of milliwatts and operation up to 80°C without coolers, something traditional GaAs or InP lasers struggled with on non-native substrates. Meanwhile, on native substrates, QD laser development continued: researchers achieved multiwavelength QD lasers (leveraging the inhomogeneous broadening to get a comb of lines, useful for WDM communications), mode-locked QD lasers generating ultrashort pulses for THz and comb applications, and vertical-cavity surface-emitting lasers (VCSELs) with QD active regions for potential low-cost optical links. By the end of the 2010s, QD lasers were no longer a laboratory novelty, they were on the cusp of widespread use in data centers and other systems.

5.3.4 2020s and Beyond

The 2020s have so far been about scaling up and new frontiers. On one hand, there have been dramatic demonstrations of manufacturing scalability. In 2020 to 2022, researchers showed that QD lasers could be grown on industry-standard large wafers. One landmark example was the first QD lasers grown on a 300-mm silicon wafer by molecular beam epitaxy, reported in 2022 [14]. These lasers, integrated into prepatterned silicon photonic waveguides, achieved continuous-wave lasing up to 60C and output powers of approximately 126 mW. The result signaled that mass production of QD lasers in conventional semiconductor fabs is feasible, opening the door to low-cost, high-volume deployment. These advances have the explicit goal of lowering costs and enabling QD lasers to be produced in large quantities for industry; for example, for the thousands of lasers needed in data center interconnects or photonic integrated circuits.

On the other hand, new frontiers are being explored. A notable beyond 2020 development is the revival of colloidal QD (CQD) lasers [15]. CQDs are nanocrystals synthesized in solution (such as PbS or CdSe QDs) that had long been studied for light emission, but achieving lasing was difficult due to losses and Auger recombination. In 2024, a Los Alamos National Lab–led team demonstrated a breakthrough by engineering a special type of CQD heterostructure (called type-(I+II) dots) that suppressed Auger recombination and achieved liquid-state lasing with tunability over a wide spectral range. This was essentially a liquid QD laser, using a solution of QDs as the gain medium in a fluidic cavity. The new QDs had compartmentalized core-shell structures that nearly eliminated the nonradiative Auger process, allowing optical gain to build up even at low QD concentrations. The result was a tunable laser (in a Littrow cavity) that operated stably without the circulation apparatus needed in traditional dye lasers. Such an achievement is not directly an electrically pumped diode laser, but it expands the scope of QD lasers into realms like optofluidics and lab-on-chip devices, with potential wavelengths spanning the visible to infrared. It also highlights the versatility of the QD concept—from epitaxial dots in solid-state diode lasers to colloidal dots in liquid lasers, the underpinning idea is the same quantum confinement and discrete levels. Looking forward from 2025, QD lasers are positioned to become workhorses in many applications. Current production-grade QD lasers (especially those on GaAs or InP for 1.3- to 1.5-μm telecom bands) are being refined for higher power and direct modulation speeds (recent devices have reached 25- to 50-Gbps modulation with low chirp) [16–20]. There is also ongoing development to extend QD lasers to new wavelength regimes, such as the > 2-μm midinfrared (for sensing) using materials like InAs/InP or InAs/GaSb dots, and the blue-green spectrum (using III–nitride QDs) [21, 22]. Each of these involves materials challenges, but the past decades have shown a consistent trend: what begins as a theoretical analogy (QDs as artificial atoms) can, through persistent innovation, become an engineering reality. The evolution of QD lasers thus reflects a broader truth in photonics—advances often spring from combining imaginative conceptual models with steady improvements in fabrication technology.

5.4 Applications Across Diverse Fields

QD lasers have matured into a technology with wide-ranging applications. Thanks to their unique properties—low threshold, temperature stability, multiwavelength emission, and compact size—QD lasers are finding use (or expected use) in many sectors. In this section we survey general application areas for QD lasers. In all these applications, the key point is that QD lasers are not confined to a single niche, their features make them versatile. They bring particular value where one or more of the following is needed: high-temperature operation, low noise, multiwavelength output, defect tolerance (for integration on various substrates), or extremely low thresholds. As manufacturing techniques mature (discussed next), we can expect to see QD laser deployment broaden, much like diode lasers based on quantum wells became ubiquitous in the 1990s and 2000s. QD lasers, born from a quantum tunneling analogy, are now poised to illuminate everything from global networks to handheld devices.

5.4.1 Optical Communications and Datacenters

This is arguably the current primary driver for QD laser development. In telecom and datacom, lasers operating at 1.3 μm and 1.55 μm are needed for fiber-optic links [16]. QD lasers at 1.3 μm have demonstrated high stability over temperature (important for uncooled transceivers) and resilience to optical feedback, which is useful in dense photonic integrated circuits. They are widely considered for on-chip light sources in silicon photonics interconnects, enabling faster optical data transfer in and between data centers. The low chirp and narrow linewidth of QD lasers also suit advanced modulation formats in coherent communications. In practical terms, QD lasers can reduce the cost of optical links by allowing simpler packaging (e.g., no isolator or cooler needed in some cases) and by being integrable with silicon CMOS processes. Major silicon photonics foundries are investigating incorporating QD laser dies or even directly growing QD lasers on silicon wafers for optical interconnects in high-performance computing.

5.4.2 Sensing and Imaging

QD lasers offer advantages for various sensing applications. For example, their stability and spectral purity can improve gas sensing and spectroscopy, where a laser's precise wavelength is tuned to detect molecular absorption lines [23–25]. It has been noted that deployable systems for gas sensing and even biomolecular sensing could benefit from QD laser sources. In the automotive sector, eye-safe lidar could employ near-infrared QD lasers that remain stable over car operating temperatures and provide short pulses (using mode-locked QD lasers for example) [26]. Additionally, the ability to fabricate QD lasers on different substrates raises the prospect of integrating lasers onto sensor chips directly. In biomedical imaging, QD lasers emitting in the red/near-IR could be useful for techniques like optical coherence tomography or fluorescence imaging, especially if their coherence and low noise improve image quality. The multiwavelength capacity of inhomogeneously broadened QD gain could be harnessed to create tunable lasers or even simultaneous multiline lasers for advanced imaging modalities. Moreover, as shown by the colloidal QD laser advances, we might see lab-on-a-chip light sources using liquid QD lasers for microfluidic sensors. These could enable portable analytical devices where the gain medium can be reconfigured or replaced easily (simply by fluid exchange).

5.4.3 Quantum Information and Cryptography

Interestingly, QD lasers can operate in regimes spanning classical coherent emission to single-photon emission, depending on how they are used. If operated with one electron–hole pair at a time in a QD, the device acts as a single-photon source rather than a conventional laser [27–30]. QD single-photon sources are a key component for quantum cryptography (quantum key distribution) and potentially for optical quantum computing, providing on-demand photons with quantum purity. Meanwhile, as lasers, QD devices can serve in quantum communication as reliable sources for quantum key distribution systems that require lasers with very particular stability and linewidth characteristics. The analogy of QDs as artificial atoms has even broader implications in quantum tech. For example, QD lasers can be used to generate entangled photon pairs via cascaded emissions or serve as the gain medium in entangled photon generation setups [31–35]. In emerging neuromorphic photonics (covered more below), mode-locked QD lasers have been used to emulate spiking neurons for photonic neural networks, bridging into the computing domain.

5.4.4 High-Speed Electronics and Microwave Photonics

The ultrafast gain dynamics of QD lasers (with bandwidth enhancements via tunneling injection or simply inherently fast carrier capture in small dots) make them attractive for microwave photonics applications. QD mode-locked lasers that produce picosecond pulses can generate broadband frequency combs, useful in frequency metrology or as multiwavelength sources for parallel radio frequency (RF) signal generation [36–38]. Their low noise and reduced mode partition noise also mean QD lasers can produce relatively low-jitter pulse trains. This has implications for optical sampling oscilloscopes, clock distribution, and photonic analog-to-digital conversion. For radio-over-fiber links (carrying wireless signals over optical fiber), QD lasers' high linearity and low relative intensity noise are advantageous. In summary, any application requiring stable, high-quality laser signals across varying conditions can potentially benefit from QD lasers.

5.4.5 Military and Aerospace

In aerospace or satellite communications, temperature extremes are common, and QD lasers' temperature robustness (with some devices exhibiting threshold stability from –50°C up to 100°C) is a major asset [39, 40]. They could be used in satellite interconnects or deep-space optical communication where reliability over a range of conditions is critical. There is also interest in midinfrared QD lasers for infrared countermeasure systems or remote sensing (detecting chemical agents, pollutants, etc.). QD lasers might also find roles in lidar for autonomous drones or vehicles due to their stability and possibly reduced speckle (if operated in certain multimode regimes). The relatively low manufacturing cost (once scaled) can also be appealing for deploying many units in distributed systems.

5.4.6 General Computing and Photonic Integration

As computing systems look beyond electrons to photons for speed, QD lasers are strong candidates for integration into photonic chips for optical computing. Their tolerance to back-reflections and ability to be integrated without isolators simplify building complex photonic circuits where multiple components (modulators, detectors, waveguides) are connected [16]. Some experimental demonstrations even use QD lasers in optical neural network accelerators, leveraging their fast dynamics to perform computations. The earlier-mentioned demonstration of a QD laser acting as an artificial neuron hints at future optical processors made of networks of such lasers. Furthermore, QD lasers on silicon open possibilities for 3D-integrated systems where lasers are directly bonded or grown above electronic circuits, providing tightly integrated electrophotonic systems for on-chip optical communication.

5.5 Technical and Production Challenges

Figure 5.5 illustrates the basic architecture of a quantum dot laser integrated on a silicon substrate—a key step toward unifying electronic and photonic technologies. The structure comprises a ridge waveguide in which the active layer contains an ensemble of self-assembled quantum dots, as highlighted in the magnified region. Each quantum dot confines carriers in all three spatial dimensions, resulting in discrete electronic energy levels and efficient radiative recombination. This confinement not only lowers threshold current densities but also enhances temperature stability compared to conventional quantum well lasers.

The surrounding cladding layers guide the optical mode along the ridge, ensuring strong overlap with the QD active region to facilitate stimulated emission. The silicon substrate serves as a mechanically stable and thermally

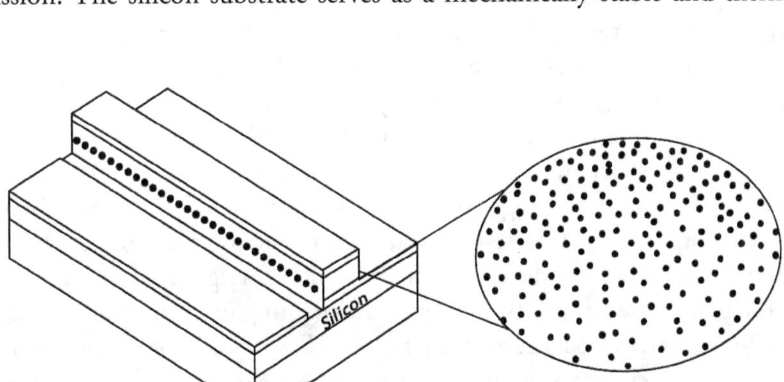

Figure 5.5 Simplified schematic of a quantum dot laser structure integrated on a silicon substrate. The device consists of a layered waveguide configuration with a quantum dot active region (magnified on the right) embedded within the gain medium. Each quantum dot acts as an individual light-emitting center, collectively enabling lasing through quantum confinement effects. The use of silicon as a base platform supports integration with photonic circuits and electronic components.

conductive foundation, while also enabling compatibility with established complementary metal–oxide–semiconductor (CMOS) fabrication processes. This integration capability is particularly valuable for silicon photonics, where on-chip light sources are required for optical interconnects and data communication systems.

The inset on the right of Figure 5.5 emphasizes the nanoscale distribution of quantum dots within the active layer. Although each dot behaves as an individual artificial atom, their collective emission forms the optical gain that drives laser action. The random yet dense arrangement of dots contributes to broad spectral tunability while maintaining the benefits of strong carrier localization. Overall, this schematic encapsulates the physical and technological principles that make quantum dot lasers central to next-generation integrated photonic platforms.

Despite the impressive progress, QD lasers face several technical and production challenges today. Overcoming these issues is crucial for QD lasers to reach their full potential in commercial and scientific domains. We outline the key challenges and difficulties below.

5.5.1 Inhomogeneous Broadening and Dot Uniformity

Unlike atomic systems, practical QD ensembles have variability in size, shape, and composition, leading to slightly different energy levels for each dot. This results in an inhomogeneously broadened gain spectrum, which is essentially a spread of gain frequencies rather than a single sharp peak [41–45]. While some broadening can be tolerated (and even leveraged for tunability), too much broadening means many QDs are off-resonance and do not contribute effectively to lasing. They instead soak up carriers and emit spontaneously, raising threshold. Reducing this inhomogeneity has been a long-standing materials challenge. It requires exquisite control of QD growth to achieve uniform sizes and compositions. Techniques like using patterned substrates to nucleate QDs at specific sites, or optimizing growth kinetics to narrow the dot size distribution, are being developed. State-of-the-art QD materials have achieved 24-meV photoluminescence linewidths at room temperature, a significant improvement indicating fairly uniform dots. However, maintaining uniformity across an entire 300-mm wafer or between production batches is nontrivial. This challenge is tied to manufacturing scale-up: while small-area MBE growth can be tightly controlled, large-scale MOCVD processes might introduce greater distributions that need to be reined in.

5.5.2 Low Dot Density Versus Gain Requirements

A single QD has a tiny volume and can only emit so much light (one or a few photons at a time in essence). For a laser, many QDs must collectively provide enough gain to overcome losses. Early QD lasers struggled because the areal density of active QDs was limited (e.g., 10^{10}–10^{11} cm^{-2}), and not all the semiconductor area contributes to gain, unlike a quantum well which is continuous. The solution has been to grow multiple layers of QDs (often 5 to 10 layers) separated by thin spacers. However, stacking layers can introduce strain and lead to nonuniform dot sizes in upper layers or the formation of elongated dots (quantum dashes on InP). Ensuring each layer is of high quality requires careful strain management. Additionally, too high a dot density per layer can cause dots to interact or coalesce. Thus, finding the right balance and repeating it layer after layer is a challenge. Innovations like using a slightly different alloy in each layer to prevent vertical alignment of dots (which can cause columnar coalescence) have been explored. This challenge is fundamentally about getting sufficient gain: QD lasers often reach a point of gain saturation at high currents because once the ground state of each dot is filled (gain clamped), additional current only goes into excited states or wasteful recombination. Having more dots or more radiative states per dot (like utilizing an excited-state transition) can increase the gain, but may compromise other benefits. Ongoing research tries to increase dot density without sacrificing uniformity, perhaps by using novel self-assembly techniques or even artificial QD placement (lithographic QDs).

5.5.3 Carrier Injection and Phonon Bottleneck

In QD lasers, injecting carriers (electrons and holes) efficiently into the dots is not straightforward [3]. Carriers are first supplied to a higher dimensional reservoir (like a quantum well called the wetting layer or separate confinement layer) and then must relax into the dot's quantized levels. If the separation between energy levels is larger than one longitudinal optical (LO) energy, carriers cannot easily shed the excess energy in one go, this is the so-called phonon bottleneck [46–49]. They may accumulate in the reservoir or excited states, leading to heating and slower modulation response. The tunneling injection approach addresses this by quantum-mechanically coupling a quantum well directly to the dot level [13]. When designed properly, an electron in the injector well can tunnel into the dot's lasing level, bypassing the need to emit a phonon (the process can be aided by phonon resonance, as one team tuned the injector-dot separation to one LO phonon energy [13]). Such designs greatly alleviate hot carrier effects and improve modulation speed, as demonstrated by small-signal modulation bandwidth enhancements. However, adding such complexity (extra layers and thin tunneling barriers) is itself a fabrication challenge. It requires precise control of layer thickness down to a few angstroms.

Figure 5.6 provides a conceptual overview of the tunneling injection mechanism in quantum-dot lasers, which plays a pivotal role in determining

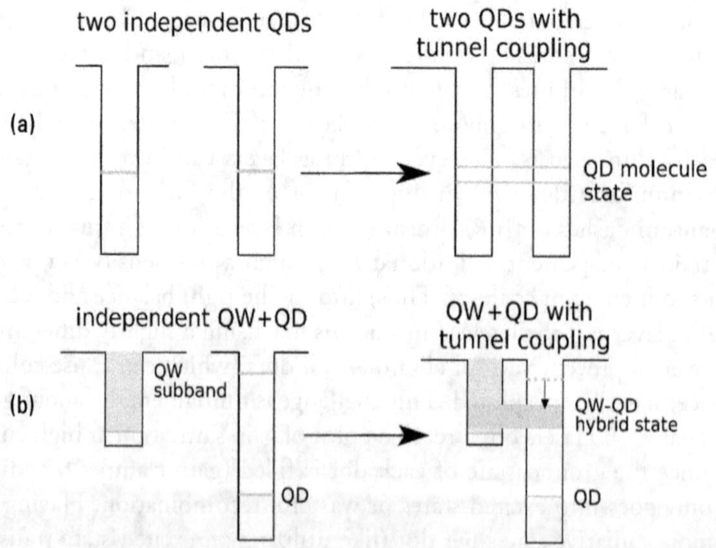

Figure 5.6 Schematic illustration of tunnel coupling mechanisms relevant to tunneling injection in QD lasers: (a) formation of a QD molecule state due to tunneling between two independent quantum dots, and (b) hybridization between a QW subband and a QD energy level when tunnel coupling is introduced, leading to an energy-lowered QW–QD hybrid state. The diagrams highlight the evolution from isolated potential wells to coupled systems, illustrating how band alignment and tunneling interactions govern carrier dynamics in tunneling injection architectures. (Image inspired by [8].)

carrier capture and recombination dynamics. As shown in part (a), when two quantum dots are spatially separated and electronically independent, their discrete energy states remain localized. Upon introducing tunnel coupling between them, however, the individual dot states hybridize to form a QD molecule state. This coupling leads to delocalization of electronic wave functions, effectively modifying the carrier transport properties within the active region.

Panel (b) extends this concept to a quantum well–quantum dot (QW–QD) system, where the QW subband acts as a carrier reservoir. In the absence of tunneling, carriers must rely on thermal or phonon-assisted relaxation to populate the QD levels. When tunnel coupling is enabled, a hybrid state forms between the QW subband edge and the QD excited (p-shell) state, resulting in a lowering of the system's overall energy. This process enhances the efficiency of carrier injection into the QD ground (s-shell) state and can improve modulation response under optimal band alignment conditions. However, as discussed in the text, structural inhomogeneity among self-assembled QDs can disrupt uniform coupling, potentially degrading laser performance. The figure thus visually captures both the physical mechanism and the potential trade-offs inherent in tunneling injection QD laser design.

Not all QD laser designs use tunneling injection; many rely on fast scattering processes (interface roughness scattering, Auger, etc.) that were found to partly circumvent the phonon bottleneck. Nonetheless, carrier dynamics engineering remains a challenge: to further increase direct modulation speeds toward 50 to 100 GHz, one must manage how quickly carriers can be captured and recycled in QDs. P-type doping of QDs is another strategy used to prefill dots with holes, reducing the hole capture bottleneck and mitigating thermal escape of holes at high temperatures [50]. But doping can introduce its own issues (like increased Auger recombination or altered band structure). Thus, balancing injection efficiency, speed, and side effects is an ongoing design challenge.

As illustrated in Figure 5.7, the electronic band profiles of InAs/GaAs quantum-dot lasers reveal the contrasting effects of undoped and codoped structures on carrier dynamics [50]. In the undoped configuration (Figure 5.7(a)), the conduction and valence bands exhibit well-defined quantum confinement, yet carrier recombination efficiency is limited by an imbalance between electron and hole populations within the InAs quantum dot. In contrast, the codoped structure (Figure 5.7(b)) introduces direct n-type doping inside the InAs QD, supplying excess electrons, while the modulation p-type doping

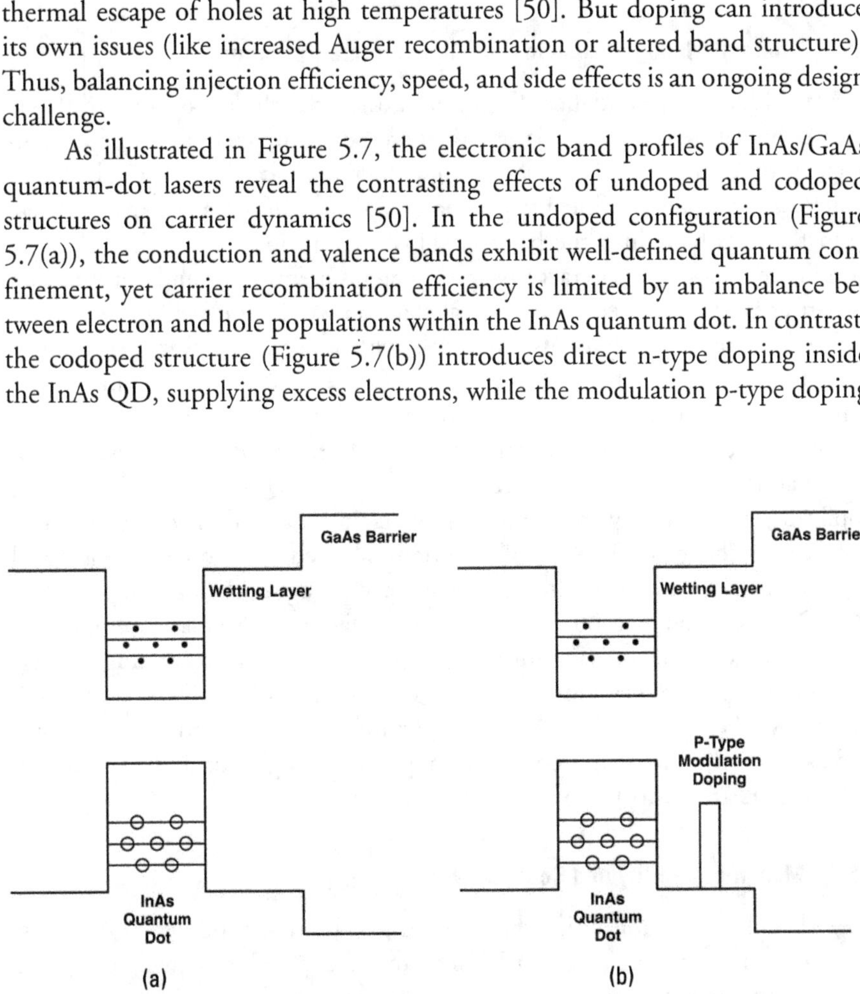

Figure 5.7 (a) Energy band diagram of an undoped InAs/GaAs quantum-dot structure showing discrete confined states in the InAs quantum dot and the InAs wetting layer, and (b) schematic illustration of the codoped configuration incorporating both direct n-type doping within the quantum dot and modulation p-type doping in the adjacent GaAs barrier region. The combined codoping technique enhances carrier confinement and radiative recombination efficiency by balancing electron and hole populations within the active region. (Image inspired by [50].)

layer positioned near the barrier region contributes additional holes that diffuse toward the active zone. This strategic doping configuration establishes a quasi-charge neutrality that strengthens radiative recombination and mitigates ground-state quenching at elevated temperatures. Consequently, the codoping scheme leads to lower threshold current densities and improved temperature stability across a wide operational range, as confirmed by both experimental and simulated results in the study.

5.5.4 Thermal Management and High-Temperature Operation

While QD lasers have higher T_0 (meaning their threshold varies more slowly with temperature) than QW lasers, they are not completely immune to temperature effects [2, 51, 52]. At very high temperatures, carriers (especially holes, which have smaller confinement in many QD systems) can thermally escape from the dots back into the barriers. This leads to loss of population inversion in the dots. To counteract this, p-doping (as mentioned) and optimization of barrier height are used, but an ideal solution would confine both electrons and holes more strongly or use materials with inherently lower carrier effective masses. Additionally, like any laser, QD lasers generate heat during operation (from nonradiative recombination and resistive heating). Their small active volume means heat can concentrate; if not removed, it will raise the dot temperature and affect performance. For high-power QD lasers (e.g., pumping applications or high-power DFB lasers), managing heat and preventing thermal rollover is a challenge. The use of radiation-balanced laser concepts (discussed later) or improved heat sinking in device packaging are being considered to keep QD lasers cool without active cooling. So far, many QD lasers still require temperature control to maintain optimal performance in demanding settings (though for moderate environments they can run uncooled). Thus, continuing to improve the intrinsic thermal stability (through band structure engineering) and extrinsic thermal management (through packaging) is critical, particularly as QD lasers are pushed to higher output powers.

5.5.5 Manufacturability and Scalability

Even though remarkable strides have been made in mass-producing QD lasers, scaling to full commercialization involves numerous practical challenges. One issue is consistency: ensuring that every chip on a wafer has similar QD size and density, and thus similar threshold and wavelength. Small fluctuations in growth temperature or flux can cause noticeable changes in QD properties. Inline monitoring and feedback control during epitaxy may be needed to tighten this consistency. Another issue is that traditional laser fabrication processes (etching, metallization, facet cleaving, or polishing) might need adaptation for QD material—for example, QD active layers might be sensitive to certain dry etch chemistries, or the facet coating process might interact with the dot layers differently than with bulk or QW layers. When moving to large wafers, thermal stresses during cool-down (in epitaxy) or subsequent wafer bonding (for heterogeneously integrating QD lasers with silicon) can introduce defects or cracks; these must be managed by careful thermal cycle design. Testing and yield for QD lasers also pose challenges: the inhomogeneous gain means that lasing wavelength might vary across a wafer more than in QW lasers, so binning or trimming (maybe via current or thermal tuning) might be necessary to get lasers on exact channel wavelengths for DWDM communications, for instance. Ultimately, cost reduction is a challenge: today QD lasers are a bit more niche and costly than equivalent DFB or Fabry–Perot QW lasers, mainly due to lower volumes and trickier growth. If yield and uniformity can be improved, and perhaps if QD structures can be grown on larger substrates (like GaAs or Si) with many devices per wafer, the cost per device can drop significantly. This is actively being pursued, since cost competitiveness will decide the extent to which QD lasers replace conventional ones in the market.

5.5.6 Integration and Packaging

Finally, integrating QD lasers into systems (particularly photonic integrated circuits) raises some challenges. For example, feedback sensitivity is low but not zero, and therefore designing circuits without isolators requires careful consideration of reflections. QD lasers have different impedance and capacitance characteristics, so driving circuits might need optimization to achieve the fastest modulation speeds. For integrating on silicon, differences in thermal expansion and refractive index have to be managed to efficiently couple light from the QD laser (III-V material) to silicon waveguides (one approach used in the 300-mm demonstration was a butt-coupled cavity design to directly couple light laterally). Also, QD material on silicon sometimes exhibits a higher threshold than on native substrates due to residual defects; improving that is a continuing effort. On the packaging side, if QD lasers are more temperature tolerant, one might be able to eliminate thermoelectric coolers, but then one must ensure the system can handle slightly varying wavelengths or output power due to ambient changes. Moreover, for high-speed use, the parasitics in packaging (bond wires, etc.) must be minimized, something true for all lasers, but every new laser structure has to be characterized for microwave performance.

In summary, while QD lasers have surmounted many scientific hurdles, these engineering and manufacturing challenges are the focus of current R&D. Each challenge, however, is being actively addressed through a combination of innovative design (like tunnel injection to fix bottlenecks) and process improvement (like better uniformity and larger scale growth). The trajectory of QD lasers so far gives confidence that these issues will be progressively resolved, much as earlier hurdles (e.g., early high thresholds, limited output) were overcome by the community.

5.6 Cross-Domain Analogies Driving Innovation

Just as tunneling and confinement analogies gave birth to QD lasers, the field continues to evolve through new rounds of analogy thinking. Each innovation reflects a fresh mapping from another discipline—biology, thermodynamics, metamaterials, or artificial intelligence—into photonics. The following sections examine how these cross-domain analogies embody the recursive cycle of analogy thinking introduced in Chapter 1, where new source domains continually feed and refine photonic technologies.

From the very beginning, analogy has played a powerful role in the development of QD lasers, the analogy to quantum tunneling and particles in a box led to their conceptual birth. Moving forward, researchers continue to look to other scientific domains and even nature for analogies that could spark new ideas or solutions. In this section, we explore how analogies from various fields, such as biology, thermodynamics, metamaterials, and artificial intelligence, might influence the next innovations in QD lasers and photonic devices.

5.6.1 Biological Inspiration

Learning from nature's light systems: Natural organisms have evolved sophisticated ways to harvest, emit, and control light. These biological systems can inspire new approaches to lasers. One striking example is the use of photosynthetic complexes (such as those in purple bacteria, particularly *Rhodobacter sphaeroides*) to create a bioinspired laser [53, 54]. Scientists have recently taken the light-harvesting antenna complexes from these bacteria, which form nanoring structures with near-perfect energy transfer efficiency, and incorporated them into a laser cavity. These rings, approximately 60 nm across, efficiently funnel absorbed sunlight energy into a reaction center. By combining them with man-made gain elements and placing them in a resonant cavity, researchers demonstrated a concept of a sunlight-powered laser that piggybacks on billions of years of evolution in energy transfer. The natural antenna acts much like an exciton funnel, an idea that could be applied to QD lasers: one could imagine an array of quantum dots being fed by an energy transfer network, similar to how the antenna ring feeds a reaction center. This might overcome injection and pumping inefficiencies, essentially a biomimetic energy wiring for lasers.

Biology also offers ideas in self-assembly and patterning. Viruses and proteins can organize nanocomponents with atomic precision. Some researchers are experimenting with virus-templated QD assemblies to achieve superlattices of QDs that might have more uniform spacing or coupling. While still in early stages, such methods could yield highly controlled QD ensembles for lasers or LEDs. Additionally, consider how biological systems manage heat and damage: the eye, for instance, has mechanisms to avoid photoreceptor saturation and uses enzymatic repair for photodamaged molecules. Analogously, one could envision self-healing laser materials or dynamic adaptation in QD laser arrays (perhaps via microfluidic cooling or replacing QD solution, etc., in extreme cases).

The lesson from biology is that novel light-emission strategies (like the single-photon generation in fireflies, random laser action in certain white beetle scales, or the distributed feedback in peacock feather iridescence) can often be reinterpreted in engineering. Already, researchers have looked at biomaterials for random lasers, using materials like cellulose or proteins as scattering matrices with QDs as the gain, creating inexpensive lasers for sensing. Moving forward, the fusion of biomolecules with semiconductor QDs could lead to hybrid devices that operate in environments or modes traditional lasers cannot. The use of biological analogies exemplifies thinking outside the box, or in this case, thinking outside the die, to improve and perhaps revolutionize photonic devices.

5.6.2 Thermodynamic and Heat Engine Analogies

A laser can be thought of as a kind of optical heat engine: it converts a higher energy fuel (pump energy) into a lower energy but more ordered form (coherent laser photons), with waste heat generated in the process. Using analogies from thermodynamics can yield insights into efficiency and new operational regimes. For instance, in thermodynamics one strives for reversible processes to maximize efficiency. In lasers, the concept of radiation-balanced lasers applies a similar idea: if one could eliminate or recycle all waste heat, one could approach a reversible, high-efficiency limit. The radiation-balanced laser uses anti-Stokes fluorescence (optical refrigeration) to literally cool the laser medium as it operates. While demonstrated in certain crystal lasers (like Yb-doped materials), one could envision applying this to semiconductor lasers. Perhaps QD lasers could be designed with built-in upconversion layers that absorb subbandgap luminescence and emit higher energy photons that carry heat away, effectively cooling the chip. This is speculative, but it shows how a thermodynamic analogy (laser as a Carnot cycle) might guide new designs focused on managing entropy and heat.

Another area is the analogy to thermal insulation. In buildings, we use multiple panes of glass with vacuum gaps to reduce heat loss (preventing conduction and convection). Analogously, one might use QD layers as energy insulation in a device, for example, using a wider bandgap barrier around QDs to confine carriers and also block thermal coupling to the lattice. Already, the use of high bandgap shells in colloidal QDs (a concept borrowed from chemistry and thermodynamics of core-shell particles) has greatly improved their efficiency by confining carriers away from nonradiative surface traps [55]. In epitaxial QDs, surrounding a dot with a larger bandgap well (creating something akin to a quantum-dot-in-well, or dot-in-a-well structure) can reduce carrier leakage at higher temperatures, akin to adding extra insulation to keep the carriers from evaporating out. Thermodynamics also teaches the importance of entropy: a laser reduces entropy by creating order (coherence) out of disorder (incoherent pump). If too much disorder (noise, temperature fluctuations) enters the system, the laser performance degrades. Thus, engineers use techniques like regenerative feedback or injection locking to keep lasers in low-entropy states. These can be seen as analogous to refrigerators or engines with feedback control to stabilize temperature. In the context of QD lasers, one could use coupled cavities or mode locking (creating a phase-stable comb) to reduce phase noise (entropy in the phase). Indeed, a recent advance achieved self-injection locking of a QD laser to achieve ultranarrow linewidth suitable for coherent communication, essentially using an external high-Q cavity to enforce order on the QD laser output. This parallels how in thermodynamics a flywheel might stabilize an engine's output. By viewing lasers through a thermodynamic lens, researchers can devise ways to push them closer to ideal efficiency and stability, identifying where entropy generation is happening (be it spontaneous emission, leakage currents, or thermal broadening) and attacking those via creative engineering.

5.6.3 Metamaterials and Photonic Analogies

Metamaterials are engineered structures designed to control waves in ways conventional materials cannot, often drawing analogies to electronic band structure, magnetic responses, and more. In photonics, photonic crystals are a prime example: they were inspired by the analogy to semiconductor crystals, creating a periodic dielectric structure that affects photons similarly to how a periodic atomic lattice affects electrons (producing bandgaps, etc.). This analogy gave rise to photonic crystal lasers, which use a 2D or 3D periodic lattice as the cavity.

As discussed in Chapter 4, metamaterials and photonic crystals represent a broader class of wave-manipulating media where electromagnetic behavior is engineered through structural design rather than intrinsic material composition. The principles introduced there, such as bandgap engineering, negative refractive index, and tailored dispersion, form the conceptual backdrop for this section. Here, we build on those ideas to explore how similar analogies extend into the design and enhancement of QD lasers.

In fact, many photonic crystal lasers have used QD emitters as the gain medium, combining two analogies in one device: a photonic crystal providing feedback and a QD providing the gain. The result are very low threshold lasers with high beam quality. Going forward, metasurfaces (2D metamaterials) might be used to create novel cavity mirrors or mode selectors for QD lasers. For example, a metasurface mirror could impart a phase profile that shapes the output beam or forces the laser into a single lateral mode, improving beam quality for applications like lidar. The use of metamaterial claddings could also help confine light more tightly to the QD active region or even allow coupling between QD lasers in unusual ways (e.g., coupling through subwavelength tunneling channels engineered by metamaterials, an analog of quantum coupling between dots, but implemented optically).

Metamaterial resonators (like split-ring resonators) might be integrated on-chip to create ultrafast modulators for QD lasers by exploiting plasmonic analogy (using plasmonic absorption to modulate the QD gain). Another cross-domain analogy is between quantum tunneling and optical tunneling: just as electrons tunnel through barriers, photons can tunnel through classically forbidden regions (for instance, evanescent waves enabling frustrated total internal reflection, which is often cited as a direct optical analog of quantum tunneling). Engineers use this optical tunneling effect in distributed Bragg reflectors and coupled cavities. In coupled QD laser arrays, evanescent (tunneling-like) coupling could synchronize phases (for coherent combining) or create Vernier effects for multiwavelength generation.

Essentially, by treating light with the same mathematical formalisms as quantum particles, one can port techniques back and forth, an approach sometimes called quantum optical analogies. For QD lasers, which are fundamentally quantum-mechanical systems producing classical light, these analogies can be especially apt. We might see future devices where a metamaterial structure is tuned to mimic a potential landscape, and QDs embedded in it experience modified spontaneous emission (Purcell effect) or new emission pathways. For example, hyperbolic metamaterials (which have high density of photonic states) could be used to enhance spontaneous emission from QDs, effectively increasing the material gain.

In summary, the rich toolset of metamaterials and analogical thinking between electronics and photonics, which were first established in Chapter 4 and extended here, promises new mechanisms to control QD laser behavior, whether through novel feedback structures, coupling mechanisms, or emission control.

5.6.4 Artificial Intelligence and Computational Analogies

The rise of AI and machine learning is not only an application area for lasers (e.g., optical neural networks) but also a tool for designing them. The analogy here is perhaps more abstract—treating a photonic device design problem as something that can be learned or optimized by an artificial brain. Already, machine learning is being applied to optimize QD growth and device performance. For instance, researchers have trained models to predict how changes in growth conditions affect QD size distribution or to optimize laser cavity designs for certain performance metrics. This is analogous to having a virtual apprentice that learns from data rather than explicit human analogies. However, AI itself can suggest analogies: generative algorithms might identify nonintuitive patterns in QD arrangements that mimic, say, a biological structure or a mathematical tiling, which human designers hadn't considered. On the application side, analogy to neural networks has directly inspired using lasers as computing elements. We discussed the demonstration of a QD mode-locked laser acting as an all-optical neuron—here the designers explicitly drew an analogy between the limit-cycle behavior of a passively mode-locked laser and the spiking of a neuron, and showed that the laser could integrate and fire optical pulses in a neural-like fashion. This cross-pollination between neuroscience and laser physics could pave the way for optical neuromorphic processors where each neuronal node is a tiny QD laser. The advantage would be the high speed (GHz spiking) and the direct compatibility with fiber-optic communication for linking nodes.

Moreover, AI can help manage complexity: an optical circuit with many QD lasers and photonic components is complex to tune, but a machine learning controller could adapt bias currents and phase shifters to maintain optimal operation (an analogy to how our brain's neural networks self-adjust synapses). Lastly, one intriguing analogy is using QDs themselves as neurons in a computational material, given their ability to have two states (like a two-level system) and optical coupling. While this borders on quantum computing, some have proposed QD networks for Ising-type computations.

In any case, the interplay of AI and photonics is growing: Not only do lasers enable optical computing for AI, but AI in turn enables better lasers. The analogy here is more of a virtuous cycle than a one-to-one mapping. Each of these analogies—biological, thermodynamic, metamaterial, AI—provides a different perspective on overcoming the challenges or expanding the capabilities of QD lasers. Biological analogies teach us about efficient light harvesting and self-assembly, which could improve how we pump and manufacture lasers. Thermodynamic analogies highlight ways to manage heat and noise, pushing lasers toward physical efficiency limits. Metamaterial analogies inspire new structures for feedback and emission control, potentially leading to lasers with exotic properties (like ultranarrow beams or simultaneous multicolor outputs). AI and neural analogies not only offer new applications (neuromorphic computing) but also help solve complex design and control problems. Historically, the analogy to quantum tunneling was a conceptual masterstroke that led to the invention of QD lasers; moving forward, continued analogical thinking will similarly be the catalyst for innovation in photonic devices. In research and development, an engineer who can think like a physicist, a biologist, and a computer scientist all at once—drawing connections between disparate phenomena—is well equipped to invent the next-generation laser.

5.7 Conclusion

The story of QD lasers is, at its core, a case study in analogy thinking. Following the framework introduced in Chapter 1, researchers moved from a well-understood source domain (quantum tunneling) to a challenging target domain (semiconductor lasers), performed mapping and abstraction to extract the idea of confinement, executed transfer through nanofabrication, refined designs via constraint matching, and achieved insight generation that transformed photonic engineering.

What began as an abstract comparison between tunneling barriers and carrier confinement gradually evolved through iterative validation into one of the most versatile and scalable laser technologies of our time. Each subsequent stage, including commercialization, integration with silicon, and expansion into new materials, represents a continued progression of the analogy-thinking cycle that generates new sources, new mappings, and new insights.

Looking forward, QD lasers remind us that analogy thinking is not just a cognitive tool but a creative discipline. Whether drawing from biology's efficiency, thermodynamics' balance, or AI's adaptability, the future of photonics will likely depend on how effectively researchers continue to think by analogy—connecting the known to the unknown to generate the next wave of innovation.

References

[1] Esaki, L., "New Phenomenon in Narrow Germanium p–n Junctions," *Physical Review*, Vol. 109, No. 2, 1958, pp. 603–604, https://doi.org/10.1103/PhysRev.109.603.

[2] Arakawa, Y., and H. Sakaki, "Multidimensional Quantum Well Laser and Temperature Dependence of Its Threshold Current," *Applied Physics Letters*, Vol. 40, No. 11, June 1982, pp. 939–941, https://doi.org/10.1063/1.92959.

[3] Akahane, K., "Introduction and History of Quantum Dot Lasers," in *Handbook of Radio and Optical Networks Convergence*, T. Kawanishi (ed.), Singapore: Springer, 2024, https://doi.org/10.1007/ 978-981-33-4999-5 49-1.

[4] Kirstaedter, N., et al., "Low Threshold, Large T_0 Injection Laser Emission from (InGa)As Quantum Dots," *Electronics Letters*, Vol. 30, No. 17, 1994, pp. 1416–1417, https://doi.org/10.1049/el:19940939.

[5] Asada, M., et al., "Gain and the Threshold of Three-Dimensional Quantum-Box Lasers," *IEEE Journal of Quantum Electronics,* Vol. 22, No. 9, 1986, pp. 1915–1921, https://doi.org/10.1109/ JQE.1986.1073149.

[6] Ekimov, A. I., et al., "Quantum Size Effect in Semiconductor Microcrystals," *Solid State Communications,* Vol. 88, No. 11–12, 1993, pp. 947– 950, https://doi.org/10.1016/0038-1098(93) 90275-R.

[7] Merz, J. P., et al., "Epitaxial Growth and Optical Properties of Quantum-Dot Structures," *Journal of Applied Physics,* Vol. 61, No. 10, 1987, pp. 4996–5001, https://doi.org/10.1063/1.338660.

[8] Khanonkin, I., et al., "On the Principle Operation of Tunneling Injection Quantum Dot Lasers," *Progress in Quantum Electronics,* Vol. 81, Jan. 2022, p. 100362, https://doi.org/10.1016/j.pquantelec.2021.100362.

[9] Lester, L. F., et al., "Optical Characteristics of 1.24 μm InAs Quantum-Dot Laser Diodes," *IEEE Photonics Technology Letters,* Vol. 11, No. 8, 1999, pp. 931–933, https://doi.org/10.1109/68.775303.

[10] Park, .G, et al., "Room-Temperature Continuous-Wave Operation of a Single-Layered 1.3 μm Quantum Dot Laser," *Applied Physics Letters,* Vol. 75, No. 21, 1999, pp. 3267–3269, https://doi.org/10.1063/1.125320.

[11] Mirin, R., et al., "Room Temperature Lasing from InGaAs Quantum Dots," *Electronics Letters,* Vol. 32, No. 18, 1996, pp. 1732–1734, https://doi.org/10.1049/el:19961147.

[12] Kim, K., et al., "Gain Dynamics and Ultrafast Spectral Hole Burning in In(Ga)As Self-Organized Quantum Dots," *Applied Physics Letters,* Vol. 81, No. 4, 2002, pp. 670–672, https://doi.org/10.1063/1.1493665.

[13] Norman, J. C., et al., "A Review of High-Performance Quantum Dot Lasers on Silicon," *IEEE Journal of Quantum Electronics,* Vol. 55, No. 2, 2019, p. 2000511, https://doi.org/10.1109/JQE.2019. 2901508.

[14] Shang, C., et al., "Electrically Pumped Quantum-Dot Lasers Grown on 300mm Patterned Si Photonic Wafers," *Light: Science & Applications,* Vol. 11, 2022, p. 299, https://doi.org/10.1038/s41377-022-00982-7.

[15] Hahm, D., et al., "Colloidal Quantum Dots Enable Tunable Liquid-State Lasers," *Nature Materials,* Vol. 24, 2025, pp. 48–55, https://doi.org/10.1038/s41563-024-02048-y.

[16] Shekhar, S., et al., "Roadmapping the Next Generation of Silicon Photonics," *Nature Communications,* Vol. 15, 2024, p. 751, https://doi.org/10.1038/s41467-024-44750-0.

[17] Wang, C., et al., "Enhanced Dynamic Performance of Quantum Dot Semiconductor Lasers Operating on the Excited State," *IEEE Journal of Quantum Electronics,* Vol. 50, No. 9, 2014, pp. 723–734, https://doi.org/10.1109/JQE.2014.2335811.

[18] Zhang, Z., et al., "25 Gb/s Directly Modulated Ground-State Operation of 1.3 μm InAs/GaAs Quantum Dot Lasers up to 75 C," *Chinese Optics Letters,* Vol. 18, No. 7, 2020, p. 071401.

[19] Koscica, R., et al., "Quantum Dot DBR Lasers Monolithically Integrated on Silicon Photonics by In-Pocket Heteroepitaxy," *Journal of Lightwave Technology,* Vol. 43, No. 12, 2025, pp. 5773–5781, https://doi.org/10.1109/JLT.2025.3555555.

[20] Zenari, M., et al., "Degradation of 1.3 μm Quantum Dot Laser Diodes for Silicon Photonics: Dependence on the Number of Dot-in-a-Well Layers," *IEEE Journal of Selected Topics in Quantum Electronics*, Vol. 31, No. 2, 2025, p. 1500308, https://doi.org/10.1109/JSTQE.2024.3430050.

[21] Raun, A., et al., "Ultralow Threshold Blue Quantum Dot Lasers: What's the True Recipe for Success?" *Nanophotonics*, Vol. 10, No. 1, 2020, pp. 97–121, https://doi.org/10.1515/nanoph-2020-0382.

[22] Lin, X., et al., "Blue Lasers Using Low-Toxicity Colloidal Quantum Dots," *Nature Nanotechnology*, Vol. 20, 2025, pp. 229–236, https://doi.org/10.1038/s41565-024-01812-0.

[23] Nevsky, A. .Y., et al., "A Narrow-Line-Width External Cavity Quantum Dot Laser for High-Resolution Spectroscopy in the Near-Infrared and Yellow Spectral Ranges," *Applied Physics B*, Vol. 92, 2008, pp. 501–507, https://doi.org/10.1007/s00340-008-3113-4.

[24] Chen, T.- L., et al., "Noise-Immune Cavity-Enhanced Optical Heterodyne Molecular Spectrometry on N2O 1.283 μm Transition Based on a Quantum-Dot External-Cavity Diode Laser," *Optics Letters*, Vol. 40, No. 18, 2015, pp. 4352–4355, https://doi.org/10.1364/OL.40.004352.

[25] Chen, T. -L., et al., "Sub-Doppler Resolution Near-Infrared Spectroscopy at 1.28 μm with the Noise-Immune Cavity-Enhanced Optical Heterodyne Molecular Spectroscopy Method," *Optics Letters*, Vol. 42, No. 13, 2017, pp. 2447–2450, https://doi.org/10.1364/OL.42.002447.

[26] Tang, H., et al., "A Review of High-Power Semiconductor Optical Amplifiers in the 1550 nm Band," *Sensors*, Vol. 23, No. 17, 2023, p. 7326, https://doi.org/10.3390/s23177326.

[27] Yuan, Z., et al., "Electrically Driven Single-Photon Source," *Science*, Vol. 295, No. 5552, 2002, pp. 102–105, https://doi.org/10.1126/science.1066790.

[28] Michler, P., et al., "A Quantum Dot Single-Photon Turnstile Device," *Science*, Vol. 290, No. 5500, 2000, pp. 2282–2285, https://doi.org/10.1126/science.290.5500.2282.

[29] Bennett, A. J., et al., "Microcavity Single-Photon-Emitting Diode," *Applied Physics Letters*, Vol. 86, No. 18, 2005, p. 181102, https://doi.org/10.1063/1.1921332.

[30] Ellis, D. J. P., et al., "Oxide-Apertured Microcavity Single-Photon Emitting Diode," *Applied Physics Letters*, Vol. 90, No. 23, 2007, p. 233514, https://doi.org/10.1063/1.2747200.

[31] Tokman, M., et al., "Generation of Entangled Photons via Parametric Down-Conversion in Semiconductor Lasers and Integrated Quantum Photonic Systems," *Physical Review A*, Vol. 105, No. 3, 2022, p. 033707, https://doi.org/10.1103/PhysRevA.105.033707.

[32] Stevenson, R. M., et al., "A Semiconductor Source of Triggered Entangled Photon Pairs," *Nature*, Vol. 439, No. 7073, 2006, pp. 179–182, https://doi.org/10.1038/nature04446.

[33] Akopian, N., et al., "Entangled Photon Pairs from Semiconductor Quantum Dots," *Physical Review Letters*, Vol. 96, No. 13, 2006, p. 130501, https://doi.org/10.1103/PhysRevLett.96.130501.

[34] Dousse, A., et al., "Ultrabright Source of Entangled Photon Pairs," *Nature*, Vol. 466, No. 7303, 2010, pp. 217–220, https://doi.org/10.1038/nature09148.

[35] Liu, J., et al., "A Solid-State Source of Strongly Entangled Photon Pairs with High Brightness and Indistinguishability," *Nature Nanotechnology*, Vol. 14, 2019, pp. 586–593, https://doi.org/10.1038/ s41565-019-0435-9.

[36] Dong, B., et al., "Broadband Quantum-Dot Frequency-Modulated Comb Laser," *Light: Science & Applications*, Vol. 12, 2023, p. 182, https://doi.org/10.1038/s41377-023-01225-z.

[37] Huang, J.- Z., et al., "Ultra-Broadband Flat-Top Quantum Dot Comb Lasers," *Photonics Research*, Vol. 10, No. 5, 2022, https://doi.org/10.1364/PRJ.446349.

[38] Liu, L., et al., "Simple and Seamless Broadband Optical Frequency Comb Generation using an InAs/InP Quantum Dot Laser," *Optics Letters*, Vol. 42, No. 6, 2017, pp. 1173–1176, https://doi.org/10.1364/OL.42.001173.

[39] Ledentsov, Jr., N., et al., "Quantum Dot 850 nm VCSELs with Extreme High Temperature Stability Operating at Bit Rates up to 25 Gbit/s at 150 C," *Solid-State Electronics*, Vol. 155, 2019, pp. 150–158, https://doi.org/10.1016/j.sse.2019.03.018.

[40] Abdollahinia, A., et al., "Temperature Stability of Static and Dynamic Properties of 1.55 μm Quantum Dot Lasers," *Optics Express*, Vol. 26, No. 5, 2018, pp. 6056–6066, https://doi.org/10.1364/OE.26.006056.

[41] Chow, W. W., et al., "Extraction of Inhomogeneous Broadening and Nonradiative Losses in InAs Quantum-Dot Lasers," *Applied Physics Letters*, Vol. 107, No. 17, 2015, p. 171106, https://doi.org/10.1063/1.4934838.

[42] Kim, J., et al., "Effect of Inhomogeneous Broadening on Gain and Phase Recovery of Quantum-Dot Semiconductor Optical Amplifiers," *IEEE Journal of Quantum Electronics*, Vol. 46, No. 11, 2010, pp. 1670–1680, https://doi.org/10.1109/JQE.2010.2058793.

[43] Dery, H., et al., "The Impact of Energy Band Diagram and Inhomogeneous Broadening on the Optical Differential Gain in Nanostructure Lasers," *IEEE Journal of Quantum Electronics*, Vol. 41, No. 1, 2005, pp. 26–35, https://doi.org/10.1109/JQE. 2004.837953.

[44] Basset, F. B., et al., "Spectral Broadening in Self-Assembled GaAs Quantum Dots with Narrow Size Distribution," *Journal of Applied Physics*, Vol. 126, No. 2, 2019, p. 024301, https://doi.org/10.1063/1.5097277.

[45] Ngo, C.- Y., et al., "Effects of Size and Shape on Electronic States of Quantum Dots," *Physical Review B*, Vol. 74, 2006, p. 245331, https://doi.org/10.1103/PhysRevB.74.245331.

[46] Sugawara, M., et al., "Effect of Phonon Bottleneck on Quantum-Dot Laser Performance," *Applied Physics Letters*, Vol. 71, No. 19, 1997, pp. 2791–2793, https://doi.org/10.1063/1.120135.

[47] Razm-Pa, M., et al., "Phonon Bottleneck and Auger Recombination Effect on the Performance of a Self-Assembled Quantum Dot Laser," *Silicon*, Vol. 10, 2018, pp. 1513–1521, https://doi.org/10.1007/s12633-017-9634-x.

[48] Huang, Y. Q., et al., "Effect of a Phonon Bottleneck on Exciton and Spin Generation in Self-Assembled Quantum Dots," *Physical Review Applied*, Vol. 9, 2018, p. 044037, https://doi.org/10.1103/PhysRevApplied.9.044037.

[49] Ikeda, K., et al., "Phonon Bottleneck Effects in InAs/GaInP Quantum Dots," *Journal of Luminescence*, Vol. 108, No. 1-4, 2004, pp. 273–276, https://doi.org/10.1016/j.jlumin.2004.01.058.

[50] Deng, H., et al., "1.3 μm InAs/GaAs Quantum-Dot Lasers with p-Type, n-Type, and Co-Doped Modulation," *Advanced Physics Research*, Vol. 3, No. 10, 2024, p. 2400045, https://doi.org/10.1002/apxr.202400045.

[51] Mukai, K., et al., "High Characteristic Temperature of Near-1.3 μm InGaAs/GaAs Quantum-Dot Lasers at Room Temperature," *Applied Physics Letters*, Vol. 76, No. 23, 2000, pp. 3349–3351, https://doi.org/10.1063/1.126644.

[52] Shchekin, O. B., et al., "1.3 μm InAs Quantum Dot Laser with T_0 = 161 K from 0 to 80°C," *Applied Physics Letters*, Vol. 80, No. 18, 2002, pp. 3277–3279, https://doi.org/10.1063/1.1476708.

[53] Mattiotti, F., et al., "Bio-Inspired Natural Sunlight-Pumped Lasers," *New Journal of Physics*, Vol. 23, No. 10, 2021, p. 103015, https://doi.org/10.1088/1367-2630/ac2852.

[54] Pullerits, T., et al., "Exciton Dynamics in the Light-Harvesting Complexes of *Rhodobacter sphaeroides*," *Chemical Physics Letters*, Vol. 224, No. 3-4, 1994, pp. 355–365, https://doi.org/10.1016/0009-2614(94)00561-3.

[55] Fang, F., et al., "Core-Shell Structure Induced Surface Reconstruction of PbS Quantum Dots Toward High-Detectivity Short-Wave Infrared Photodetectors," *Inorganic Chemistry Frontiers*, 2025, https://doi.org/10.1002/inc2.70003.

6

Mirasol Display Technology: From Butterfly Wings to MEMS Innovation

After reading this chapter, you should be able to understand:

- How biomimicry of butterfly wings enabled Mirasol display innovation;
- How manufacturing challenges and competition limited commercial success;
- How Mirasol's legacy continues to influence reflective display research.

6.1 Introduction

As outlined in Chapter 1, this case exemplifies the role of analogy thinking as a structuring principle within bioinspired design. The Mirasol display's origin, rooted in observations of butterfly wing iridescence, can be seen as a paradigmatic instance of what Chapter 1 described as the transition from natural inspiration to technological embodiment.

Long before human engineers dreamed of colorful low-power screens, nature had already solved the problem of vivid color without pigments. The wings of butterflies, especially the tropical Blue Morpho, display brilliant iridescent blues that do not come from any dye. Instead, their color arises from microscopic structure. In the Morpho's case, tiny tree-like nanostructures on its wing scales interact with light waves to amplify some colors and cancel others, a phenomenon now known as structural coloration. Early naturalists in the nineteenth century, such as Goureau in 1842 [1], noted these iridescent hues and correctly suspected they were caused by physical structures rather than pigments. However, it wasn't until the advent of electron microscopy in the mid-twentieth century that scientists could actually see the nanoscale ridges and layers on butterfly scales and confirm how they produced color. Researchers like A. G. Anderson and A. G. Richards, in the 1940s [2], documented that the Morpho's dazzling blue comes from multilayer interference: the wing's surface is covered in overlapping scales, each bearing incredibly fine ridges and microribs. When white light strikes those nanostructures, certain wavelengths (blue, in this case) constructively interfere and reflect intensely, while other wavelengths cancel out. The result is a vivid blue that never fades because it's generated by geometry rather than a dye.

Such natural optical engineering captured the imagination of scientists and engineers. If a butterfly wing—made of biological material—could manipulate light so precisely, perhaps human-made structures could do the same. Throughout the late-twentieth century, as micro- and nanofabrication techniques improved, the Morpho idea moved from scientific curiosity into practical photonics. The Morpho butterfly offered a powerful analogy: by mimicking the layered nanostructures found in its wing scales, one could create brilliant colors or optical effects in an artificial device. By the 1990s, researchers in optics and materials science were actively trying to recreate structural color in the lab. The stage was set for a device that would use this principle to solve a pressing engineering problem: how to make a display screen that is as bright and color-rich as an LCD, yet as energy-efficient and sunlight-visible as a butterfly's wing.

The *Morpho didius* male butterfly demonstrates a remarkable dichotomy in wing coloration, as illustrated in Figure 6.1. The dorsal (upper) surface of the wings displays a vivid, metallic blue sheen—an optical effect produced by structural coloration, where nanoscale ridges on the scales interfere with incident light to create intense iridescence. In contrast, the ventral (lower) surface

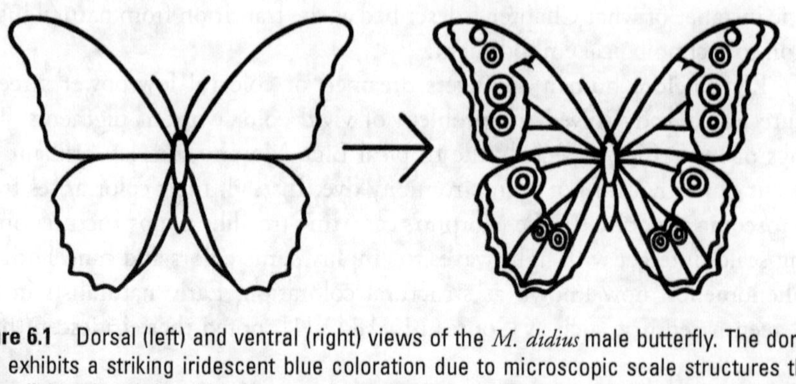

Figure 6.1 Dorsal (left) and ventral (right) views of the *M. didius* male butterfly. The dorsal side exhibits a striking iridescent blue coloration due to microscopic scale structures that reflect light selectively, while the ventral side displays a muted brown hue with multiple ocelli (eye spots) that provide camouflage and predator deterrence.

adopts a cryptic brown tone adorned with concentric eyespots. This duality serves as an adaptive mechanism: the dorsal brilliance is used in territorial or mating displays, while the subdued ventral side provides camouflage when the butterfly is at rest with wings closed. Such contrasting pigmentation is a defining feature of the Morpho genus, exemplifying the intersection of optical physics and evolutionary adaptation in Lepidoptera.

6.2 From Butterfly Wings to Interferometric Modulators

The scientific principle behind the Morpho's color is multilayer interference in a periodic structure. In simple terms, when light waves reflect off multiple closely spaced surfaces, the reflections can interfere with each other. If the spacing is tuned just right (on the order of the wavelength of light), certain colors will add up (constructive interference) while others cancel out (destructive interference). The Morpho's wing scales contain slender, branched nan-ridges (400 nm in size) stacked in layers–often likened to a Christmas tree structure–that act as a natural multilayer interference filter [3–7]. Only specific wavelengths (e.g., blue) are reflected back strongly to the viewer, producing a pure, intense color. This is why the Morpho's blue color shifts not only in brightness but also in hue with changes in viewing angle and lighting, an effect known as iridescence.

The hierarchical nanostructure of the Morpho butterfly's wings, as depicted in Figure 6.2, serves as the foundation for its vivid blue iridescence and optical sensitivity. Each scale contains a series of lamellae stacked periodically along ridges, forming a photonic lattice that interacts strongly with visible light through multilayer interference and diffraction. The geometry—defined by the ridge width, lamella width, and alternating layers of dielectric (l_1) and air gaps (l_2)—determines the reflected wavelength and color purity. Such a configuration

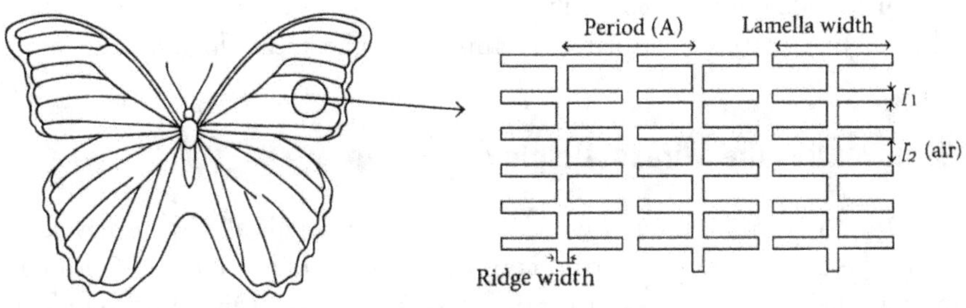

Figure 6.2 Schematic representation of the Morpho butterfly wing structure showing the hierarchical arrangement of ridge-lamella photonic nanostructures. The figure illustrates the periodic ridge width, lamella width, air gap (l_2), and dielectric layer thickness (l_1), which together define the characteristic period (A) responsible for the butterfly's structural coloration through multilayer interference and diffraction effects.

produces high reflectance in the blue spectral range while suppressing other wavelengths, resulting in the butterfly's distinctive hue. Mimicking this architecture enables engineered photonic structures with tunable optical responses for sensing, display, and vapor-detection applications.

Translating this to human technology, one can engineer an optical cavity or thin-film stack that selectively reflects a desired color. In optics, a well-known example is the Fabry–Perot interferometer, which consists of two parallel reflecting surfaces forming a resonant cavity. Light of certain wavelengths will resonate in the cavity and emerge constructively. By adjusting the cavity thickness, you can tune the color it reflects. The key insight for a display is that if you could dynamically adjust the cavity spacing electrically, you could switch colors on and off. This idea, a tunable interferometric reflector, is what bridges the butterfly wing to a digital screen. By the late 1990s, engineers realized that advances in microelectromechanical systems (MEMS) made it feasible to build such tiny tunable mirrors on a chip. Instead of a static stack of layers like the Morpho wing, one could create an array of microcavities with movable mirrors [8–12]. Each microscopic cavity would act like an on-demand Fabry–Perot etalon: in one position it reflects a particular color, and in another position it turns dark. An array of millions of these could form pixels on a display. This concept became known as the interferometric modulator (IMOD), a MEMS device that modulates light via interference [13–21].

As illustrated in Figure 6.3, the Mirasol display operates on the principle of interferometric modulation, where the perceived color arises from constructive interference within optical cavities of varying thicknesses. Each pixel is composed of stacked reflective membranes that define independent air gaps corresponding to red, green, and blue subpixels. These cavities act as resonant optical filters—when the membrane spacing is adjusted electrostatically, the interference conditions change, and the reflected wavelength shifts accordingly. This mechanism allows the display to produce vivid colors without the need for traditional backlighting or color filters, significantly enhancing power efficiency and sunlight readability compared to emissive display technologies.

6.3 Inventing the Mirasol Display

The inspiration from butterfly wings coalesced into a radical new display concept in the early 2000s. A startup company called Iridigm Display Corporation, founded in 2000, set out to commercialize this idea. One of its founders, electrical engineer Mark W. Miles, is credited with inventing a MEMS-based interferometric modulator inspired by the Blue Morpho's wing [13, 22]. The goal was ambitious: build an electronic screen that reflects light to show color (just as a butterfly reflects only certain wavelengths) instead of using materi-

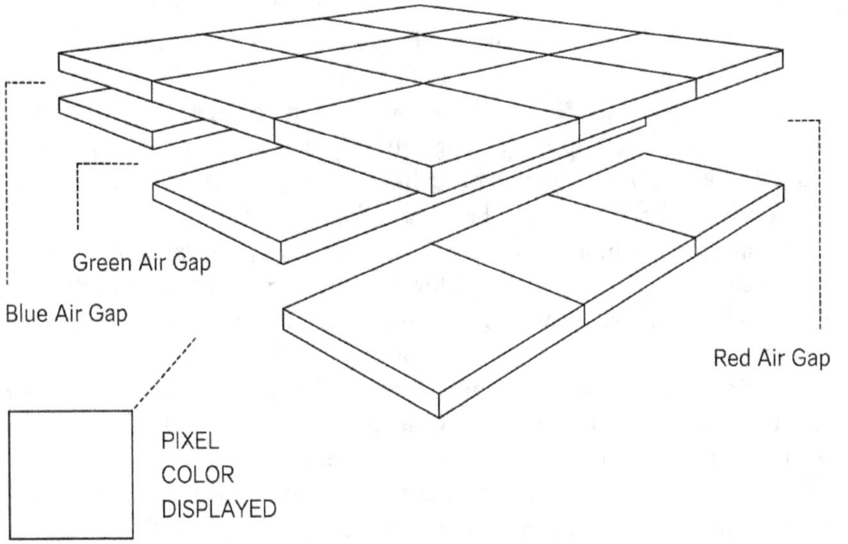

Figure 6.3 Schematic illustration of a single Mirasol display pixel structure. Each pixel consists of multiple reflective membranes separated by air gaps tuned to specific optical resonances corresponding to red, green, and blue wavelengths. By varying the spacing of these air gaps, the interferometric modulation mechanism selectively reflects different colors, resulting in the pixel color displayed.

als with wavelength-selective optical properties or backlights. Such a screen, in theory, would be easily viewable in bright sunlight and consume far less power than an LCD, since it wouldn't need a backlight. Miles and his colleagues at Iridigm developed tiny MEMS devices, which were essentially miniature mirrors with adjustable spacing, that could mimic the multilayer interference effect of Morpho scales and produce vibrant colors electronically.

The concept was elegant. Rather than copying a butterfly's biological material in detail, the engineers asked: what is the principle behind the butterfly's color, and how can we achieve it with modern materials? In nature, the wing's structures are fixed, yielding one static color, but an electronic display needs to switch colors rapidly to show moving images. Iridigm's solution was to make each pixel a tiny MEMS structure: two parallel reflecting surfaces (a conductive mirror and a thin-film stack) that can move relative to each other. By applying a voltage, one surface (a flexible membrane) can be pulled closer to the other, changing the gap between them. This gap determines which wavelength of light they reinforce or cancel. In effect, each little device is a tunable mirror that can select a specific reflected color depending on the gap. When the gap is open to a certain size, a specific color (red, green, or blue) is reflected back; when the gap is collapsed (mirror pulled in), the device reflects no visible light and appears dark. By 2003, Iridigm had demonstrated small reflective displays using these interferometric modulators, branding the technology Digital Paper [22].

The promise of a low-power, sunlight-visible color screen caught the attention of industry. In 2004, telecommunications giant Qualcomm (better known for its mobile chipsets) acquired Iridigm for $170 million. Qualcomm rebranded the IMOD technology as Mirasol™ (a name evoking "mirror" and the Spanish word for sun, *sol*). The Mirasol logo even featured a butterfly motif as homage to the biological muse. Under Qualcomm MEMS Technologies (QMT), engineers refined the design for mass production. Essentially, they were building a butterfly wing on a chip, using standard semiconductor fabrication processes to make millions of tiny movable mirrors on silicon. Each Mirasol pixel consisted of an array of subpixel cavities. A typical implementation used three subpixels tuned to reflect red, green, and blue, respectively. Each subpixel is a MEMS interferometric modulator that is bistable, it has two stable states: an open state (reflecting its designated color), and a closed state (membrane pulled in, appearing black). By driving each subpixel on or off, the pixel as a whole can display a full range of colors via additive mixing. For example, if the red and green subpixels are on while blue is off, the pixel appears yellow. If all three are on, the pixel shows white, and so forth.

Unlike an LCD, which continuously consumes power to modulate light and requires a backlight, the Mirasol IMOD needs power only to switch states; it draws virtually no power to hold an image because the two stable states don't require a sustained voltage (much like E-ink in this regard). And since it reflects ambient light, its brightness actually increases in brighter environments, the opposite of emissive displays. This meant that a Mirasol screen would be perfectly readable outdoors in sunlight, where traditional LCDs or OLEDs wash out. The core pixel structure of Mirasol thus comprised a conductive thin-film stack on a glass substrate (which acted as a partial mirror and also an absorber in the closed state), a thin air gap, and a reflective movable membrane above the gap. The movable membrane was usually made of a metallic film (like aluminum alloy) on a deformable mechanical support. When a voltage is applied between the membrane and the superstrate, electrostatic attraction pulls the membrane up until it nearly contacts the substrate, eliminating the air cavity. In that state, the optical cavity no longer supports the interference of visible light, and the pixel appears black (or whatever underlying nonreflective color the stack has). Release the voltage, and the membrane springs back to its designed gap (a few hundred nanometers), restoring the reflective color. This fast switch can happen in microseconds, enabling video-capable refresh rates. Qualcomm's engineers published detailed descriptions of this MEMS modulator design and its optical performance in technical papers, validating that the approach could yield a high-reflectivity, bistable display element.

As illustrated in Figure 6.4, a Mirasol pixel comprises a glass superstrate, a partially reflective top electrode, and a reflective bottom electrode that together define an addressable air cavity. By electrostatically forming or collapsing this cavity, the device toggles the optical path length between the reflecting interfaces, thereby turning a narrowband interferential reflection on or off. Optically, this is the same thin-film interference observed in the bioinspiration from the Morpho butterfly: its wing scales contain multilayer nanostructures that achieve vivid structural colors by the very same interference mechanism [19–21].

6.4 Manufacturing Challenges and Industry Partnerships

Turning the Mirasol concept into a commercial product was a formidable challenge. Making a few prototype displays in a lab is very different from mass-producing high-resolution panels. Each Mirasol pixel contained moving parts and nanometer-scale gaps that had to be uniform across the entire screen. MEMS devices are notoriously difficult to manufacture at high yield–tiny variations or particles can cause a mirror to stick or a cavity to seal improperly [23, 24]. In the mid-2000s, Qualcomm poured significant investment into building production capacity for Mirasol. They constructed dedicated fabrication lines for the MEMS displays, including a new plant in Taiwan. In fact, by 2011 Qualcomm committed roughly \$1 billion to a Mirasol manufacturing facility in

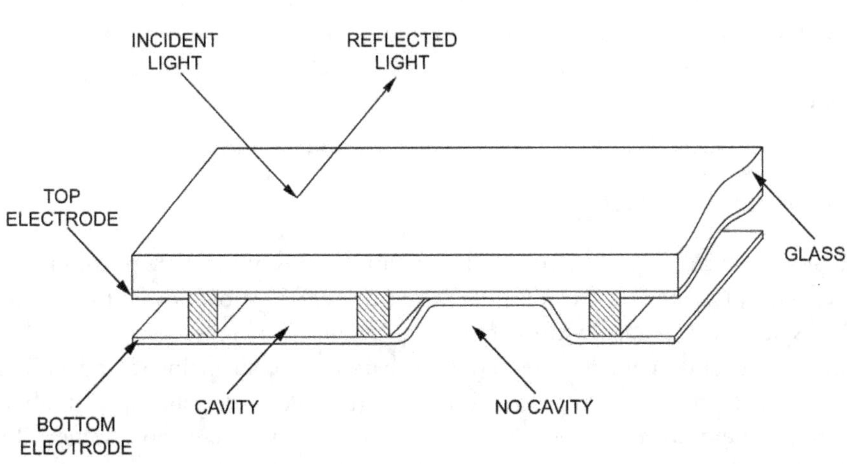

Figure 6.4 Interferometric modulator (Mirasol) pixel operation. Incident light reflects from a partially reflective top electrode and a reflective bottom electrode separated by an addressable cavity. When the cavity is present (left), the optical path difference selects a reflected color via thin-film interference; when the cavity is collapsed (right), the resonance is suppressed and the pixel appears dark. The optical behavior is analogous to the multilayer nanostructures of Morpho butterfly scales: both systems exploit thin-film interference to generate structural color [19–21].

Taiwan, underscoring the company's gamble on this technology. This bet was to enable volume production of Mirasol panels for e-readers, phones, and more. To address expertise gaps in display manufacturing, Qualcomm also sought partnerships. In December 2012, Qualcomm announced a collaboration with Sharp Corporation, a veteran LCD manufacturer, to advance next-generation Mirasol displays. As part of the deal, Qualcomm invested in Sharp (which was developing IGZO thin-film transistor technology) to leverage Sharp's display fabrication know-how. The idea was that Sharp's facilities and experience, combined with Qualcomm's MEMS design, could solve the scaling issues. This partnership was also aimed at hybridizing Mirasol with efficient backplanes (the driving circuitry) to improve resolution and yield. For a time, Sharp and Qualcomm worked together on pilot production, and Sharp's involvement lent credibility that Mirasol might finally reach mass-market devices.

Building the Mirasol display involved several technological hurdles. The MEMS fabrication had to integrate with conventional glass panel processes. Tiny spacers and support posts for the MEMS membranes had to be formed with semiconductor lithography techniques. The air cavity for each subpixel had to be created (often by depositing a sacrificial layer and then etching it away to leave an air gap). All of this had to remain stable over the device's lifetime, meaning the MEMS couldn't deform or fatigue after millions of flexes. Qualcomm's team made numerous engineering tweaks: for example, refining the materials for the moving membrane to prevent stiction (the tendency of microscale parts to stick together due to surface forces) and optimizing the cavity design to maximize reflectance for a given gap. By 2010, they were showing off production prototypes, indicating that these challenges were at least partially overcome.

6.5 Market Deployment

As prototype color displays emerged, excitement grew around the possibility of a reflective, low-power screen for consumer devices. Demos showed that Mirasol displays were readable in direct sunlight, could play video with reasonable frame rates, and enabled extremely long battery life, all major selling points for mobile electronics. Qualcomm initially targeted e-readers and small mobile devices, where power efficiency and outdoor visibility were most valued. In late 2011, the first commercial product with a Mirasol display was launched: a color e-reader by Kyobo, South Korea's largest bookseller. This Kyobo eReader featured a 5.7-inch Mirasol screen with XGA resolution (1024 × 768, 223 ppi) and touch input. It could display vivid color content and even play video (Qualcomm demonstrated up to 30 fps on that display) while still offering weeks of use on a single charge. The Kyobo eReader, powered by a Qualcomm

Snapdragon S2 processor, was introduced at a price of around $300 in the Korean market. Its launch was significant—it proved that Mirasol was not just a lab experiment but a manufacturable product. Following Kyobo, similar Mirasol-based e-readers appeared in China and Europe in 2012. For instance, the Chinese firm Shanda released the Bambook Sunflower, an e-reader using the same 5.7-inch Mirasol panel, and a company named Koobe introduced a color e-reader in certain markets. These devices all capitalized on the ability to show color illustrations and even video (something e-ink could not do), while retaining an "always readable, always on" screen. They were particularly pitched for educational content, (e-textbooks, magazines, etc.), where color is useful but a full tablet (with its backlit LCD) would have poor battery life or outdoor performance.

Qualcomm also explored the use of Mirasol in wearables. In 2013, they unveiled the Qualcomm Toq smartwatch as a proof-of-concept for original equipment manufacturers (OEMs). The Toq featured a 1.55-inch color Mirasol display that was always-on, sunlight-visible, and power-efficient, yielding multiple days of runtime on a charge, a stark contrast to early OLED-based smartwatches that often struggled to last a single day. The Toq's Mirasol screen allowed users to view notifications and even small videos outdoors without draining the battery quickly. It was meant to showcase how the technology could enable new classes of devices (like wearables) where conventional displays were a weak link. Consumers and tech media were intrigued by these products. A color e-reader that you could use on the beach in bright sun, or a smartwatch you didn't have to charge nightly, were compelling promises. Videos of the Kyobo eReader in action showed smooth enough video playback and acceptable color graphics. The reflective screen did not have the same saturation as an LCD, but it was reported to be comfortable on the eyes and usable in a wide range of lighting. Industry analysts wondered if Mirasol might carve out a niche between e-ink (great in sunlight but monochrome and slow) and LCD/OLED (full color and fast, but power hungry). Qualcomm even hinted at plans for Mirasol in smartphones and tablets eventually, once resolution and size could be scaled up. By 2012–2013, Mirasol was positioned as a potentially disruptive display technology in an industry hungry for improvements in power and outdoor readability.

6.6 The Fall: Why Mirasol Did Not Achieve Widespread Success

Despite the strong technical merits and the initial deployments, Mirasol never achieved wide adoption [25]. By 2014, Qualcomm announced it would stop direct development of Mirasol—each phase representing what Chapter 1 termed an incremental analogical iteration, where successive refinements sustained continuity with the original biological principle. Displays for consumer products, pivoted to a strategy of licensing the technology instead. The dedicated Mirasol factory in Taiwan was scaled back, and plans for larger panels or new devices were shelved. Even the partnership with Sharp did not yield a sustainable manufacturing pipeline; Sharp moved on to focus on its IGZO LCD and OLED efforts as Qualcomm retreated. In essence, after investing nearly a decade and significant capital, Qualcomm wound down its Mirasol business unit. Several key factors contributed to Mirasol's commercial disappointment, which are discussed next.

6.6.1 Manufacturing Complexity and Yield Problems

Scaling MEMS to high-resolution, large displays proved extremely challenging. Each pixel's mechanical movement had to be uniform and reliable; even a tiny defect could result in a pixel that stayed stuck on or off (reflective or black). Achieving high yield (percentage of good devices on a wafer/panel) was difficult, which drove up costs. Qualcomm struggled with issues like particles causing stiction, variations in cavity gap leading to color nonuniformity, and membrane materials wearing out. While they did produce functional screens, the yields were likely far lower than competing LCD/OLED, making Mirasol panels expensive. Building a brand-new display manufacturing ecosystem from scratch (as opposed to using existing LCD fabs) meant high up-front costs and learning curve delays. These issues limited the sizes and quantities of displays Qualcomm could produce, and thus limited the market reach.

6.6.2 Color and Performance Limitations

Although Mirasol produced color, the color saturation was lower than contemporary LCD or OLED screens. The colors tended to appear more muted or pastel, especially under typical indoor lighting. This is partly inherent to reflective displays that rely on ambient light—in dim indoor light, Mirasol colors looked dull (though one could activate a front-light in some e-reader implementations). Additionally, the viewing angles were not as wide as LCD/OLED; the interference effect could be highly angle-dependent. As a result, the image could shift in color or brightness when viewed off-center. For consumers used to the bright, vibrant screens of modern smartphones, Mirasol's image quality didn't quite measure up except in niche situations (like direct sunlight). The video was demonstrated at 30 frame per second, but the initial versions had some motion artifacts and flicker under certain conditions, which made them less ideal for multimedia-heavy use. These performance gaps made device manufacturers hesitant to adopt the technology for mass-market products where user expectations were set by LCD/OLED quality.

6.6.3 Lack of OEM Adoption and Ecosystem Support

Getting device manufacturers (OEMs) on board proved difficult. Using Mirasol in a product was not a simple drop-in replacement for an LCD. It required custom driving electronics, tailored software (e.g., color processing for the IMOD's color space), and likely modifications to the device's design (since front-lighting might be needed for low-light usage, and touch layers had to be added without hurting reflectivity). Many OEMs were already invested in existing display supply chains and were not willing to risk using an unproven technology that might face production delays or post-sale support issues. The big players in e-readers (Amazon, Barnes & Noble) never adopted Mirasol, nor did major phone or tablet makers. Without a strong ecosystem, Qualcomm was essentially trying to do everything from materials to display integration to content adaptation. This made it hard to reach economies of scale. When the initial devices launched to only lukewarm reception, other potential partners grew even more cautious. In the end, only a handful of niche products ever used Mirasol, which wasn't enough to sustain its development.

6.6.4 Rapid Improvements in Competing Display

While Qualcomm was refining Mirasol, the mainstream display technologies were not standing still. OLED displays in particular made great strides in the early 2010s, improving their power efficiency and outdoor visibility (with pixel designs to boost brightness) and being implemented in small devices like watches. LCDs also got better with transflective coatings, brighter backlights, and lower power modes. The relative power advantage of Mirasol, one of its chief selling points, diminished as LCD/OLED power consumption was reduced with each generation. By 2013, an LCD could be engineered (with a reflective layer or adaptive brightness) to be somewhat usable outdoors and might only consume moderately more power than Mirasol for typical use. Moreover, high-resolution LCDs and OLEDs delivered high DPI with perfect color and video, something Mirasol had yet to achieve. In wearables, for instance, some devices opted for transflective LCDs (like certain Garmin watches) as a compromise for outdoor viewability. Mirasol was no longer the only game in town for low-power displays, and the gap in overall user experience between it and conventional displays remained large.

6.6.5 Strategic Shift by Qualcomm

Lastly, there was an internal strategic reality: Qualcomm is fundamentally a semiconductor and IP company, not a display manufacturer. Running a display fabrication business (with its capital-intensive fabs, supply chain management, and need to secure customers in consumer electronics) was far outside Qualcomm's core competencies. The company likely reassessed the opportunity cost as the difficulties mounted. By 2014, Qualcomm took a $142 million charge to write down the Mirasol display business, and began looking to licensing or sale of the technology instead. They even sold their Taiwan Mirasol plant (reports later emerged that Apple Inc. took over the facility, possibly for its own R&D). In essence, the combination of technical hurdles, slow adoption, and competition led Qualcomm to pull back. It was a classic case of innovator's dilemma in hardware: an impressive innovation that could not cross the chasm to broad commercial viability in time. Industry commentators analyzed these factors in postmortem articles. Display experts noted that while Mirasol was a breakthrough in concept, it was "a solution looking for a problem" at a time when LCDs and OLEDs were rapidly improving. The technical advantages it offered (low power, sunlight readability) were somewhat narrow in appeal, and many consumers prioritized the high color fidelity and responsiveness of their existing screens. As one commentary put it, innovation alone doesn't guarantee success—the technology also needs a supportive ecosystem and timing on its side.

6.7 Legacy of Mirasol and Future Directions in Biomimetic Displays

Although Qualcomm's Mirasol initiative wound down without conquering the market, it left an important legacy in the display and photonics community. It demonstrated that biomimicry in optical engineering can yield viable, if challenging, technological solutions. Mirasol proved that it was possible to create an electronic color display that uses no incandescent or emissive light source, only ambient light and interference. This had never been commercialized before at such a scale. The very existence of Mirasol displays in the wild (even if few in number) has inspired further research into reflective and nonemissive display technologies. For instance, the concept of MEMS-based optical modulators continues to be explored. Some startups and research groups have looked at variations like electrowetting displays, photonic crystal displays, and other MEMS mirrors for color generation. The lessons learned from Mirasol (materials for reliability, the need for optical efficiency, etc.) inform these efforts.

In academic circles, Qualcomm's work spurred interest in photonic crystals and structured reflectors for low-power color generation; even outside of displays, applications such as color-changing sensors, tunable filters, and anticounterfeiting devices have drawn on the ideas pioneered with Mirasol [26–32]. Mirasol also underscored the importance of matching technology to market. The quest for ultralow-power displays hasn't gone away. In the era of ubiquitous mobile devices and now the Internet of Things (IoT), there is renewed interest in screens that can operate for long periods on minimal energy. E-paper (electrophoretic displays) remains a strong player for reading devices, and companies have been working on color e-paper (some using filters on E-ink, others using electrofluidic or electrowetting approaches). These are direct alternatives in the space Mirasol targeted. It's possible that future breakthroughs in nanofabrication could revisit the interferometric modulator idea—perhaps using new materials (like graphene MEMS mirrors, which have been experimented with to achieve higher switching speeds and better optical range graphenea.com), or improved manufacturing techniques.

The concept of a bistable, reflective-color pixel is still highly attractive for many applications. In a broader sense, Mirasol's story is a case study in the challenges of biologically inspired design. It's not enough to copy nature; engineers must also deal with constraints of scalability and integration. The Morpho butterfly did not have to manufacturability-test its wing scales on silicon wafers, but Qualcomm did. The effort expanded the toolkit of MEMS fabrication (for example, developing processes for large-area optical MEMS), and will likely benefit other projects down the line. Finally, Mirasol forced the display industry to consider use-cases that were long neglected by mainstream technologies. Sunlight readability and power frugality are now selling points for specialized devices. We see, for example, memory LCDs and low-refresh displays in smartwatches for outdoor use, and dual-layer LCDs with reflective modes.

The pursuit of a color reflective display continues. As of the mid-2020s, companies like E Ink have developed improved color e-paper (using e-ink pigments), and some are exploring Cholesteric LCD panels that reflect specific wavelengths. The dream that Mirasol chased—a full-color, video-capable digital paper—is still alive, and Mirasol proved it was at least partially attainable. In retrospect, the Mirasol journey—from butterfly wings, to a startup's lab, to billion-dollar fabrication lines, to an abrupt retreat—illustrates the complex interplay between innovation and market forces. It showed that a technology can be brilliant in concept and execution, yet still falter due to external factors. Nonetheless, Qualcomm's Mirasol should be remembered not as a failure, but as a bold exploration at the frontiers of display engineering. It expanded our notion of what's possible by applying lessons from nature. As display researchers continue to seek the next leap in low-power screens, the pioneering work on interferometric modulators will surely inform and inspire them. The butterfly may not have taken flight in the market, but its influence flutters on in labs and ideas around the world.

6.8 Conclusion

The story of the Mirasol display is a vivid reminder that even when an invention draws on brilliant analogy thinking—when it captures the essence of nature's designs with elegant engineering—success in the marketplace is never guaranteed. Mirasol's core concept, inspired by the Morpho butterfly's shimmering wings, was a technical marvel: a sunlight-readable, low-power, color-reflective display built with MEMS ingenuity. Yet despite its innovation, Mirasol faltered not because the invention itself was fundamentally flawed, but because the broader ecosystem failed to align.

At its heart, Mirasol's downfall was not purely a technical story. It was the ecosystem—the intricate web of manufacturing maturity, cost structure, OEM adoption, content optimization, and consumer expectations—that ultimately determined its fate. Qualcomm faced steep yield challenges in scaling MEMS to high-resolution panels, but even if yields had improved, the uphill battle remained: overcoming the dominance of LCD and the rising momentum of OLED, both fueled by powerful economies of scale. In parallel, OLED technology continued its relentless march forward, rapidly improving brightness, color, and efficiency just as Mirasol was trying to carve a niche. The bar for new display entrants was rising faster than Mirasol could clear it.

This history naturally invites a pointed question: Will microLED display, with each phase representing what Chapter 1 termed an incremental analogical iteration where successive refinements sustained continuity with the original biological principles, face a similar fate?

MicroLEDs, like Mirasol, promise transformative benefits: exceptional brightness, long lifetimes, low power consumption, and high color purity. But they, too, confront formidable challenges, particularly around manufacturing yield and ecosystem readiness. Just as MEMS-based Mirasol displays struggled with uniformity, stiction, and tiny mechanical tolerances, microLEDs must grapple with the daunting task of transferring, aligning, and bonding millions of microscopic emitters with extreme precision [33–37]. Yield losses at such small scales can be crippling, both for cost and for consistency across large panels. The technical ambition is breathtaking, but technical achievement alone may not be enough.

Even more critically, microLEDs enter a world where OLED displays are not static targets [38–42]. OLEDs have steadily improved in efficiency, peak brightness, and burn-in resistance, thanks to billions of dollars of cumulative investment and the sheer scale of manufacturing. OLED has built an ecosystem—from materials suppliers to equipment makers to panel fabs to device OEMs—that is resilient, mature, and cost-optimized. MicroLEDs, like Mirasol before them, must create not just better pixels but an entire infrastructure that makes better pixels accessible, affordable, and reliable at scale.

Thus, the cautionary lesson from Mirasol remains sharply relevant: analogy-driven innovation can yield extraordinary ideas, but adoption demands far more than brilliance in the lab. It demands an ecosystem synchronized in timing, economics, and supply chain readiness. It demands not just technical proof but proof of manufacturability, reliability, and economic viability at volumes that move markets. And even then, it requires navigating a landscape where incumbent technologies improve faster than newcomers can establish a beachhead.

Mirasol did not fail for lack of vision. It failed because translating vision into market-shifting reality is a different kind of engineering—one that goes beyond the device itself and into the messy, unforgiving terrain of cost curves, risk aversion, customer habits, and timing. As microLEDs race toward commercialization today, the shadows of Mirasol's experience loom large. The butterfly took flight, if only briefly. Whether microLEDs can avoid the same fate remains an open and pressing question.

References

[1] Goureau, C. C., "Note sur la transparence et le lustre des ailes de certains insectes," *Annales de la Societe Entomologique de France*, Vol. 10, 1842, pp. 385–392.

[2] Anderson, T. F., and A. G. Richards, Jr., "An Electron Microscope Study of Some Structural Colors of Insects," *Journal of Applied Physics*, Vol. 13, No. 12, 1942, pp. 748–758, https://doi.org/10.1063/1.1714827.

[3] Vukusic, P., et al., "Quantified Interference and Diffraction in Single Morpho Butterfly Scales," *Proceedings of the Royal Society of London. Series B: Biological Sciences*, Vol. 266, No. 1427, 1999, pp. 1403–1411, https://doi.org/10.1098/rspb.1999.0794.

[4] Kinoshita, S., et al., "Mechanisms of Structural Colour in the Morpho Butterfly: Cooperation of Regularity and Irregularity in an Iridescent Scale," *Proceedings of the Royal Society of London. Series B: Biological Sciences*, Vol. 269, No. 1499, 2002, pp. 1417–1421, https://doi.org/10.1098/rspb.2002.2019.

[5] Siddique, R. H., et al., "Theoretical and Experimental Analysis of the Structural Pattern Responsible for the Iridescence of Morpho Butterflies," *Optics Express*, Vol. 21, No. 12, 2013, pp. 14 351–14 361, https://doi.org/10.1364/OE.21.014351.

[6] Thome, M., et al., "Light Guidance in Photonic Structures of Morpho Butterfly Wing Scales," *Applied Physics A*, Vol. 126, 2020, https://doi.org/10.1007/s00339-020-03948-x.

[7] Giraldo, M. A., et al., "Colouration Mechanisms and Phylogeny of Morpho Butterflies," *Journal of Experimental Biology*, Vol. 219, No. 24, 2016, pp. 3936–3944, https://doi.org/10.1242/jeb.148726.

[8] Patterson, J. D., et al., "Micro-Mechanical Voltage Tunable Fabry–Perot Filters Formed in (111) Silicon," NASA Langley Research Center, *Tech. Rep. NASA-TP-3702*, 1997, https://ntrs.nasa.gov/api/citations/19970037592/downloads/19970037592.pdf.

[9] Moon, J. S., et al., "Performance Limits of Micromachined Tunable-Cavity Filter," in *TechConnect Briefs*, 2004, https://briefs.techconnect.org/papers/performance-limits-of-micromachined-tunable-cavity-filter/.

[10] Hirokubo, N., et al., "Wideband Visible Wavelength Range MEMS Fabry–Perot Tunable Filter with Highly Accurate Calibration System," *IEEE Sensors Journal*, Vol. 13, No. 8, 2013, pp. 2930–2936, https://doi.org/10.1109/JSEN.2013.2264458.

[11] Brunne, J., et al., "Tunable MEMS Axicon Mirror Arrays," *Optics Letters*, Vol. 38, No. 11, 2013, pp. 1939–1941, https://doi.org/10.1364/OL.38.001939.

[12] Mao, H., et al., "MEMS-Based Tunable Fabry–Perot Filters for Adaptive Multispectral Thermal Imaging," *IEEE Journal of Selected Topics in Quantum Electronics*, Vol. 22, No. 3, 2016, pp. 1–8, https://doi.org/10.1109/JSTQE.2016.2535114.

[13] Miles, M., 2000, *Interferometric Modulation*, US Patent 6055090.

[14] McMichael, I., et al., 2001, *Reflective Display Apparatus Using Interferometric Modulators*, US Patent 6271897.

[15] Mounier, E., et al., "New Emerging MEMS Applications," in *Micromachining Technology for Micro-Optics and Nano-Optics V*, SPIE, Vol. 6462, 2007, pp. 97–103.

[16] Felhofer, D., et al., "Device Physics of Capacitive MEMS," *Microelectronic Engineering*, Vol. 84, No. 9–10, 2007, pp. 2158–2164, https://doi.org/10.1016/j.mee.2007.04.118.

[17] Ma, J., et al., "Advanced MEMS-Based Technologies and Displays," *Displays*, Vol. 37, 2015, pp. 2–10, https://doi.org/10. 1016/j.displa.2014.10.003.

[18] Liao, C.- D., et al., "The Evolution of MEMS Displays," *IEEE Transactions on Industrial Electronics*, Vol. 56, No. 4, 2009, pp. 1057–1065, https://doi.org/10.1109/TIE.2008.2005684.

[19] Lan, J.- H., 2007, *High Aperture-Ratio Top-Reflective AM-IMod Displays*, US Patent 7643199B2.

[20] Natarajan, B. R., and S. Ganti, 2007, *MEMS Devices and Processes for Packaging Such Devices*, US Patent 7746537B2, 2007.

[21] Lin, W.- J., et al., 2003, *Optical Interference Display Panel and Manufacturing Method Thereof*, US Patent 8004736B2.

[22] Miles, M., et al., "Digital Paper™ for Reflective Displays," *Journal of the Society for Information Display*, Vol. 11, No. 1, 2012, pp. 209–215, https://doi.org/10.1889/1.1831708.

[23] Larsson, E., et al., "Manufacturing Issues of BEOL CMOS-MEMS Devices," *IEEE Access*, Vol. 9, 2021, pp. 83149–83162, https://doi.org/10.1109/ACCESS.2021.3086867.

[24] Gao, L. L., et al., "Statistical Analysis And Yield Enhancement Of MEMS Devices By Considering Multi-Process Variations," *Key Engineering Materials*, Vol. 645-646, 2015, pp. 589–594, https://doi.org/10.4028/www.scientific.net/KEM.645-646.589.

[25] Kozlowski, M., "The Rise and Fall of Qualcomm Mirasol e-Readers," 2018, https://goodereader.com/blog/electronic-readers/the-rise-and-fall-of-qualcomm-mirasol-e-readers.

[26] Wang, Danyan, et al., "Structural Color Generation: From Layered Thin Films to Optical Metasurfaces," *Nanophotonics*, Vol. 12, No. 6, 2023, pp. 1189–1210, https://doi.org/10.1515/nanoph-2022-0063.

[27] Wang, H., et al., "Photonic Crystal Structures with Tunable Structure Color as Colorimetric Sensors," *Sensors*, Vol. 13, No. 4, 2013, pp. 4192–4213, https://doi.org/10.3390/s130404192.

[28] Hu, H., et al., "Photonic Anti-Counterfeiting using Structural Colors Derived from Magnetic-Responsive Photonic Crystals with Double Photonic Bandgap Heterostructures," *Journal of Materials Chemistry*, Vol. 22, No. 22, 2012, pp. 10926–10932, https://doi.org/10.1039/c2jm30169e.

[29] Liu, W., et al., "Broadly Tunable Filter Based On A Graphene MEMS-Photonic Crystal Composite Structure And Its Application In Single-Pixel Full-Color Displays," *Journal of Materials Chemistry C*, Vol. 12, No. 18, 2024, pp. 6588–6595, https://doi.org/10.1039/d3tc03646d.

[30] Minh, N. H., et al., "Anti-Counterfeiting Labels of Photonic Crystals with Versatile Structural Colors," *Nanoscale Advances*, Vol. 6, No. 23, 2024, pp. 5853–5860, https://doi.org/10.1039/d4na00814f.

[31] Meng, Z., et al., "Photonic Crystal-Based Structural Color Switches," *Materials Horizons*, Vol. 20, 2025 https://doi.org/10.1039/d5mh00768b.

[32] Xie, X., et al., "Structured Color Filters Based on Fabry–Perot Resonance for High Performance RGB Transmission," *Applied Physics A*, Vol. 131, 2025, p. 527, https://doi.org/10.1007/ s00339-025-08654-0.

[33] Chen, F., et al., "Mass Transfer Techniques for Large-Scale and High-Density MicroLED Arrays," *International Journal of Extreme Manufacturing*, Vol. 4, No. 4, 2022, p. 042005, https://doi.org/10.1088/2631-7990/ac92ee.

[34] Ozeki, H., et al., "Enabling Next-Generation MicroLED Displays Through ACF-Based Bonding Solutions," *Information Display*, Vol. 41, No. 5, 2025, pp. 12–16, https://doi.org/10.1002/msid.1608.

[35] Zhang, S., et al., "Research Progress of Micro-LED Display Technology: Chip, Mass Transfer, Full-Color Display, Bonding and Driving," *Crystals*, Vol. 13, No. 7, 2023, p. 1001, https://doi.org/10.3390/cryst13071001.

[36] Xie, Y., et al., "High-Yield Micro-LED Laser Transfer Accomplished Using an Ablation-Type Release Material," *APL Materials*, Vol. 12, No. 10, 2024, p. 101111, https://doi.org/10.1063/5.0232745.

[37] Zhu, G., et al., "Mass Transfer, Detection and Repair Technologies in Micro-LED Displays," *Science China Materials*, Vol. 65, No. 6, 2022, pp. 2128–2153, https://doi.org/10.1007/s40843-022-2110-2.

[38] Ahmed, K., *MicroLED Devices and Systems*, Norwood, MA: Artech House, 2024.

[39] Huang, Y., et al., "Mini-LED, Micro-LED and OLED Displays: Present Status and Future Perspectives," *Light: Science & Applications*, Vol. 9, No. 1, 2020, p. 105, https://doi.org/10.1038/s41377-020-0341-9.

[40] Hsiang, E.- L., et al., "Prospects and Challenges of Mini-LED, OLED, and Micro-LED Displays," *Journal of the Society for Information Display*, Vol. 29, No. 6, 2021, pp. 446–465, https://doi.org/10.1002/jsid.1058.

[41] Zhu, B., et al., "Research Progress of Micro-LED Display Technology," *Crystals*, Vol. 13, No. 7, 2023, p. 1001, https://doi.org/10.3390/cryst13071001.

[42] Gao, H., et al., "Advances in Pixel Driving Technology for Micro-LED Displays," *Nanoscale*, Vol. 15, No. 43, 2023, pp. 19085–19105, https://doi.org/10.1039/d3nr01649h.

7

Moth-Eye to Manufacture: Analogy Thinking in Antireflective Coatings

After reading this chapter you should be able to:

- Understand the biological principles behind moth-eye structures and how they suppress optical reflection through graded-index effects, framing the moth eye as a source domain and modern AR challenges as target domains within an analogy-thinking process.
- Explain how moth-eye-inspired nanostructures are fabricated using top-down and bottom-up techniques, and how these methods have evolved into scalable, manufacturable processes—highlighting the mapping, abstraction, and transfer steps from biological template to engineered surface.
- Identify the broad range of applications—including solar panels, optical systems, sensors, and displays—where moth-eye coatings provide performance advantages, including antireflective (AR), self-cleaning, and antifogging properties, and evaluate them via constraint matching and validation loops that refine designs and generate insight.

7.1 Introduction

The reduction of surface reflections has long been a challenge in optics and photonics, whether in telescopes, solar panels, camera lenses, or display technologies [1]. Traditional AR coatings, primarily based on thin-film interference, are often limited by their narrow wavelength or angle-specific performance. A revolution began, however, with an unexpected source of inspiration: the compound eye of the nocturnal moth. In the late 1960s, researchers using electron microscopy discovered that the surface of a moth's eye is covered with nanoscale protuberances [2–4]. These structures, smaller than the wavelength of visible light, create a gradual transition in refractive index from air to tissue, significantly reducing reflection. This chapter explores how that biological insight evolved into a suite of engineered coatings, now transforming optical design across industries.

In fact, the nipple-array concept is not restricted to moth eyes but also appears on insect wings. For example, certain cicadas, dragonflies, and damselflies display subwavelength protuberance arrays on their wing membranes that act as antireflective or antiglare surfaces. These wing structures serve not only optical functions but also hydrophobic or self-cleaning roles, suggesting a rich diversity of natural source domains. Recognizing this broader biological occurrence reinforces the value of our analogy-thinking journey—from source to target—and invites engineers to widen their scanning of candidate domains beyond the corneal surfaces of moths [5–8].

Recent investigations have revealed that not only highly ordered nipple-array nanostructures, but also partially disordered or irregular arrays can yield greater optical transparency than strictly periodic systems [9–12]. In particular, insect wings with so-called nipple protuberances, sometimes exhibit randomness in height, spacing or packing, and these subtle deviations from perfect order have been shown to broaden the effective refractive-index gradient and suppress reflection across a wider angular and spectral range. By embracing a modest degree of disorder, nature appears to optimize the graded-index mapping from air into the wing membrane, thereby enhancing light transmission under real-world illumination. This insight suggests a refined mapping in our analogy-thinking framework: the source domain (insect nipple arrays) not only provides an ordered prototype, but also invites abstraction of structured disorder as a design lever. The constraint-matching step therefore moves beyond simply copying geometry, toward embracing controlled irregularity as a mechanism for AR [13–16].

Figure 7.1 illustrates the geometric configuration of the moth-eye absorber [17]. The model consists of a periodic array of semiellipsoidal nanocones arranged in a square lattice of period P and height h on a thick substrate of thickness t. The interplay between these parameters allows the structure to maintain high absorptance (> 94%) without requiring extreme aspect ratios.

Beyond suppressing Fresnel losses, insect wing nipple arrays often couple optical transparency with liquid repellence. A useful mechanistic picture emerges when the composite protuberance is decomposed into a truncated conical base capped by a hemispherical dome: the cone establishes a depth-varying fill

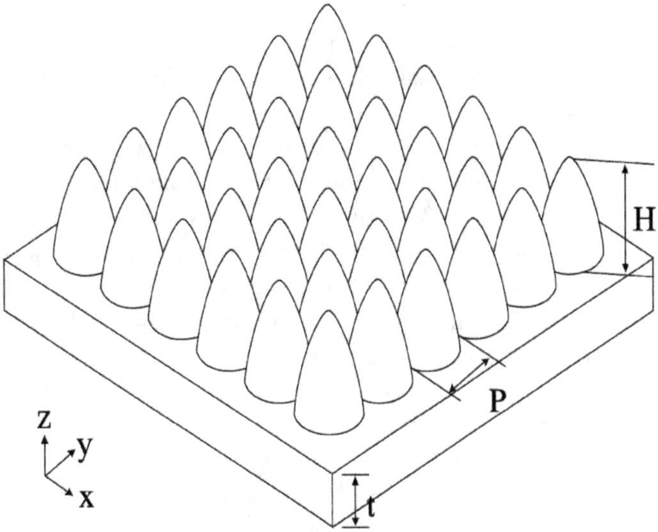

Figure 7.1 Three-dimensional schematic of a periodic moth-eye nanostructure.

fraction that realizes the graded-index transition responsible for broadband antireflection, whereas the dome locally maximizes curvature and air entrapment, elevating apparent contact angles and reducing pinning [14]. This two-level division of labor, cones for impedance matching, domes for hydrophobicity, has been resolved on cicada wings by combined SEM/optical/wetting studies and reproduced in biomimetic replicas that retain both high transmission and water repellence. Thus, in the analogy-thinking frame used here, the same biological source domain provides a multifunctional mapping: a single nanoscale motif simultaneously addresses optical and interfacial constraints, guiding engineered surfaces that target glare suppression and self-cleaning in one texture [18–21].

In addition to the engineered introduction of photocatalytic materials for antimicrobial functionality, it is important to note that insect tissues themselves often exhibit intrinsic antibacterial behavior, particularly in the wing and cuticle surfaces of various species. Studies have shown that the nanopillarrich surfaces of, for example, cicada or dragonfly wings not only suppress fouling via hydrophobicity or antiglare performance, but also mechanically disrupt bacterial cells upon contact, thus providing a biological antimicrobial mechanism that predates any added agent. The presence of such functionality in the natural source domain strengthens our analogy-thinking framework: the biological surface is not a blank optical texture, but a multifunctional template—optical, hydrophobic, and antimicrobial all at once. Highlighting this means our abstraction phase should explicitly recognize that the cuticular nanostructure contributes both graded-index AR behavior and intrinsic cell-rupturing or antiadhesive bacterial defense. That recognition enriches the mapping step and opens the possibility of bioinspired coatings that integrate optical AR performance and passive antimicrobial ridges without relying solely on exogenous photocatalyst [22–25].

Table 7.1 shows representative measurements of facet diameters, nipple distance, and nipple height for a number of butterflies. The facet diameters see variations from 17 μm to 33 μm; the nipple distance does not show large variability. The nipple height shows much larger variability amongst butterflies.

The nipple distance and nipple height are sub-wavelength for visible light spectrum (400 nm to 700 nm) what results in unique light-matter interaction that produce unique scattering of visible light.

The purpose of this chapter is to extend the discussion presented in Chapter 1, particularly emphasizing the role of analogy thinking as a key methodological approach. As introduced earlier, analogy thinking provides a structured process for transferring insights across domains, fostering creative problem solving and conceptual expansion. This chapter builds upon those foundations by illustrating how the same analogical reasoning framework can be used to interpret and apply the core ideas presented here.

Analogy-thinking lens for this chapter: We will explicitly treat the moth eye as the source domain and engineered AR coatings as the target domain. The biological microtexture-to-optics correspondence constitutes the mapping; the

Table 7.1
Dimensions of the Corneal Nipple Array of Butterflies

	D (μm)	d (nm)	h (nm)
Papilionidae			
Graphium sarpedon	28	230	30
Papilio memnon	31	n.d.	10
Papilio protenor	33	240	20
Papilio xuthus	25	230	20
Pachliopta aristolochiae	26	235	20
Pieridae			
Pieris rapae	22	210	210
Anthocharis cardamines	24	215	170
Lycaenidae			
Everes argiades	17	215	140
Pseudozizeeria maha	21	180	120
Narathura japonica	17	200	90

Average values of measurements by SEM, TEM and AFM. D, facet diameter; d, nipple distance; h, nipple height; n.d., not determined. Errors: $\Delta D = 3\ \mu$m, $\Delta d = 10$ nm, $\Delta h = 10$ nm. (*From:* [15].)

graded-index principle distilled from geometry is the abstraction; its implementation in lithography, etching, colloids, and nanoimprinting is the transfer. Throughout, we will use constraint matching (materials, scale, durability, manufacturability) and validation (optical/environmental testing) to close the loop and drive insight generation for next iterations.

7.2 Biological Origins and Early Engineering Efforts

The study of antireflective phenomena in nature represents one of the earliest and most compelling examples of how analogy thinking bridges biology and engineering. Long before nanofabrication matured as a discipline, biologists had already observed that certain insects possessed optical structures capable of suppressing reflection without relying on pigments or coatings. These naturally evolved systems hinted at physical principles that could be abstracted, mapped, and ultimately transferred into human-made technologies. Within this context, the moth's eye emerged as a particularly rich source domain for analogical reasoning in optics. Its surface morphology, characterized by nanoscale periodicity, offered not only aesthetic intrigue but also a functional model of refractive-index grading that minimized optical impedance mismatches. This discovery provided a conceptual springboard for engineers seeking new target domains—artificial surfaces capable of similar broadband and angle-independent antireflective performance. The early decades that followed transformed the biological insight into a reproducible engineering framework. Through iterative cycles of mapping, abstraction, and constraint matching, researchers began to formalize the physics underlying moth-eye optics and adapt it to glass, silicon, and polymer substrates. The result was a pioneering shift from observing biological phenomena to harnessing them as engineered principles, a hallmark of analogy thinking in photonic innovation.

As illustrated in Figure 7.2, the moth's compound eye comprises a dense matrix of ommatidia arranged in a convex hemispherical configuration. Each ommatidium acts as a discrete photoreceptive channel, oriented at a slightly different angle to maximize spatial coverage and light capture. This intricate microstructural organization grants the moth an exceptionally wide-angle field of view, enabling effective detection of motion and contrast even under extremely low-light conditions. The structural precision evident in the figure also reflects an evolutionary optimization toward nocturnal activity, where enhanced sensitivity rather than color discrimination is critical for survival and navigation. The artistic rendering captures not only the geometric regularity of the ommatidia but also the subtle surface curvature that contributes to minimizing optical aberrations, thus offering both scientific clarity and visual fidelity.

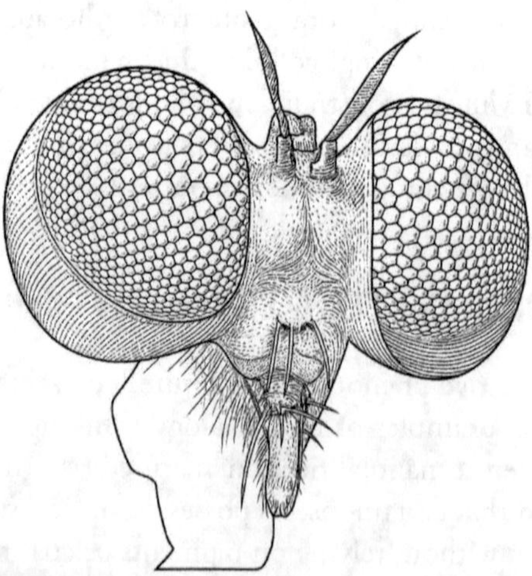

Figure 7.2 An artistic rendering of a moth's compound eye showing the dense and orderly arrangement of ommatidia across the hemispherical surface. Each ommatidium functions as an individual photoreceptive unit, collectively contributing to the moth's broad field of vision and heightened sensitivity to dim light, an adaptive trait for nocturnal navigation.

7.2.1 The Moth Eye's Microstructure

In the late 1960s, a quiet revolution in bio-optics began with the work of G. D. Bernhard [26, 27], who made a striking discovery using electron microscopy: the surface of a moth's eye was not smooth as once believed, but rather blanketed with an exquisitely ordered array of nanoscale protrusions. These tiny structures, often referred to as nipple arrays, were approximately 200-nm tall and spaced about the same distance apart. At the time, this was a revelation. The structures were so small that they fell well below the wavelength of visible light, meaning they didn't scatter light like larger surface irregularities. Instead, they played a more subtle and profound role—manipulating how light transitioned from air into the biological tissue beneath. These structures suppress reflections by acting as a graded-index medium. The surface refractive index gradually shifts from that of air $n \sim 1.0$ to that of the biological tissue $n \sim 1.4$, dramatically reducing Fresnel reflections across a wide spectral range [28–32].

What Bernhard and others quickly realized was that these structures weren't random; they were evolutionary adaptations with a very specific optical function. The surface of the moth eye, with its textured nanoscale landscape, acted as a kind of optical bridge. In a traditional flat interface between air and another medium—say glass or biological tissue—a sharp jump in refractive index causes a portion of the incident light to reflect back, a phenomenon governed by Fresnel's equations. But with the moth's eye, something different happened. The nanoscale protrusions created a gradient, a smooth transition of refractive index from the air, where the index is about 1.0, to the eye tissue, with an index closer to 1.4. Light entering the eye did not encounter an abrupt change in optical impedance. Instead, it experienced a tapering path inward, like a slope rather than a cliff.

This graded-index effect is crucial. Because the transition is gradual, the amount of light that reflects is significantly reduced. Across a broad range of wavelengths, effectively spanning the visible spectrum, the moth's eye appears nonreflective. This has practical implications in the insect's natural habitat. Moths are nocturnal and often vulnerable to predators that detect reflections from eyes, especially in dim or moonlit conditions. By suppressing glint, the textured surface of their eyes enhances their chances of survival. But the implications go far beyond biology.

Once this phenomenon was understood, scientists and engineers began to realize its potential in human-made optical systems. If nature could solve the problem of unwanted reflection without multilayer coatings, perhaps those principles could be translated to lenses, sensors, or solar cells. The core idea—that surface geometry alone, when designed at the right scale, could manipulate light in powerful ways—opened up a new frontier. Bernhard's initial observation thus became the seed of a new branch of optical engineering, one that would eventually marry biology with nanotechnology and transform how we think about controlling light at surfaces. It was a reminder that sometimes, evolution's solutions—honed over millions of years—can surpass even our best industrial efforts, and by paying attention to them, we can unlock new directions in science and technology.

Analogy-thinking note: Here the moth eye serves as the source domain. The functional correspondence between tapered feature density and effective refractive index is the key mapping that later enables engineering transfer to human-made substrates.

7.2.2 First Replications in the Laboratory

In 1973, a pivotal moment in biomimetic optics unfolded when researchers Peter Clapham and Michael Hutley succeeded in creating the first artificial moth-eye surface on a glass substrate [33]. Motivated by the earlier biological observations of the moth cornea, they set out to replicate nature's elegant solution to the problem of optical reflection. Utilizing the emerging techniques of photolithography and subsequent etching, they patterned a series of subwavelength structures onto glass, mimicking the nanoscale topography found in the moth's eye. What they achieved was remarkable for its time: a surface that exhibited significantly reduced reflectivity not only at a specific wavelength or angle, but across a broad spectrum and wide range of incident angles. This was a major departure from conventional AR coatings, which typically relied on multiple stacked dielectric layers tuned for specific optical paths.

Yet their breakthrough came with a major limitation—fabrication scale. In the early 1970s, nanofabrication was in its infancy. The tools available lacked the precision, repeatability, and throughput necessary for manufacturing large-area or commercial-scale coatings. As a result, Clapham and Hutley's prototypes remained confined to small test surfaces, more a proof-of-principle than a viable product. Nonetheless, their work laid the foundation for decades of innovation, marking the beginning of engineered, nature-inspired optics.

Analogy-thinking note: This phase operationalizes transfer from biology to engineered glass while revealing constraint mismatches (throughput, uniformity) that require iterative validation and process refinement.

7.3 Optical Principles and Photonic Mechanisms

The translation of biological insight into engineered optical design depends on a clear understanding of the physical principles that govern light-matter interaction at structured interfaces. Once the functional behavior of the moth eye was recognized as a natural antireflective system, the challenge shifted from observation to explanation: how could nanoscale surface geometry produce broadband suppression of reflection without the aid of multilayer coatings? Addressing this question required a framework capable of connecting geometry, material composition, and optical response within a unified description. The field of photonics provided the necessary conceptual tools, allowing researchers to reinterpret biological textures through the lens of wave theory and electromagnetic boundary conditions. By abstracting the problem from the specifics of biology, scientists identified a generalized mechanism—the gradual modulation of refractive index—that could be modeled, optimized, and ultimately fabricated. This transition from empirical mimicry to analytical understanding represents a key moment of analogy thinking: the movement from source (biology) to target (optical engineering) through structural and functional mapping. The subsections that follow introduce the theoretical underpinnings of this process, beginning with effective medium theory (EMT), which formalizes how subwavelength structures collectively behave as continuous optical media with graded refractive indices.

7.3.1 Effective Medium Theory

When surface features are crafted at dimensions significantly smaller than the wavelength of incident light, something fascinating occurs. Rather than interacting with the individual structures as discrete scattering centers, light perceives the surface as a smooth, continuous medium whose optical properties vary with depth. This is where the framework of EMT becomes essential [34, 35]. Instead of viewing the surface as a collection of separate ridges, cones, or bumps, EMT treats the entire textured region as a homogenized layer whose refractive index changes gradually from one side to the other. This simplification is powerful because it captures the essence of how light behaves in the presence of intricate subwavelength architecture without having to resolve every nanoscale detail.

At the heart of this approach is the idea of fill fraction—the ratio of solid material to air within any infinitesimally thin slice of the textured layer. Near the top of the structure, where the features taper off, the volume is mostly air, and thus the effective refractive index remains close to that of air, around 1.0 [36]. As one moves deeper into the layer, the proportion of solid material increases, causing the effective index to rise accordingly, approaching the value of the underlying substrate. The gradual increase in optical density mimics a soft transition rather than an abrupt boundary, reducing the likelihood of reflection due to index mismatch.

This graded-index profile serves as a gentle optical ramp, allowing light to move from one medium to another with minimal impedance mismatch. Because the transition is smooth and continuous over a subwavelength scale, reflection is suppressed not at a single wavelength or incident angle, but broadly across a wide spectral and angular range. It is this subtle control over light's momentum and phase at the interface that makes EMT so crucial in the design and understanding of moth-eye-inspired optical coatings.

As illustrated in Figure 7.3, the fundamental optical distinction between thin-film coatings and moth-eye nanostructures lies in how each system manages light reflection at material interfaces. Panel (a) represents a conventional thin-film system, where light encounters abrupt refractive index discontinuities between air, the film, and the substrate. These discrete interfaces cause partial reflections at each boundary, and the reflected beams interfere with one another depending on their phase difference and optical path length. This phenomenon, known as thin-film interference, produces wavelength-dependent reflection and transmission spectra, often manifested as colorful fringes on coated surfaces such as soap bubbles or antireflective glass layers.

Panel (b), in contrast, demonstrates a moth-eye-inspired surface characterized by an array of subwavelength conical protuberances that form a graded refractive-index profile. Instead of abrupt boundaries, the effective refractive index (n) gradually transitions from that of air (nair) to that of the substrate material (nsi). According to effective medium theory, when the structural period is much smaller than the wavelength of incident light, the medium behaves as a homogeneous gradient-index layer. This continuous refractive-index variation eliminates Fresnel reflections, minimizing interference and maximizing broadband antireflection performance. Such biomimetic nanostructures, which are modeled after the corneal nipple arrays of moth eyes, achieve near-zero reflectivity over a wide spectral and angular range, offering superior optical performance compared with discrete thin-film coatings.

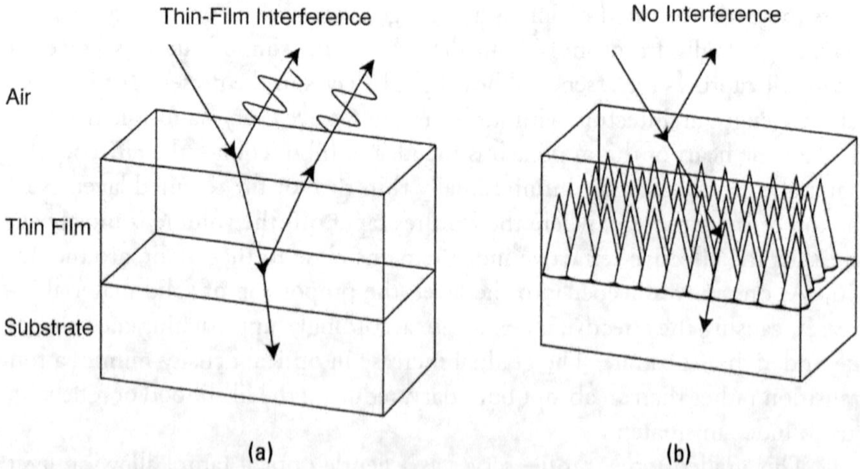

Figure 7.3 Comparison between (a) a conventional thin-film coating exhibiting interference effects, and (b) a moth-eye-inspired graded-index surface structure. In (a), discrete interfaces between air, the thin film, and the substrate cause multiple reflections that interfere constructively or destructively, resulting in thin-film interference. In (b), the conical nanostructures gradually change the refractive index from air (n_{air}) to the substrate (n_{si}), effectively suppressing reflections and eliminating interference by providing a smooth optical impedance transition.

As illustrated in Figure 7.4, the optical response of moth-eye (ME) textures transitions continuously with feature size *d*. When $d \ll \lambda$, the subwavelength cones establish a smooth graded-refractive-index (GRIN) profile that

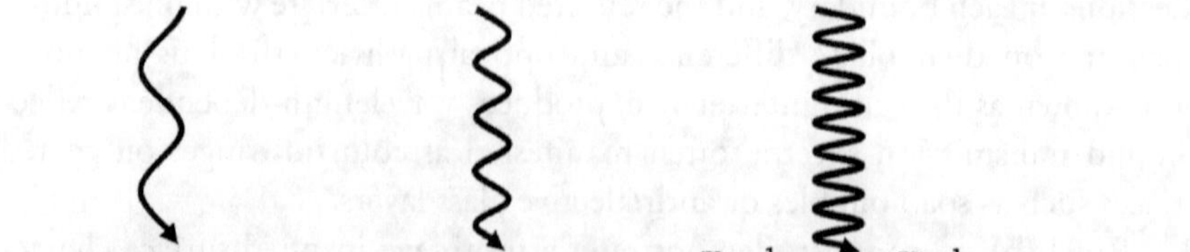

Figure 7.4 Schematic of light–surface interaction regimes for moth-eye nanostructures as the structural length scale *d* varies relative to the wavelength λ. For $d \ll \lambda$ (left), the array behaves as an effective medium with only the zeroth diffraction order ($m = 0$); for $d \sim \lambda$ (middle), diffraction and scattering appear and ± 1 orders can radiate; for $d \gg \lambda$ (right), ray-optics/Fresnel behavior dominates.

suppresses Fresnel reflections and eliminates diffracted orders, so only the specular ($m = 0$) channel remains. As d approaches λ, the periodicity supports additional momentum, allowing ± 1 diffraction orders and diffuse scattering that redirect energy away from the specular beam; this regime explains the onset of angle-dependent color and transmission roll-off reported for ME arrays with pitches near the operating band. For $d \gg \lambda$, the surface effectively becomes optically flat, and propagation is well described by ray optics and standard Fresnel relations. This regime map, reproduced from the review on ME-inspired antireflective surfaces, underpins design choices for broadband AR in the IR and for light-extraction engineering in LEDs [37].

Analogy-thinking note: EMT represents the abstraction step: it strips away biological specifics and isolates the transferable principle, continuous index grading via varying fill fraction.

7.3.2 Broadband and Omnidirectional Properties

Traditional AR coatings have long relied on the principle of destructive interference, using carefully designed stacks of thin dielectric layers—often referred to as quarter-wave coatings. These are engineered so that reflected light waves from different interfaces cancel each other out at a targeted wavelength and angle of incidence. While highly effective under specific conditions, this approach is inherently narrowband. It tends to work optimally for a single wavelength of light and for rays striking the surface at a certain angle, typically normal or near-normal incidence. As soon as the light shifts in wavelength or arrives at a steeper angle, the delicate interference balance is disrupted, and reflectance increases.

In stark contrast, moth-eye-inspired surfaces adopt a fundamentally different approach. Rather than relying on interference effects, they reduce reflection by controlling how light perceives the change in optical impedance between air and the underlying material. By constructing a surface textured with features far smaller than the wavelength of light, typically a few hundred nanometers or less, one can engineer a smooth and gradual transition in refractive index. This is not a series of discrete layers, but a continuously varying medium that guides the light gently into the substrate.

One of the more striking demonstrations of this principle can be seen in nanostructures fabricated from gallium phosphide (GaP), a semiconductor with a high intrinsic refractive index of about 3.4 in the visible range. By creating a forest of tapered nanostructures on the surface of a GaP substrate, the region closest to the air is composed mostly of voids—essentially air—while the region deeper into the material becomes increasingly solid. The result is a refractive index gradient that spans the full range from 1.0 to 3.4 over just a few hundred nanometers of depth. Light approaching this surface experiences a gentle optical slope, not an abrupt jump [38].

Because this gradient is continuous and subwavelength in nature, it doesn't give rise to reflections that would otherwise occur at discrete boundaries. Instead, the impedance mismatch is so well-managed that reflections are suppressed across virtually the entire visible spectrum and over a broad range of incident angles. This wideband, omnidirectional AR behavior is not only more robust but also more versatile than conventional coatings. It allows devices to perform reliably under changing lighting conditions, with various viewing angles, and across a full range of wavelengths, making moth-eye surfaces an elegant and powerful alternative to traditional AR technologies.

Analogy-thinking note: The broadband, angle-robust performance is an insight emerging from validated mapping (tapered geometry → graded index) under realistic constraints (visible wavelengths, high-index substrates).

7.4 Advances in Fabrication Technologies

The translation of moth-eye principles into practical optical coatings depended not only on theoretical understanding but also on the evolution of fabrication technologies capable of realizing subwavelength features at scale [39, 40]. Early laboratory demonstrations confirmed that graded-index nanostructures could drastically suppress reflection, yet their widespread application remained limited by the precision, cost, and throughput of existing manufacturing tools. As photonic design matured, a new challenge emerged: how to move from conceptual analogies and small prototypes to reproducible, large-area surfaces suitable for real-world deployment. The subsequent decades witnessed a parallel revolution in nanofabrication—spanning lithography, self-assembly, and imprinting—that made it possible to sculpt and replicate nature-inspired textures with unprecedented fidelity. These advances transformed the analogy between the moth eye and engineered optics from a metaphor into a manufacturable paradigm. Each fabrication strategy reflected a distinct balance between control and scalability, from the atomic precision of top-down etching to the adaptability of bottom-up growth. Collectively, these technologies represent the transfer phase of analogy thinking: the stage where abstracted principles of graded-index behavior are translated into material form. The subsections that follow trace this evolution, beginning with bottom-up nanostructures and culminating in high-throughput imprinting methods that bridge scientific insight and industrial implementation.

As illustrated in Figure 7.5, the evolution of lithographic techniques has been characterized by a continuous decrease in exposure wavelength to achieve finer patterning resolutions. The transition from mercury i-line (365 nm) to KrF (248 nm) and ArF (193 nm) excimer lasers marked major milestones in the 1990s and early 2000s, respectively. This wavelength reduction directly

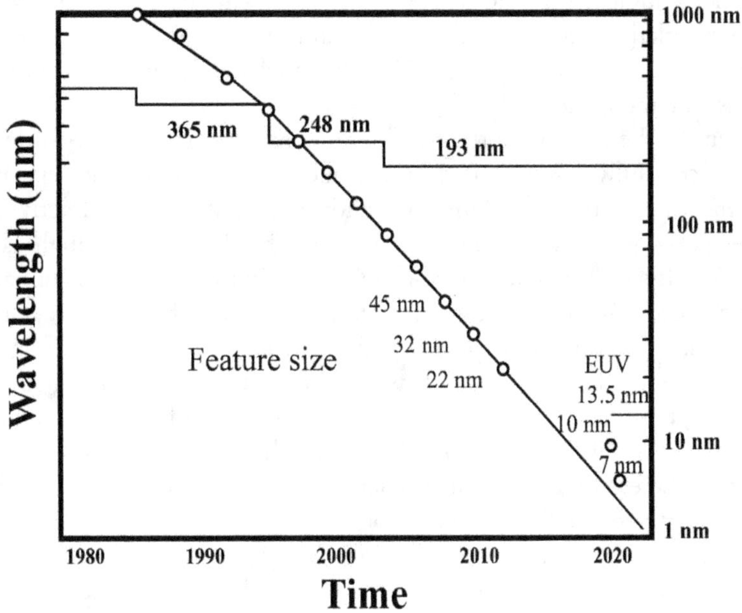

Figure 7.5 Evolution of lithographic wavelength and corresponding feature size over time. The plot illustrates the progressive reduction of exposure wavelength—from early i-line lithography at 365 nm in the 1980s, to KrF (248 nm), ArF (193 nm), and ultimately EUV lithography at 13.5 nm—enabling continued device miniaturization. The decreasing feature size, shown alongside, reflects the advancement of optical lithography techniques and scaling trends in semiconductor manufacturing. (Adapted from [39].)

influenced the achievable feature sizes, scaling from hundreds of nanometers down to tens of nanometers. The introduction of extreme ultraviolet (EUV) lithography at 13.5 nm represents the latest paradigm shift, enabling sub-10-nm features critical for advanced semiconductor nodes. This trend underscores the essential relationship between optical wavelength and device scaling, as shorter wavelengths yield higher resolution but require increasingly sophisticated optics and resist materials.

7.4.1 Bottom-Up Nanostructures

By the late 1990s, the pursuit of biomimetic AR coatings had begun to take a new direction, driven by advances in nanotechnology and materials chemistry (see Figure 7.5). While earlier efforts had focused heavily on top-down fabrication methods such as photolithography and etching—techniques that, though precise, were inherently limited in scale and adaptability—researchers began exploring bottom-up strategies that could sidestep some of these challenges. Instead of carving features into a substrate, the idea was to build the structure molecule by molecule, or particle by particle, allowing nature-inspired surfaces to be assembled rather than sculpted. This shift marked an important evolution in how moth-eye-like coatings could be fabricated and deployed across different surfaces and applications.

One of the more important breakthroughs came in 2007, when research groups successfully demonstrated the use of colloidal nanoparticles, specifically silica and titania, to create thin films with gradually varying refractive index profiles [41–44]. These coatings didn't require harsh etching or high-vacuum deposition. Instead, they were formed by depositing a mixture of nanoparticles in a controlled way so that the density and composition of the material changed with depth. At the microscopic level, this gradient in particle packing created a continuous shift in optical density from air to the underlying substrate, mimicking the function of natural moth-eye textures without needing discrete physical features like cones or pillars. The result was a coating that appeared smooth to the eye, but which manipulated light in sophisticated ways through subwavelength-scale refractive-index control.

A key advantage of this method was its compatibility with flexible and irregular surfaces. Because the nanoparticles could be applied in solution, often via dip-coating, spin-coating, or spray techniques, the films could conform to curved or nonplanar geometries with ease—something that had been prohibitively difficult using earlier lithographic approaches. Moreover, the process was more scalable, offering a potential route to cover large areas without relying on slow and expensive clean room fabrication steps. It also allowed researchers to fine-tune optical properties by adjusting parameters such as particle size, refractive-index contrast, and layer thickness. This new generation of moth-eye coatings, built from the bottom up, opened the door to a broader range of commercial uses, from antiglare films for displays to high-transparency protective layers on solar cells. In this way, the vision of mimicking nature's optical finesse was no longer restricted to rigid substrates or laboratory-scale experiments—it was becoming increasingly adaptable, accessible, and manufacturable

Analogy-thinking note: Bottom-up films implement the transfer of the abstracted graded-index principle while addressing constraint matching for large-area, curved, and flexible targets.

7.4.2 Nanowire Forests and Metamaterials

In 2009, an important milestone in the evolution of biomimetic optics was reached when a research team led by Jaime Gómez Rivas fabricated dense arrays of gallium phosphide (GaP) nanowires designed to replicate the AR performance observed in the eyes of moths [45]. These nanowire structures represented a deliberate departure from traditional layered thin films or even etched cones and domes, offering instead a three-dimensional Strong Geometrical Dependence of the Absorption of Light in Arrays of Semiconductor Nanowires forest of high-refractive-index material that could manipulate light in a highly controlled, yet naturalistic, fashion. GaP, known for its excellent transparency in the visible range and high refractive index of around 3.4, provided an ideal platform for exploring the limits of graded-index behavior.

What made this work especially compelling was not just the structural resemblance to the moth-eye effect but the degree to which it could reduce reflectance and enhance light transmission. The nanowires, grown vertically from a substrate and spaced at subwavelength distances, created an optical environment in which light encountered a gradual increase in refractive index, starting from nearly 1.0 at the tips and rising smoothly toward the dense material of the GaP base. This continuous transition allowed incoming photons to couple more efficiently into the material, with minimal reflection and negligible backscattering.

Simulations predicted that these structures could theoretically reach transmission efficiencies exceeding 99%. Experimental measurements, while slightly lower due to fabrication imperfections and slight scattering, confirmed a dramatic reduction in reflectance across a wide spectral and angular range. Importantly, the researchers showed that the reduction in reflectivity was not the result of destructive interference or surface scattering, but a true graded-index effect that funneled light into the material with exceptional efficiency. This demonstration helped cement the idea that engineered nanostructures could match, or even exceed, the optical performance of their natural counterparts, provided the material, geometry, and spatial arrangement were precisely controlled.

Figure 7.6 presents both a structural and conceptual comparison between step-index and graded-index antireflection mechanisms. Panel (a) shows a scanning electron micrograph of a GaP nanowire array, which functions as a graded-index layer through a smooth, nanoscale variation of effective refractive index. This morphology mimics the natural moth-eye effect, where subwavelength features suppress reflections by eliminating abrupt optical impedance mismatches. Panel (b) depicts the conventional step-index interface, in which incident light at an angle θ in encounters a sharp boundary, leading to partial Fresnel reflections at each surface. Interference between these reflected waves can partially cancel reflection, but not eliminate it entirely. Conversely, panel (c) illustrates the graded-index case, where the refractive index changes continuously, allowing the electromagnetic wave to bend progressively into the substrate. This smooth transition effectively matches the optical impedance between air and the medium, resulting in near-zero reflectivity and improved light trapping—an essential feature for high-efficiency photovoltaic and optoelectronic devices.

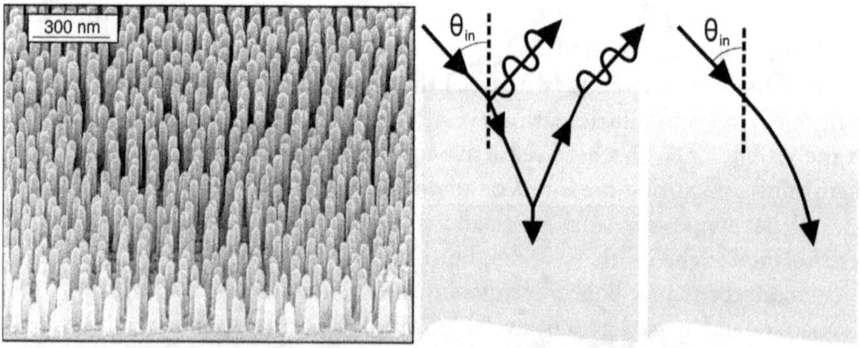

(a) (b) (c)

Figure 7.6 (a) Artistic rendering of a gallium phosphide (GaP) nanowire array demonstrating a dense, vertically aligned structure designed for antireflective behavior, (b) schematic representation of a conventional step-index interface, where discrete refractive index discontinuities cause partial reflections and interference between reflected waves, and (c) illustration of a graded-index interface that provides a gradual refractive-index transition from air to substrate, thereby minimizing Fresnel reflections and enabling enhanced optical-coupling efficiency. (Inspired by [45].)

Analogy-thinking note: GaP nanowire forests exemplify precise mapping (tapered fill fraction → effective index) and rigorous validation (simulation ↔ experiment) to confirm the abstraction.

7.4.3 Roll-to-Roll and Imprint Lithography

By the early 2010s, the development of moth-eye inspired AR coatings was no longer confined to laboratory demonstrations or small-scale fabrication experiments. A significant transformation was underway, driven by the maturation of scalable nanofabrication technologies that could finally translate the delicate subwavelength structures of bioinspired designs into manufacturable products. Among the most promising of these techniques were anodic aluminum oxide (AAO) templating and ultraviolet (UV) nanoimprint lithography—methods that, while very different in their underlying physics, shared a common advantage: the ability to pattern large areas quickly, with nanoscale precision, and at a cost that could support commercial rollout.

The AAO method gained traction because of its elegant simplicity and self-assembly nature. By electrochemically oxidizing aluminum in an acidic solution, researchers could generate highly ordered nanoporous films with near-perfect hexagonal symmetry. These porous membranes served as effective templates for transferring moth-eye textures onto other materials. When combined with a roll-to-roll process, this approach became a practical means of covering wide polymer films or glass sheets with arrays of nanoscale features that behaved optically like those on a moth's cornea. Mitsubishi Chemical was among the first to leverage this pathway at scale, culminating in the development of its MOSMITE™ film [46]. By using a mold derived from an AAO template and applying UV-curable resins onto transparent acrylic substrates, the company succeeded in replicating a consistent pattern of nanoscale bumps, each roughly 200-nm tall and spaced only 100 nm apart, over surfaces measured in square meters. The result was a film that reduced surface reflections so dramatically that untreated acrylic, which normally transmitted around 92% of visible light, could achieve over 99% transmission when both surfaces were coated.

Meanwhile, Dexerials Corporation pursued a different but equally compelling route. Drawing from their expertise in optical disc mastering, they employed high-precision laser lithography to fabricate master molds with tunable nanoscale textures. These molds, once perfected, enabled the replication of complex moth-eye patterns using nanoimprint stamping, a process not unlike the one used in mass-producing DVDs [47]. This offered even greater flexibility in controlling the shape, spacing, and depth of the nanostructures, allowing optical engineers to fine-tune the AR behavior across specific spectral ranges or angular profiles. The convergence of these manufacturing strategies marked a turning point: moth-eye coatings were no longer scientific curiosities, but viable components of next-generation optical systems, ready to be integrated into consumer devices, solar panels, display screens, and more.

Analogy-thinking note: Roll-to-roll and imprint routes are large-scale transfer mechanisms that explicitly manage constraints of cost, area, curvature, and replication fidelity; production testing provides continuous validation feeding the next design cycle.

7.5 Applications Across Industries

The practical influence of moth-eye-inspired coatings extends far beyond the laboratory. As fabrication methods matured and optical performance metrics became well characterized, researchers began translating these biologically derived designs into a wide range of industrial domains. Each application represents a distinct target domain where the same abstracted principles—graded refractive index, structural control at the nanoscale, and constraint matching to environmental conditions—manifest in new functional forms. From energy conversion and imaging to consumer displays and environmental sensing, these coatings exemplify how analogy thinking can guide innovation across disciplines. By mapping the biological mechanism of reflection suppression onto specific engineering needs, scientists have created solutions that are both elegant and practical. Moreover, as each field imposes its own material, thermal, or economic constraints, the analogical framework evolves through validation and adaptation, enriching the original insight. The following subsections trace this expansion of the moth-eye concept from natural observation to technological ubiquity. They reveal how a single biological template, once confined to the study of insect vision, has become a universal model for manipulating light, improving efficiency, and even enabling multifunctional surfaces that clean themselves, resist fogging, and enhance the clarity of the devices that define modern life.

7.5.1 Solar Energy

By the early 2010s, the promise of moth-eye nanostructures had moved well beyond theory and early prototypes, finding a particularly compelling application in the realm of solar energy. Multiple demonstrations were published on how these bioinspired surfaces significantly boost the performance of photovoltaic devices [44, 48, 49]. Their focus was on crystalline silicon solar cells, which, despite their widespread commercial use, suffer from a well-known limitation: reflection losses. Even with conventional AR coatings, a fraction of sunlight is inevitably bounced off the surface rather than absorbed and converted into electricity. This inefficiency becomes especially pronounced during early morning and late afternoon hours when the sun's rays strike the panel at steep angles.

To address this, the researchers engineered a surface coating composed of densely packed nanostructures, directly etched into the silicon itself, forming a gradient-index layer that mimicked the architecture of a moth's eye. The geometry of these features—each about 300 nm in diameter and several hundred nanometers tall—was carefully tuned to create a gradual refractive-index transition from air to silicon, thereby minimizing Fresnel reflections across the entire solar spectrum. The impact on optical performance was immediate and measurable. Solar cells treated with this nanostructured layer exhibited markedly increased light absorption, translating into a significant boost in photocurrent and overall energy conversion efficiency.

Equally important was the durability of these coatings. Recognizing that real-world deployment would expose solar modules to harsh environmental conditions, researchers subjected their nanostructured surfaces to a battery of stress tests. These included prolonged exposure to high humidity, ultraviolet radiation, thermal cycling, and even gamma irradiation. The results were striking. Not only did the moth-eye coatings maintain their structural integrity, but they also preserved their optical performance under conditions that would degrade or destroy many traditional coatings.

What made this development particularly promising was its manufacturability. The researchers devised a wet-etching method compatible with industrial-scale silicon processing, offering a path toward integrating moth-eye textures directly into solar cell production lines. This combination of optical efficacy, environmental robustness, and scalability signaled a critical advancement. It showed that lessons borrowed from nature, when combined with thoughtful engineering, could result in a practical solution for enhancing renewable energy technologies—not just in controlled lab settings but in the demanding conditions of everyday use.

Analogy-thinking note: Here the target domain is PV efficiency under variable illumination; environmental testing closes the validation loop and informs constraint matching for field reliability and process compatibility.

7.5.2 Optical and Imaging Systems

In the world of precision optics, even the smallest reflection can introduce distortion, reduce contrast, or produce unwanted visual artifacts. This becomes especially problematic in devices such as camera lenses, virtual and augmented reality headsets, and high-performance microscopes, where clarity, fidelity, and brightness are paramount. Traditionally, these challenges have been addressed using multilayer dielectric coatings, alternating thin films of materials with different refractive indices stacked carefully on the lens surface. While effective to an extent, these coatings tend to exhibit narrowband performance and often impart subtle but noticeable color casts due to wavelength-selective interference. A familiar example is the faint green or purple sheen visible on some coated lenses, which results from partial reflections being suppressed more effectively at certain wavelengths than others.

Moth-eye-inspired nanostructures offer an elegant alternative. Instead of relying on interference effects, they work by creating a smooth, gradual refractive-index transition between air and the optical substrate. This allows them to reduce reflections uniformly across a broad spectral range without favoring one color over another. The effect is visually neutral and exceptionally effective. For optics that must operate across wide fields of view or under varying lighting conditions, this is a decisive advantage. In augmented reality systems, for example, the combination of light from both the real world and digital overlays must be transmitted and managed with as little distortion or loss as possible. Any reflection at the interface can cause ghosting or dimming of the image, problems that are significantly mitigated by applying a moth-eye surface to the transparent elements in the optical path.

Until recently, one of the challenges in deploying such nanostructures was the difficulty of applying them to curved or nonplanar surfaces [50]. Unlike flat glass panes, the complex geometries of modern lenses or eyepieces require fabrication techniques that can conform to curvature without compromising precision. Advances in plasma etching and nanoimprint lithography have addressed this issue [51–54]. Researchers have demonstrated methods to uniformly texture the surface of curved optical elements, such as hemispherical lenses or convex eyepieces, with subwavelength features [55]. These methods allow for the same broadband, angle-independent AR performance previously limited to planar surfaces. As a result, the visual experience in devices like headsets, camera systems, and advanced imaging tools is noticeably improved, with richer contrast, reduced flare, and better transmission across the visible spectrum. These advancements mark a significant step toward realizing optics that are not only higher performing but also more visually immersive and functionally versatile.

As illustrated in Figure 7.7, the fabrication of moth-eye-inspired nanostructures on curved polymer substrates involves a UV-assisted nanoimprint lithography process using a flexible PDMS stamp. Initially, a thin UV-curable resist layer is applied to a PMMA sheet, which can either possess a concave or convex curvature. The PDMS stamp, carrying the inverse of the desired nanostructure pattern, is brought into conformal contact with the resist layer. Owing to the flexibility of PDMS, intimate contact is achieved even on nonplanar surfaces. Upon UV exposure, the resist is cross-linked and the nanostructure pattern is transferred accurately to the curved substrate. Following demolding, the resulting surfaces exhibit well-defined moth-eye nanostructures on both

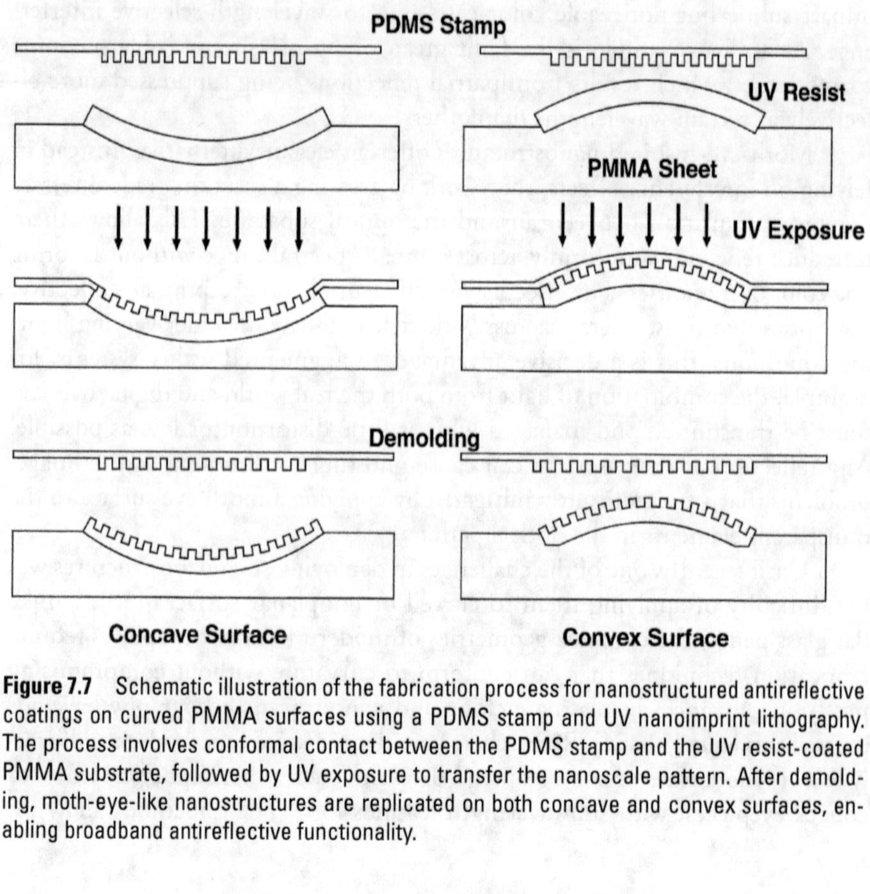

Figure 7.7 Schematic illustration of the fabrication process for nanostructured antireflective coatings on curved PMMA surfaces using a PDMS stamp and UV nanoimprint lithography. The process involves conformal contact between the PDMS stamp and the UV resist-coated PMMA substrate, followed by UV exposure to transfer the nanoscale pattern. After demolding, moth-eye-like nanostructures are replicated on both concave and convex surfaces, enabling broadband antireflective functionality.

concave and convex geometries, which significantly reduce surface reflectance by creating a graded refractive index transition between air and the substrate. This technique provides an efficient and scalable route for fabricating antireflective coatings on optically curved components such as lenses or domes.

Analogy-thinking note: Applying textures to curved optics is a transfer challenge; manufacturable conformity and optical neutrality are key constraints that shape the mapping from flat to freeform elements.

7.5.3 Sensors and Detectors

As optical sensing technologies have advanced, devices like lidar systems, thermal imaging cameras, and photodetectors have become essential across a wide range of industries, from autonomous vehicles and aerospace to medical diagnostics and industrial inspection. The performance of these systems often hinges on the ability to capture as much light as possible, particularly when operating under challenging conditions or within specific spectral bands. Yet, a persistent problem lies at the very first interface the light encounters: the window or lens. Even small reflective losses at these surfaces can lead to significant reductions in signal strength, especially when compounded across multiple optical elements. This issue becomes especially critical in the infrared spectrum, particularly the long-wave infrared (LWIR) range from 8 to 12 μm, where traditional thin-film AR coatings tend to perform poorly or become prohibitively complex and expensive to implement.

It is in this context that moth-eye-inspired nanostructures have shown tremendous promise. Unlike conventional coatings that rely on material stacks designed for specific wavelengths, these nanostructured surfaces reduce reflections through a gradual transition in refractive index—an effect that is inherently broadband and less dependent on precise layer thicknesses or materials. When applied to materials like germanium or chalcogenide glasses, which are commonly used in IR optics due to their transparency in the LWIR range, moth-eye textures can dramatically reduce surface reflections that would otherwise lead to signal loss. The improvement in light throughput has direct consequences for the sensitivity and efficiency of the sensing device. For a lidar system, more incoming light means better range and higher accuracy in object detection. In thermal cameras, especially those used for night vision or industrial diagnostics, higher transmission allows for clearer images with finer thermal resolution. And in photodetectors, particularly those tasked with capturing faint or fast-changing signals, reducing optical loss at the input surface can make the difference between a usable signal and background noise.

Recent advancements in fabrication techniques, including direct etching with focused ion beams or nanoimprint lithography, have made it feasible to apply moth-eye patterns onto curved or nonstandard surfaces often used in IR optics. These processes preserve the precise geometry required to maintain the graded-index effect while allowing scalability and robustness. The result is an optical element that, although inspired by biology, is thoroughly modern in its performance—capable of withstanding thermal cycling, mechanical wear, and even harsh environmental exposure while still providing a measurable boost in optical efficiency.

Analogy-thinking note: In IR sensing, constraint matching (materials with high LWIR transmission, thermal robustness) governs which mappings remain valid, and field testing provides validation for system-level gains.

7.5.4 Displays and Consumer Electronics

In an age where screens are our constant companions—embedded in smartphones, tablets, laptops, and even wearable devices—the visibility and readability of displays under varying lighting conditions has become more than a matter of convenience; it is a fundamental aspect of usability. One of the most persistent challenges has been screen glare, particularly in bright environments where ambient light reflects off the display surface, making it difficult or even impossible to read the content. While matte coatings and polarizing filters have been used to mitigate this issue, they often come at a cost: reduced brightness, altered color fidelity, or a muted visual texture that undermines the crispness of high-resolution displays, especially those using OLED technology known for its vivid contrast and deep blacks.

Enter moth-eye-inspired coatings, which is a concept rooted in biology and realized through modern nanofabrication. These surfaces are patterned with subwavelength textures that mimic the moth's compound eye, creating a smooth refractive-index transition from air to the screen material. This graded interface minimizes reflections without producing interference colors or altering display characteristics. One notable implementation is Mitsubishi Chemical's MOSMITE film [46], applied to transparent acrylic or glass to enhance light transmission and suppress glare. As a result, screens coated with such films remain clear, bright, and color-accurate even under direct sunlight.

The key advantage of these coatings lies in their neutrality. Unlike traditional AR treatments that often impart a green or purple hue due to narrowband interference effects, moth-eye surfaces do not distort the screen's natural color profile. Moreover, they preserve the display's original brightness by minimizing reflection losses without absorbing light. This has made them particularly attractive for next-generation microLED displays [56], where maintaining both brightness and contrast is essential for performance. As fabrication methods like UV nanoimprint lithography and roll-to-roll processing have matured, it has become increasingly feasible to integrate these coatings into mass-produced consumer electronics. The result is a user experience that feels more natural and immersive, where the barrier between the digital image and the real-world environment is subtly diminished—achieved not through more pixels or higher resolution, but through mastering light at the nanometer scale.

Analogy-thinking note: Consumer displays illustrate a clean transfer of the abstraction with strict constraints (cost, touch durability, color neutrality) that shape pattern geometry and replication method.

7.5.5 Functional Surfaces

While moth-eye structures are most celebrated for their remarkable ability to suppress reflections and enhance optical transmission, their benefits extend well beyond the realm of photonics. At the nanoscale, the very same surface textures that manipulate light also influence how water and other contaminants interact with the material. Specifically, these subwavelength structures confer hydrophobic properties, turning the surface into one that naturally resists wetting and fouling. When water comes into contact with a moth-eye-textured film, it does not spread or soak into the surface as it might on untreated glass or plastic. Instead, it beads up into nearly spherical droplets, perching delicately atop the nanoscale features, with minimal contact area between the liquid and the solid beneath.

This behavior is reminiscent of the lotus leaf effect, where hierarchical micro- and nanostructures create superhydrophobicity in nature. On moth-eye surfaces, the water droplets are easily displaced by gravity or a gentle breeze, often carrying away dust, oils, or other particles in the process. The result is a self-cleaning effect, where the surface stays clearer for longer periods, reducing the need for manual maintenance or chemical cleaning agents. In outdoor environments, such properties are especially valuable. Whether on solar panels exposed to rain and pollution, on camera lenses used in fieldwork, or on vehicle windshields, a surface that repels water and grime without degrading its optical function offers significant practical advantages.

Equally important is the role of moth-eye textures in mitigating fogging [57, 58]. Fog forms when condensation occurs on a surface, typically as fine microdroplets that scatter light and obscure visibility. On a hydrophilic surface, these droplets adhere and spread, creating a cloudy film. But on a hydrophobic, nanostructured surface, condensation is discouraged from settling evenly. Instead, any moisture that does form tends to appear as isolated beads that scatter far less light and evaporate more quickly. This antifogging property is invaluable in applications such as eyewear, camera enclosures, medical optics, and even greenhouse panels, where visibility and light transmission are critical.

The multifunctionality of moth-eye surfaces, combining AR performance with hydrophobic and antifogging traits, demonstrates the power of nanoscale engineering. It also underscores the broader lesson from nature: that elegant structures can often serve more than one purpose, providing layered functionality that arises not from added complexity but from intelligent, integrated design.

The wetting behavior of a surface is primarily characterized by the static contact angle, which reflects the balance between adhesive and cohesive forces acting at the liquid–solid–vapor interface. As shown in Figure 7.8, the classification ranges from superhydrophilic (θ < 10°) to superhydrophobic (θ > 150°) regimes. Superhydrophilic surfaces promote complete spreading of water droplets, facilitating rapid absorption or evaporation, while superhydrophobic surfaces exhibit extreme water repellence, causing droplets to roll off easily due to minimal solid–liquid adhesion.

The schematic also highlights three classical models that describe surface wetting phenomena. The Young model defines the equilibrium contact angle on a smooth, ideal surface based on interfacial tensions between the solid, liquid, and vapor phases. The Wenzel model extends this concept to rough surfaces, where liquid fully penetrates surface asperities, amplifying the intrinsic wettability of the substrate. Conversely, the Cassie–Baxter model describes heterogeneous wetting, in which air pockets trapped beneath the liquid droplet reduce the solid–liquid contact fraction, leading to enhanced hydrophobicity and

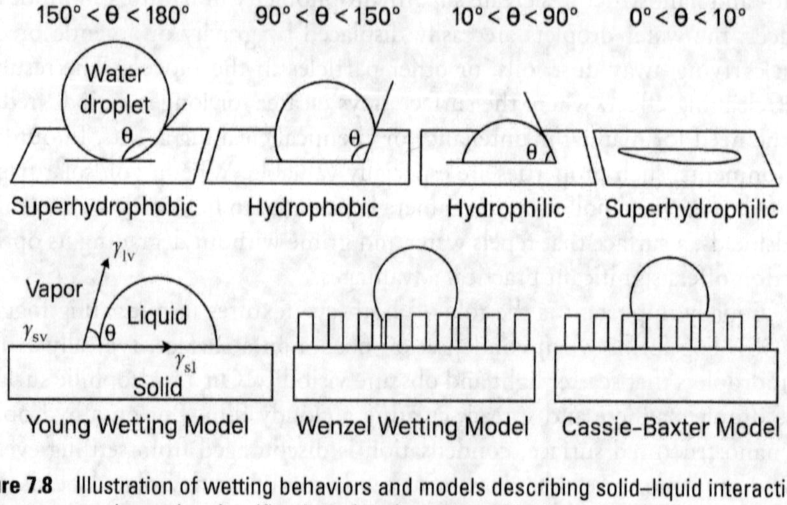

Figure 7.8 Illustration of wetting behaviors and models describing solid–liquid interactions. The upper row shows the classification of surface wettability based on the contact angle θ: superhydrophobic (150° < θ < 180°), hydrophobic (90° < θ < 150°), hydrophilic (10° < θ < 90°), and superhydrophilic (0° < θ < 10°). The lower row presents three models describing the wetting mechanisms: the Young model (ideal smooth surface), the Wenzel model (homogeneous contact on rough surfaces), and the Cassie–Baxter model (heterogeneous contact with trapped air pockets).

water repellency. Together, these models form the theoretical foundation for designing and engineering functional surfaces with tailored wetting characteristics.

Analogy-thinking note: The emergence of hydrophobic and antifog benefits constitutes insight generation beyond the original AR target; new targets can be defined and iterated using the same source-domain mapping.

7.6 Commercialization and Future Directions

As the science of moth-eye-inspired coatings has evolved from a biological curiosity to a cornerstone of advanced optical engineering, a new phase is emerging—one defined not just by performance but by practicality. Fabrication techniques that once existed only within the confines of specialized laboratories have matured into robust, repeatable processes suited for industrial-scale production. Central to this transformation is the development of roll-to-roll nanoimprint lithography, a technique that enables the continuous printing of nanostructures onto flexible substrates at high speed and low cost. This approach has unlocked the ability to apply moth-eye textures over large areas, measured not in square centimeters but in meters, making it feasible to integrate these coatings into commercial products such as display films, architectural glass, and solar modules.

The adaptability of these patterning methods has also been a significant breakthrough. Flat surfaces are no longer the sole beneficiaries of AR nanostructures. Advances in conformal imprinting and plasma-assisted etching now allow the transfer of intricate moth-eye geometries onto three-dimensional surfaces such as camera lenses, optical domes, and wearable electronics. This capability expands the applicability of moth-eye coatings into areas where optical performance is crucial but geometrical complexity previously posed a barrier. For instance, automotive sensors embedded in curved housings or optical systems in drones and autonomous vehicles can now enjoy the same reflection suppression and light transmission benefits as planar optics.

But the promise of these coatings does not end with light management. Increasingly, researchers are integrating additional functionalities into the same nanostructured layers. The natural hydrophobicity imparted by the nanoscale texture can be enhanced or tailored further with thin hydrophobic coatings, producing surfaces that are not only AR but also self-cleaning and resistant to fogging. In some cases, antimicrobial properties have been introduced through the incorporation of photocatalytic materials or specialized nanoparticle infusions, creating surfaces that not only remain clear but also hygienic—applications that are of particular interest in medical devices, public touchscreens, and food preparation environments.

Looking further ahead, a new class of moth-eye coatings is beginning to take shape, one that moves beyond static performance and into the realm of adaptability. By embedding responsive materials into the nanostructure, such as liquid crystals, phase-change compounds, or electro-optic polymers, researchers are exploring ways to create coatings with tunable reflection. These active moth-eye surfaces could dynamically adjust their optical properties in response to environmental cues like light intensity, temperature, or applied voltage. Such coatings might one day be used to create adaptive camera lenses that change their optical characteristics in real time, or display screens that optimize visibility based on ambient lighting conditions without relying on electronic dimming. This fusion of form, function, and responsiveness suggests that the future of moth-eye technology lies not just in mimicking nature, but in extending its principles into realms nature never reached: where human ingenuity builds upon biological wisdom to craft surfaces that are not only smarter but also deeply aligned with the demands of modern life.

Analogy-thinking note: Commercialization foregrounds constraint matching (cost, throughput, reliability) as a design driver; emerging tunable coatings represent new target domains seeded by prior insights, completing a recursive validation → redesign cycle.

7.7 Conclusion

The journey of moth-eye AR coatings is a testament to how deeply human innovation can be shaped by nature's quiet genius. What began with the simple act of peering through an electron microscope at the surface of a moth's eye evolved into a transformative force in photonics and materials science. From the earliest understanding of graded-index optical transitions, to the replication of subwavelength textures in glass, silicon, and polymer films, each step along this path reflects a growing mastery over light itself. At every turn, engineers and scientists have built upon the biological principle that a textured interface, if designed at the nanoscale, can guide light with precision, efficiency, and minimal loss. That principle, simple yet profound, has been applied in contexts far beyond its original environment, from solar panels on rooftops to displays in our pockets, from the delicate optics of microscopes to the robust sensors used in autonomous navigation.

And this progress has not only been about suppressing glare or boosting transmission. It has also been about redefining the very surfaces we interact with. These coatings have shown the ability to self-clean, to repel water, to resist fogging, and in the near future, to adapt their optical properties in response to stimuli. Such multifunctionality is not the product of added layers or coatings but of structure alone—proof that geometry, when understood at the scale of light's wavelength, becomes a tool as powerful as chemistry. The surface, once considered a passive boundary between media, has become an active participant in shaping optical performance.

Even more remarkable is how these principles have transitioned from lab-scale breakthroughs to manufacturable reality. The development of roll-to-roll nanoimprint lithography, conformal patterning, and scalable etching processes has allowed moth-eye coatings to leap into the world of production. They are no longer confined to journal articles or research centers; they are now appearing on consumer screens, high-efficiency solar modules, precision sensors, and biomedical optics. The future promises even more. Adaptive coatings, tunable in real time, are beginning to emerge, signaling a shift from static to dynamic optics—surfaces that respond, adjust, and optimize themselves on demand.

In reflecting on this trajectory, it becomes clear that the value of the moth-eye paradigm lies not just in what it does, but in what it reveals: that nature's designs, when understood deeply, can serve as a gateway to technological solutions that are both elegant and enduring. Through decades of research, iteration, and application, a biological survival mechanism has been reimagined as a tool for precision engineering. The result is not merely a better coating, but a new way of thinking about how light interacts with matter, and how small, intentional structures can change the world one photon at a time.

Analogy-thinking summary: Source → target (moth eye → AR surfaces); mapping (tapered geometry ↔ index gradient); abstraction (EMT); transfer (etching, colloids, imprint); constraint matching (materials, scale, cost); iterative validation (optical/environmental tests) → insight generation (broadband AR, multifunctionality, tunability). This recursive loop links Chapter 1's framework directly to the case in this chapter.

References

[1] Raut, H. K., et al., "Anti-Reflective Coatings: A Critical, In-Depth Review," *Energy & Environmental Science*, Vol. 4, No. 10, 2011, pp. 3779–3804, https://doi.org/10.1039/C1EE01297E.

[2] Gemne, G., "Ultrastructural Ontogenesis of Cornea and Corneal Nipples in the Compound Eye of Insects," *Acta Physiologica Scandinavica*, Vol. 66, No. 4, 1966, pp. 511–512, https://doi.org/10.1111/j.1748-1716.1966.tb03232.x.

[3] Miller ,W. H., et al., "The Optics of Insect Compound Eyes," *Science*, Vol. 162, No. 3855, 1968, pp. 760–767, https://doi.org/10.1126/science.162.3855.760.

[4] Bernhard, C. G., et al., "Comparative Ultrastructure of Corneal Surface Topography in Insects with Aspects on Phylogenesis and Function," *Zeitschrift für vergleichende Physiologie*, Vol. 67, 1970, pp. 1–25, https://doi.org/10.1007/BF00298117.

[5] Aideo, S. N., et al., "Limiting Hydrophobic Behavior and Reflectance Response of Dragonfly and Damselfly Wings," *Applied Surface Science*, Vol. 387, 2016, pp. 609–616, https://doi.org/10.1016/j.apsusc.2016.06.049.

[6] Stuhr, S., et al., "Structure and Chemical Organization in Damselfly *Calopteryx haemorrhoidalis* Wings: A Spatially Resolved FTIR and XRF Analysis with Synchrotron Radiation," *Scientific Reports*, Vol. 8, 2018, p. 8413, 2https://doi.org/10.1038/s41598-018-26563-6.

[7] Yoshida, A., et al., "Nanoprotuberance Array in the Transparent Wing of a Hawkmoth, *Cephonodes hylas*," *Zoological Science*, Vol. 13, No. 4, 1996, pp. 525–526.

[8] Spalding, A., et al., "Optical Modeling and Phylogenetic Analysis Provide Clues to the Likely Function of Corneal Nipple Arrays in Butterflies and Moths," *Insects*, Vol. 10, No. 9, 2019, p. 262, https://doi.org/10.3390/insects10090262.

[9] Vynck, K., et al., "Photon Management in Two-Dimensional Disordered Media," *Nature Materials*, Vol. 11, No. 12, 2012, pp. 1017–1022, https://doi.org/10.1038/nmat3442.

[10] Martins, E. R., et al., "Deterministic Quasi-Random Nanostructures for Photon Control," *Nature Communications*, Vol. 4, 2013, p. 2665, https://doi.org/10.1038/ncomms3665.

[11] Stavroulakis, P. I., et al., "Suppression of Backscattered Diffraction from Sub-Wavelength 'Moth-Eye' Arrays," *Optics Express*, Vol. 21, No. 1, 2013, pp. 1–11, https://doi.org/10.1364/OE. 21.000001.

[12] Ji, S., et al., "Optimal Moth Eye Nanostructure Array on Transparent Glass Towards Broadband Antireflection," *ACS Applied Materials & Interfaces*, Vol. 5, No. 21, 2013, pp. 10 731–10 737, https://doi.org/10.1021/am402881x.

[13] Kumar, P., et al., "Optical Probing of Long-Range Spatial Correlation and Symmetry in Complex Biophotonic Architectures on Transparent Insect Wings," *Laser Physics Letters*, Vol. 12, No. 2, 2015, p. 025901, https://doi.org/10.1088/1612-2011/12/2/025901.

[14] Dellieu, L., et al., "A Two-in-One Superhydrophobic and Anti-Reflective Nanodevice in the Grey Cicada *Cicada orni* (*Hemiptera*)," *Journal of Applied Physics*, Vol. 116, No. 2, 2014, p. 024701, https://doi.org/10.1063/1.4889849.

[15] Stavenga, D. G., et al., "Light on The Moth-Eye Corneal Nipple Array of Butterflies," *Proceedings of the Royal Society B: Biological Sciences*, Vol. 273, No. 1587, 2006, pp. 661–667, https://doi.org/10.1098/rspb.2005.3369.

[16] Finet, C., et al., "Multi-Scale Dissection of Wing Transparency in the Clearwing Butterfly *Phanus vitreus*," *Journal of the Royal Society Interface*, Vol. 20, No. 203, 2023, p. 20230135, https://doi.org/10.1098/rsif.2023.0135.

[17] Zhang, Z., et al., "Absorption Properties and Mechanisms of Metallic Moth-Eye Structures," *Optics Communications*, Vol. 540, 2023, p. 129487, https://doi.org/10.1016/j.optcom.2023.129487.

[18] Sun, M., et al., "Wetting Properties on Nanostructured Surfaces of Cicada Wings," *Journal of Experimental Biology*, Vol. 212, No. 19, 2009, pp. 3148–3155, https://doi.org/10.1242/jeb.033373.

[19] Sun, M., et al., "Influence of Cuticle Nanostructuring on the Wetting Behaviour/States on Cicada Wings," *PLoS ONE*, Vol. 7, No. 4, 2012, p. e35056, https://doi.org/10.1371/journal.pone.0035056.

[20] Yan, L., et al., "Integrated Super-Hydrophobic and Antireflective PDMS Bio-Templated from Nano-Conical Structures of Cicada Wings," *RSC Advances*, Vol. 6, No. 110, 2016, pp. 108 974–108 980, https://doi.org/10.1039/C6RA23811D.

[21] Xie, H., et al., "Rapid Fabrication of Bio-Inspired Nanostructure with Hydrophobicity and Antireflectivity on Polystyrene Surface Replicating from Cicada Wings," *Nanoscale*, Vol. 9, No. 33, 2017, pp. 11 951–11 958, https://doi.org/10.1039/C7NR04176D.

[22] Wang, W., et al., "Progress in Construction of Bio-Inspired Physico-Antimicrobial Surfaces," *Nanotechnology Reviews*, Vol. 9, No. 1, 2020, pp. 114–133, https://doi.org/10.1515/ntrev-2020-0089.

[23] van Nieuwenhoven, R. W., et al., "Verifying Antibacterial Properties of Nanopillars on Cicada Wings," *Appl. Nanoscience*, Vol. 14, 2024, pp. 531–541, https://doi.org/10.1007/s13204-024-03030-5.

[24] Ivanova, E. P., et al., "Antifungal Versus Antibacterial Defence of Insect Wings," *Journal of Colloid and Interface Science*, Vol. 603, 2021, pp. 886–897, https://doi.org/10.1016/j.jcis.2021.06.093.

[25] Ivanova, E., et al., "Natural Bactericidal Surfaces: Mechanical Rupture of *Pseudomonas aeruginosa* Cells by Cicada Wings," *Small*, Vol. 8, No. 16, 2012, pp. 2489–2494, https://doi.org/10.1002/smll.201200528.

[26] Bernard, G. D., W. Miller, and J. Allen, *Physical Optics of Invertebrate Eyes*, Cambridge, MA: MIT Press, 1968.

[27] Miller, W. H., et al., "The Optics of Insect Compound Eyes," *Science*, Vol. 162, No. 3855, 1968, pp. 760–767, https://doi.org/10.1126/science.162.3855.760.

[28] Jacques, S. L., "Optical Properties of Biological Tissues: A Review," *Physics in Medicine & Biology*, Vol. 58, No. 11, 2013, pp. R37–R61, 2 https://doi.org/10.1088/0031-9155/58/11/R37.

[29] Bezugla, N., et al., "Fundamentals of Determination of the Biological Tissue Refractive Index by Ellipsoidal Reflector Method," *Photonics*, Vol. 11, No. 9, 2024, p. 828, https://doi.org/10.3390/photonics11090828.

[30] Cairós, C., et al., "Refractive Index Estimation in Biological Tissues by Quantitative Phase Imaging," *Optical Materials*, Vol. 142, 2023, p. 114087, https://doi.org/10.1016/j.optmat.2023.114087.

[31] Lai, J., et al., "Experimental Measurement of the Refractive Index of Biological Tissues by Total Internal Reflection," *Applied Optics*, Vol. 44, No. 10, 2005, pp. 1845–1849, https://doi.org/10.1364/AO.44.001845.

[32] Dirckx, J. J., et al., "Refractive Index of Tissue Measured with Confocal Microscopy," *Journal of Biomedical Optics*, Vol. 10, No. 4, 2005, p. 044014, https://doi.org/10.1117/1.1993487.

[33] Clapham, P. B., and M. C. Hutley, "Reduction of Lens Reflexion by the 'Moth Eye' Principle," *Nature*, Vol. 244, 1973, pp. 281–282, https://doi.org/10.1038/244281a0.

[34] Qin, F., et al., "Electromagnetic Composites: From Effective Medium Theories to Metamaterials," *Journal of Applied Physics*, Vol. 132, No. 10, 2022, p. 101101, https://doi.org/10.1063/5.0099072.

[35] Mei, R., et al., "Breakdown of Effective-Medium Theory in Dielectric Composites Containing Epsilon-Near-Zero Constituents," *Physical Review B*, Vol. 109, No. 4, 2024, p. 045104, https://doi.org/10.1103/PhysRevB.109.045104.

[36] Lakhtakia, A., and R. Messier, *Sculptured Thin Films: Nanoengineered Morphology and Optics*, SPIE press, 2005, Vol. 143.

[37] Chan, L. W., et al., "Moth Eye-Inspired Anti-Reflective Surfaces for Improved IR Optical Systems & Visible LEDs Fabricated with Colloidal Lithography and Etching," *Bioinspiration & Biomimetics*, Vol. 13, No. 4, May 2018, p. 041001, https://doi.org/10.1088/1748-3190/aab738.

[38] MacLeod, H. Angus, *Thin-Film Optical Filters, 4th ed.*, Boca Raton: CRC Press, 2010, ttps://doi.org/10.1201/9781420073034.

[39] Khonina, S. N., et al., "Grayscale Lithography and a Brief Introduction to other Widely Used Lithographic Methods: A State-of-the-Art Review," *Micromachines*, Vol. 15, No. 11, Oct. 30, 2024, p. 1321, https://doi.org/10.3390/mi15111321.

[40] Kazazis, D., et al., "Extreme Ultraviolet Lithography," *Nature Reviews Methods Primers*, Vol. 4, 2024, p. Article 84, https://doi.org/10.1038/s43586-024-00361-z.

[41] Chen, M., et al., "Design of Optical Path for Wide-Angle Gradient-Index Antireflection Coatings," *Applied Optics*, Vol. 46, No. 26, 2007, pp. 6533–6538, https://doi.org/10.1364/AO.46.006533.

[42] Chau, Joseph L. H., et al., "Transparent High Refractive Index Nanocomposite Thin Films," *Materials Letters*, Vol. 61, 2007, pp. 2908–2910, https://doi.org/10.1016/j.matlet.2007.01.088.

[43] Chhajed, S., et al., "Nanostructured Multilayer Graded-Index Antireflection Coating for Si Solar Cells with Broadband and Omnidirectional Characteristics," *Applied Physics Letters*, Vol. 93, No. 25, 2008, p. 251108, https://doi.org/10.1063/1.3050463.

[44] Prevo, B. G., et al., "Assembly and Characterization of Colloid-Based Antireflective Coatings on Multicrystalline Silicon Solar Cells," *Journal of Materials Chemistry*, Vol. 17, 2007, pp. 791–799, https://doi.org/10.1039/B612734G.

[45] Diedenhofen, S., et al., "Strong Geometrical Dependence of the Absorption of Light in Arrays of Semiconductor Nanowires," *ACS Nano*, Vol. 5, 2011, pp. 2316–2323, https://doi.org/10.1021/nn103596n.

[46] Mitsubishi Chemical Group Corporation, "MOSMITE™ Antireflective Film Using Nanoimprint Technology," Online, 2020, accessed: 2025-10-19, https://www.mcgc.com/english/.

[47] Dexerials Corporation, 2021, https://www.dexerials.jp/en/.

[48] Han, Kang-Soo, et al., "Enhanced Performance of Solar Cells with Anti-Reflection Layer Fabricated by Nano-Imprint Lithography," *Solar Energy Materials and Solar Cells*, Vol. 95, No. 1, 2011, pp. 288–291, https://doi.org/10.1016/j.solmat.2010.07.018.

[49] Yamada, N., et al., "Characterization of Antireflection Moth-Eye Film on Crystalline Silicon Photovoltaic Module," *Optics Express*, Vol. 19, No. S2, 2011, pp. A118–A125, https://doi.org/10.1364/OE.19.00A118.

[50] Okabe, T., et al., "Polyimide Moth-Eye Nanostructures Formed by Oxygen Ion Beam Etching for Anti-Reflection Layers," *Microelectronic Engineering*, Vol. 242–243, 2021, p. 111559, https://doi.org/10.1016/j.mee.2021.111559.

[51] Ge, H., et al., "Hybrid Nanoimprint-Soft Lithography for Highly Curved Surface with Sub-15 nm Resolution," *Lab-on-Fiber Technology*, Vol. 56, 2014, p. 91–109, https://doi.org/10.1007/978-3-319-06998-2 5.

[52] Verschuuren, M. A., et al., "Large Area Nanoimprint by Substrate Conformal Imprint Lithography (SCIL)," *Advanced Optical Technologies*, Vol. 6, No. 3-4, 2017, p. 243–264, https://doi.org/10. 1515/aot-2017-0022.

[53] Supreeti, S., et al., "Development and Implementation of a Rotating Nanoimprint Lithography Tool for Orthogonal Imprinting on Edges of Curved Surfaces," *Nanomanufacturing and Metrology*, Vol. 4, 2021, pp. 175–180, https://doi.org/10.1007/s41871-021-00114-6.

[54] Wang, C., et al., "Discretely-Supported Nanoimprint Lithography for Patterning the High-Spatial-Frequency Stepped Surface," *Nano Research*, Vol. 14, 2021, pp. 2606–2612, https://doi.org/10.1007/s12274-020-3261-3.

[55] Scia Systems GmbH, "Nano-Texturing of Curved Optics by Plasma Etching," Online, 2024, accessed: 2025-10-19,https://www.scia-systems.com/learning/anti-reflective-nanostructures-with-moth-eye-effect.

[56] Ahmed, K., *MicroLED Devices and Systems*, Norwood, MA: Artech House, 2024.

[57] Chu, J., et al., "Recent Advances in Prevailing Antifogging Surfaces: Structures, Materials, Durability, and Beyond," *Nanoscale*, Vol. 15, No. 27, 2023, pp. 11 366–11 402, https://doi.org/10.1039/d3nr01767b.

[58] Guo, Y., et al., "Preparation and Study of Moth-Eye-Like Micro-Nano Structure Optical Anti-Reflection Coating," *Polymer Science, Series B*, Vol. 66, No. 4, 2024, pp. 182–192, https://doi.org/10.1134/S1560090424600517.

8

Biomimetic X-Ray Imaging

After reading this chapter you should be able to:

- Explain the biological inspiration and optical principles behind lobster-eye X-ray imaging systems, highlighting the role of grazing-incidence reflection in achieving wide-field sensitivity.
- Describe the engineering translation of lobster-eye geometry into microchannel plate (MCP) devices and evaluate their performance across different applications, from space telescopes to medical and industrial imaging.
- Assess the advantages, limitations, and future directions of lobster-eye optics in expanding the capabilities of X-ray imaging technologies.

8.1 Introduction

In the ever-evolving field of photonics, some of the most profound innovations have sprung not solely from high-performance computing or theoretical abstraction, but from careful observation of nature's design. Yet, both high-performance computation and theoretical modeling play a crucial enabling role in analogy thinking: They allow researchers to simulate, abstract, and generalize complex natural phenomena into transferable design principles. High-performance computing provides the numerical precision to explore multiscale optical interactions and emergent behaviors that would be impossible to isolate experimentally, while theoretical abstraction offers the conceptual scaffolding needed to translate those physical insights into generalized, domain-independent representations. Within this integrated framework, analogy thinking operates as a bridge—linking natural observation, theoretical reasoning, and computational exploration into a coherent design methodology.

One particularly striking example of this synthesis is the lobster-eye X-ray optic [1–10], a technology that captures the essence of biomimicry in its purest and most impactful form. Inspired by the unique visual system of lobsters, whose eyes are composed not of lenses but of an intricate array of square-shaped tubes, this optical architecture enables wide-field vision in dim and murky underwater environments. Unlike the human eye, which relies on refractive lenses to focus light, the lobster eye uses reflection, with each square channel directing light onto the retina from slightly different angles to produce a panoramic image (see Figure 8.1).

Traditional X-ray optics suffer from inherent limitations due to the nature of X-rays themselves, which are difficult to focus using conventional lenses or mirrors [11–14]. Lobster-eye designs, however, sidestep this challenge by guiding X-rays along reflective paths within microchannel structures, gently steering them toward a common focal plane [3, 15–18]. These structures are typically fabricated using high-aspect-ratio silicon molds or glass capillary arrays, with inner surfaces coated in reflective metals like iridium or gold to maximize efficiency. The result is a lightweight, low-power, and wide-angle imaging system that can be deployed in space telescopes, security scanners, and medical instruments. Its utility in space-based astronomy is particularly transformative, allowing satellites to scan vast portions of the sky for transient X-ray sources with minimal energy consumption. In blending evolutionary biology with optical physics and material engineering, lobster-eye optics exemplify how biomimetic thinking can reframe persistent technical limitations and enable new paradigms in imaging science.

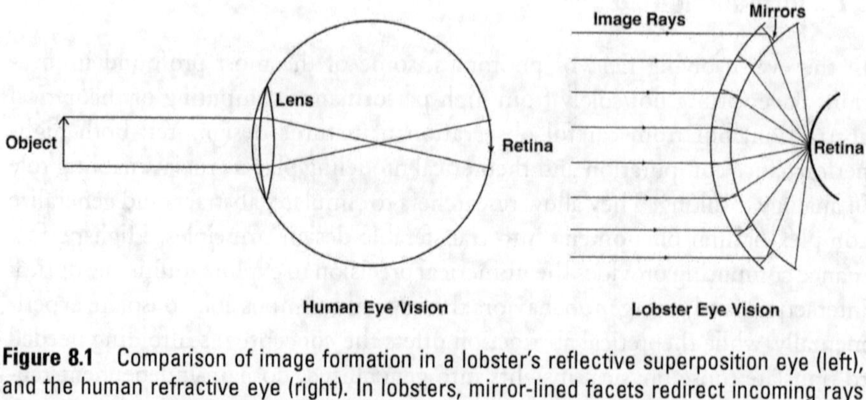

Figure 8.1 Comparison of image formation in a lobster's reflective superposition eye (left), and the human refractive eye (right). In lobsters, mirror-lined facets redirect incoming rays through a clear zone so that they converge on a single point on the retina, enhancing sensitivity at low-light levels. In humans, light is bent primarily at the cornea and then at the lens to focus an image on the retina. Arrows indicate ray directions and the retinal image location.

Analogy-thinking frame: Throughout this chapter we will explicitly use the terminology from Chapter 1 to make the cross-domain reasoning transparent: the lobster's eye serves as the source domain, wide-field X-ray imaging is the target domain, and their correspondence relies on deliberate mapping (square reflective channels → microchannels), abstraction (reflection-based superposition independent of biology), and transfer (engineering the concept into MCP optics). We will also surface constraint matching (e.g., grazing-incidence and coating limits) and show how iterative validation produces actionable insight generation for instruments and applications.

8.2 The Biological Blueprint: Structure of the Lobster Eye

Among the vast diversity of optical systems found in nature, the eye of the lobster stands out as one of the most geometrically elegant and functionally distinct. Unlike most terrestrial and aquatic animals that rely on lenses to refract light, the lobster has evolved an entirely different strategy—turning reflection into a precise imaging mechanism. Its eye presents an intricate surface of perfectly square facets, often described as graph paper wrapped across a dome. Beneath each facet lies a tube (a microscopic corridor with mirror-like interior walls) aligned along the eye's curvature. These tubes act as waveguides for incoming light via grazing-incidence reflections, steering rays toward a common retinal focal region (Figure 8.1).

Abstraction and constraint matching: The abstraction we carry into the target domain is wide-field imaging by orthogonal, grazing-incidence reflections in square pores. Biological tolerances (surface smoothness, aspect ratio) foreshadow the engineering constraint matching later required for MCPs.

For this optical strategy to work efficiently, the proportions of the tubes matter as much as their shape. The length-to-width ratio must support two controlled reflections; if the tubes are too short, light scatters, while overly long tubes suffer intensity loss due to absorption or angular mismatch. This balance—shape, proportion, and surface finish—creates a wide-angle, low-aberration imaging system ideal for dim underwater environments.

In bright conditions, lobsters migrate pigment granules to mask off-axis rays, acting as a dynamic aperture control that preserves contrast and prevents saturating glare.

Validation cue: This adaptive masking is a natural instance of validation that preserves performance as conditions change, echoed later in engineered calibration of MCP instruments.

As illustrated in Figure 8.1, lobsters form images by reflection rather than refraction. Parallel rays entering the eye strike square, mirror-like facets and are steered across a clear zone to a common focal point on the retina; the superposition of many such rays yields a bright image advantageous in dim environments. The human eye, by contrast, relies on refraction at the cornea and lens.

The comparison in Table 8.1 highlights the fundamental divergence between biological and artificial imaging objectives. Lobster vision has evolved to prioritize sensitivity and coverage over resolution. Operating in dim, bluish underwater environments, lobsters benefit from an optical system capable of detecting motion and silhouettes across a wide field of view, even under extremely low photon flux. This adaptation, achieved through reflective microchannel optics, allows for efficient light collection but sacrifices image sharpness and fine spatial detail.

In contrast, visible and infrared (IR) cameras are engineered for environments where illumination is either abundant or controllable. Their design emphasizes precision and image fidelity, necessitating high-resolution optics and minimal aberrations. Consequently, whereas blurring and diffuse contrast are acceptable compromises for biological survival, they are detrimental to photographic or analytical imaging systems, which demand sharp focus and fine-grained reconstruction of spatial features.

Overall, the table underscores a key trade-off between sensitivity and resolution: lobster eyes represent a biologically optimized, low-light sensing solution, while visible/IR cameras exemplify a technologically optimized, high-precision imaging system. This distinction explains why biomimetic designs inspired by lobster optics are successful in X-ray astronomy—where sensitivity and wide field of view are crucial—but unsuitable for visible or IR applications that require diffraction-limited performance.

Mapping note: In the language of Chapter 1, Figure 8.1 makes explicit the mapping between reflective superposition (source) and refractive imaging (contrast case), clarifying the structural (not superficial) correspondence.

Michael Land's seminal studies in the 1970s [1] revealed this reflective microchannel architecture and its panoramic sensitivity—insights that later inspired engineered lobster-eye optics for X-rays. If high-energy photons could be

Table 8.1
Comparison Between Lobster Vision and Visible/IR Camera Imaging Requirements

Goal	Lobster	IR Camera
Function	See silhouettes, detect motion in dim light	Capture sharp, detailed images
Resolution	Low	Very high
Light conditions	Dark, bluish ocean	Well-lit or controlled
Accepts blurring	Yes	No

guided via shallow-angle reflections inside square channels, one could achieve compact, wide-field nonrefractive imaging where lenses fail.

As shown in Figure 8.2, the lobster-eye configuration can also operate in reverse to generate a collimated beam from a point X-ray source: reflective microchannel surfaces guide X-rays through sequential reflections along two orthogonal planes, maintaining beam coherence without traditional lenses. This reflective principle, evolved for underwater vision, has been adapted for wide-field X-ray telescopes and imaging devices where refraction is impractical.

Source → target transfer: In Chapter 1 terms, this section narrates a full transfer: from biological source (lobster microchannels) to X-ray target, mediated by abstraction (grazing-incidence guidance), and realized through material and geometric mapping.

8.3 The Optical Physics: Grazing-Incidence Reflection

The challenge of focusing X-rays has long stood as one of the great obstacles in high-energy imaging. Unlike visible light, X-rays are poorly refracted by conventional materials. One of the few viable strategies is reflection at extremely shallow angles—grazing incidence—on atomically smooth, high-Z surfaces (e.g., Au, Ir, Pt). Under such conditions, X-rays can glance off a surface like a skipping stone, provided roughness and angle constraints are met.

This principle lies at the heart of lobster-eye optics. An orderly lattice of long, narrow square channels guides incoming photons via shallow-angle reflec-

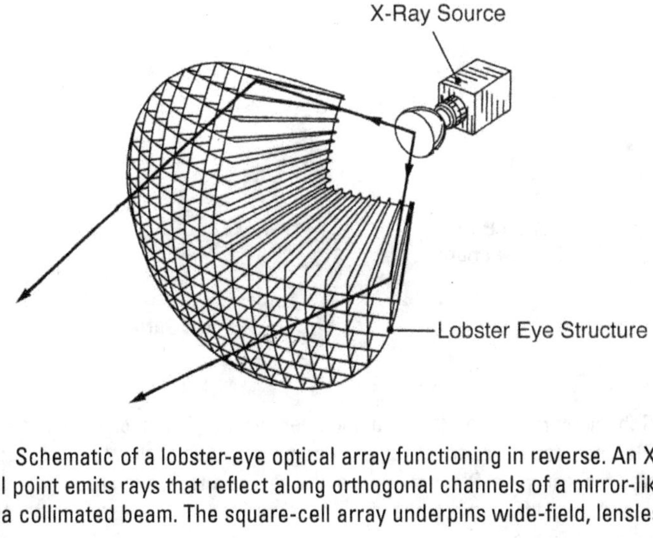

Figure 8.2 Schematic of a lobster-eye optical array functioning in reverse. An X-ray source at the focal point emits rays that reflect along orthogonal channels of a mirror-like structure, producing a collimated beam. The square-cell array underpins wide-field, lensless X-ray imaging.

tions toward a common focal plane. The internal channel walls are coated with thin high-Z layers to maintain reflectivity.

Constraint matching lens: Grazing-incidence angle, surface roughness, and coating selection dominate constraint matching and are prerequisites for credible validation.

Compared with traditional Wolter optics—composed of nested, concentric paraboloidal and hyperboloidal mirror shells that focus X-rays through double grazing-incidence reflections—lobster-eye arrays represent a fundamentally different imaging philosophy. Wolter telescopes achieve subarc-second angular resolution, but their precision-polished mirrors and long focal lengths constrain the field of view to only a few arc-minutes, demanding large, massive structures and intricate alignment. In contrast, lobster-eye systems employ dense arrays of square microchannels that redirect X-rays by internal reflection within each pore. This geometry enables modular, lightweight, and inherently wide-angle imaging at the expense of fine angular resolution [2, 6, 19–21]. Moreover, the square-pore architecture naturally lends itself to microfabrication, allowing the creation of tiled, scalable optical systems capable of monitoring vast regions of the sky or scanning extensive areas for security and industrial applications.

The unique optical design of the lobster eye allows for wide-angle vision using an array of square microchannels that function as reflective tubes. As illustrated in Figure 8.3, light rays entering these channels strike their inner mirror-like surfaces at very shallow (grazing) angles, similar to how a stone skips across water. Each reflection redirects the light toward a common focal surface, enabling efficient image formation across a broad field of view.

Different portions of the incoming light, such as those from Area A and Area B of the source image, are focused onto corresponding regions of the focal

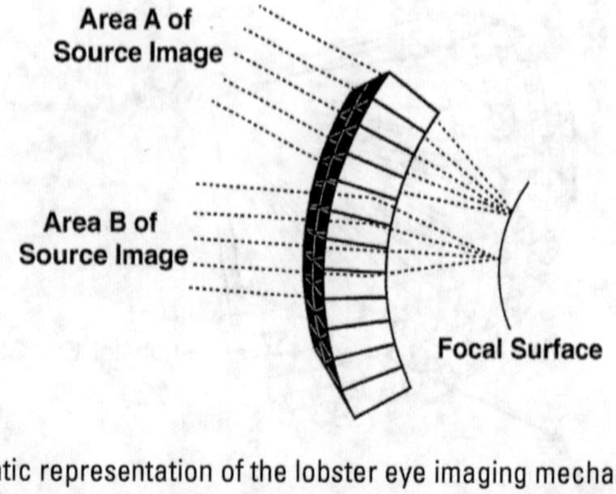

Figure 8.3 Schematic representation of the lobster eye imaging mechanism. The mirror-like surfaces inside each square channel reflect incoming light at shallow angles, redirecting it toward a curved focal surface. Distinct areas of the source image, such as Area A and Area B, are thus projected onto different regions of the focal surface, allowing the lobster to perceive a complete composite image.

surface. By integrating information from all the microchannels, the system constructs a continuous, wide-angle image of the observed object. This configuration mimics the natural compound eye of a lobster, and has inspired the development of wide-field X-ray optics in astronomical instruments.

8.4 Device Engineering: From Biology to MCPs

This section explores the engineering pathway from the biological archetype to its synthetic realization in MCPs, the fundamental building blocks of lobster-eye X-ray imaging systems. Through meticulous control of pore geometry, surface coating, and assembly curvature, these engineered analogues achieve wide-field imaging with exceptional sensitivity. We further distinguish between one-dimensional and two-dimensional MCP architectures, whose differing geometrical mappings yield distinct optical behaviors and data reconstruction challenges. Together, these elements form a coherent narrative of device translation—from natural model to functional technology—where design, fabrication, and computation jointly reproduce the power of biological vision in the high-energy domain.

8.4.1 MCP Design

At the core of the synthetic lobster-eye X-ray imaging system lies a deceptively simple yet powerful structure: the MCP. These thin slabs of glass or ceramic are densely perforated with microscopic square or rectangular channels [22]. Each channel acts as an individual waveguide for incoming X-rays; together, millions of pores form a unified optical element.

Transfer in practice: Fabrication and coating of MCPs constitute the hands-on transfer of the abstracted principle into engineered form, preserving mapped geometry while meeting manufacturability constraints.

Channel diameters typically range from 20 to 100 μm, with plate thickness many times greater, yielding aspect ratios that can exceed 40:1, which is essential for multiple grazing-incidence reflections.

Fabrication approaches include glass fiber drawing (bundle, draw, fuse, slice, and polish) and lithographic etching (e.g., DRIE in silicon or glass). Each offers trade-offs in uniformity, cost, and compatibility. Regardless of method, internal walls are coated (ALD, sputtering) with high-Z materials (Ir, Au, Pt); even minor nonuniformity degrades efficiency and angular performance.

Assembled MCPs are mounted on curved (cylindrical or spherical) carriers to emulate the natural configuration and deliver broad field of view, stitching a wide-angle image from many directional channels. The result is a sensitive, lightweight, compact system for faint X-ray signals over expansive fields.

Validation thread: Uniformity checks, roughness metrology, and coating thickness control operationalize validation under real fabrication tolerances.

8.4.2 1D Versus 2D Designs

One-dimensional lobster slab optics align channels in a single direction, focusing to a line. They are simpler and useful for slit-like geometries (e.g., linear scanning) but require motion or complementary optics for full images.

Two-dimensional designs employ square pores enabling reflections in both planes and true point focusing on a flat focal plane. Their point spread function (PSF) comprises a bright central core plus cross-shaped arms: doubly reflected photons form the core; singly reflected ones produce the orthogonal arms.

Mapping and deconvolution: The PSF shape follows directly from the square-pore mapping. Deconvolution is a computational transfer that restores detail while respecting the mapped physics.

8.5 Performance and Advantages

The translation of the lobster-eye concept into engineered microchannel systems yields profound practical benefits for X-ray astronomy and related sensing domains. By combining large solid-angle coverage with compact, lightweight construction, these optics redefine the balance between resolution, sensitivity, and field of view. The result is a new observational paradigm, which is capable not only of surveying vast regions of the sky instantaneously, but also of detecting fleeting, energetic phenomena in real time. This section examines the principal performance strengths of such systems, emphasizing their ultrawide fields of view (FOV), responsiveness to transient events, and suitability for spaceborne, resource-constrained missions.

8.5.1 Ultrawide Field of View

Wolter telescopes deliver superb resolution but very narrow FOV [23– 26]. Lobster-eye optics invert this trade: square-pore arrays accept photons from wide angles, enabling surveys across thousands of square degrees (1 sr = $(180/\pi)^2$ deg^2 ≈ 3282.806 deg^2) in a single exposure.

The Einstein Probe (launched 2024) is a landmark lobster-eye mission [22], capable of observing nearly one-tenth of the sky at once and acting as an early warning system for transient X-ray sources.

Insight generation: Wide-field vigilance is the concrete insight validated at scale.

8.5.2 Event Monitoring and Transient Detection

Transient phenomena—gamma-ray bursts (GRBs), supernovae, tidal disruption events—demand timing and coverage [27–29]. Traditional observatories excel at deep-targeted study but miss off-axis events. Lobster-eye systems provide panoramic awareness and quick localization for multiwavelength follow-up, synergizing with gravitational-wave alerts by shrinking the error region rapidly. This vigilance transforms time-domain high-energy astrophysics.

Validation loop: Rapid alerts and successful follow-ups serve as system-level validation.

8.5.3 Lightweight and Space Compatible

Replacing nested mirrors with flat, tiled MCPs slashes mass and alignment complexity, enabling CubeSats [30–32] and microsatellites [33, 34] to carry capable X-ray imagers. Constellations of small, low-cost platforms can provide persistent, global surveillance of transient activity.

Insight to architecture: Once the analogy is transferred and validated, it reshapes mission architectures.

8.6 Historical Development and Key Milestones

Angel's 1979 proposal [2], inspired by Land's biology [1], envisioned a lattice of square channels guiding X-rays via shallow reflections. Prototypes in the 1980s leveraged MCPs. The University of Leicester (Fraser and colleagues) advanced coatings, metrology, PSF characterization, and mission concepts [35–37]. NASA's 2012 suborbital test [38], China's 2022 module [38], and the full Einstein Probe (2024) [39, 40] mark maturation from biology to orbit.

Process visibility: Milestones trace repeated cycles of mapping, transfer, and validation, with constraint matching gating progress.

8.7 Applications Beyond Astronomy

While conceived for cosmic observation, the lobster-eye X-ray imaging paradigm extends far beyond astronomy. Its combination of wide-angle coverage, compact form factor, and high sensitivity lends itself naturally to diverse terrestrial and applied domains. From national security screening to biomedical diagnostics and industrial metrology, the same optical principles that once served to survey the heavens now enable safer, faster, and more efficient imaging here on Earth. This section surveys how the biologically inspired architecture, optimized for faint, diffuse sources, translates into transformative tools across security, medicine, and materials science, uniting disparate fields under a common geometrical framework.

8.7.1 Homeland Security

Wide FOV and compact architecture enable noninvasive, panoramic scanning for threat detection (weapons, shielded materials), improving throughput and reliability in airports, borders, and seaports [41–43].

Target reframing: The abstraction transfers from astrophysics to security without altering the core mapping.

8.7.2 Medical and Biomedical Imaging

Emerging work explores lobster-eye endoscopes for soft X-ray/EUV imaging at low dose, enabling sensitive, wide-angle views in compact, flexible form factors suited to early detection and intraoperative guidance [44–46].

Constraint awareness: Dose limits and biocompatibility are domain-specific constraints.

As shown in Figure 8.4, the lobster-eye optical configuration also inspires curved image formation on spherical/curved detectors, efficiently collecting over large solid angles.

8.7.3 Industrial and Materials Science

Wide-area, single-exposure X-ray fluorescence mapping and elemental analysis accelerate quality control of thin films, coatings, and wafers, integrating into inline, nondestructive inspection workflows.

Insight reuse: Industrial inspection leverages the same insight: wide-area, low-mass imaging.

8.8 Technical Challenges and Trade-Offs

Lobster-eye optics trade fine angular resolution for FOV and simplicity [48, 49]. Typical resolutions are several arcminutes (3 to 10 arcmin), versus arcseconds for Wolter systems. Manufacturing demands include pore uniformity, reflective coating quality, and precise curvature. Efficiency peaks for soft X-rays (0.1 to 10 keV) and drops at higher energies as the critical angle shrinks; multilayers help but add complexity. The cross-shaped PSF complicates deconvolution and source localization, increasing computational load, especially for crowded fields and time-sensitive alerts.

Constraint matching in focus: This section catalogs constraint matching pressures and engineering responses.

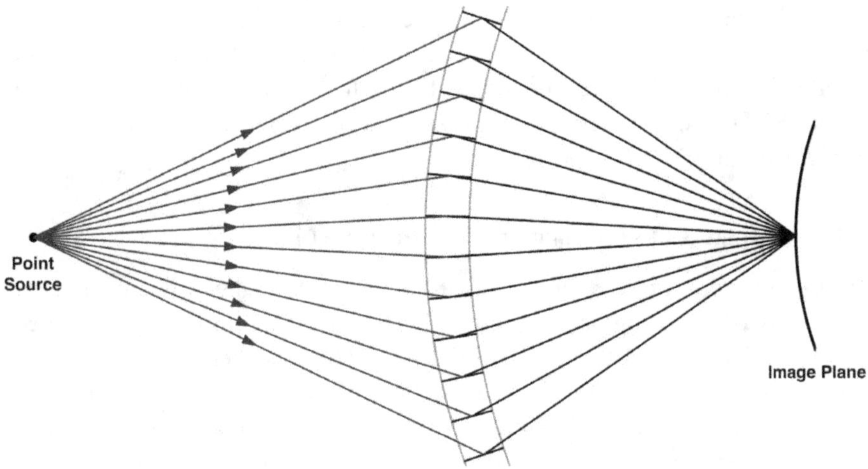

Figure 8.4 Schematic of lobster-eye optics adapted for curved image formation. Incident rays reflect within microchannel walls near the critical angle θ_c and converge on a curved image surface, enabling wide-field imaging on nonplanar detectors [47].

8.9 Future Directions

Distributed sensors on small satellites can triangulate transients in real time; mobile radiation-mapping drones benefit from panoramic awareness without moving parts; miniaturized medical probes promise real-time internal X-ray views.

AI closes the loop on PSF complexity: CNNs improve detection and classification; U-Nets/autoencoders enable deconvolution in photon-limited regimes; onboard accelerators support low-latency triggering; reinforcement learning adapts calibration and acquisition; multimodal fusion enhances localization [18, 50, 51]. Advances in ALD coatings, higher aspect-ratio MCPs, and multispectral detectors further expand capability.

The cycle continues: Each avenue restarts the Chapter 1 cycle: identify a fresh target domain, remap the abstraction, respect new constraints, and drive insight via validation.

8.10 Feasibility of Lobster-Eye Optics in the Infrared Regime

The lobster-eye optical principle is rooted in the use of grazing-incidence reflection within square or rectangular microchannels. This geometry was originally devised to address a problem unique to high-energy photons: X-rays cannot be refracted or focused by ordinary lenses because their refractive indices are close to unity and absorption is strong. By steering X-rays along shallow reflective paths, lobster-eye optics achieve efficient wide-field imaging where conventional optical designs fail. A natural question arises: C ould this concept, so effective in the X-ray domain, also be applied to longer wavelengths such as the IR? The answer is partly yes in concept, but largely no in direct physical implementation.

8.10.1 Why the X-Ray Principle Does Not Translate Directly

Infrared photons interact with materials in a fundamentally different way than X-rays. At IR wavelengths (typically 0.75 μm to several hundred μm), reflection and refraction both operate efficiently. There is therefore no need for grazing-incidence reflection. In fact, the lobster-eye geometry would be inefficient at such wavelengths because the channel dimensions are too large relative to the wavelength, leading to diffraction and scattering instead of controlled reflection.

The differences can be summarized through several key points:

1. *Interaction regime.* X-rays require extremely shallow reflection angles to avoid absorption, whereas infrared light reflects well at ordinary incidence angles.
2. *Scale mismatch.* In X-ray systems, the microchannel width (~20–100 μm) is many orders of magnitude larger than the wavelength (~0.1–10 nm), ensuring purely geometric reflection. In the IR, with λ comparable to the channel size, the geometry enters a waveguide regime dominated by diffraction.
3. *Material response.* High-Z metals (Au, Ir, Pt) used in X-ray optics for their electron density have no special advantage in the IR, where dielectric and metallic mirrors already provide high reflectivity without grazing incidence.
4. *Optical necessity.* IR optics can rely on conventional mirrors or lenses made of silicon, germanium, or chalcogenide glass; the lobster-eye approach adds complexity without benefit.

8.10.2 Conceptual Extensions and Biomimetic Inspiration

Although the literal microchannel geometry is unsuitable for infrared imaging, the conceptual framework of the lobster-eye—that of a distributed, compound reflective array providing a panoramic field of view—remains valuable. Engineers can reinterpret the principle through modern optical technologies such as micromirror arrays, metasurfaces, and diffractive optics. These systems mimic

Table 8.2
Comparison of Optical Mechanisms in Lobster Eyes, X-ray Telescopes, and Visible/IR Cameras

Feature	Lobster Eyes	X-Ray Telescopes	Visible/IR Cameras
Optical mechanism	Grazing reflection	Grazing reflection	Reflection/refraction
Works best for	Dim visible light	X-rays	Visible/IR wavelengths
Key advantage	Wide FOV, sensitivity	Focusing of X-rays	High-resolution imaging
Not suitable for	Precise visible/IR imaging	—	—

the mapping and wide-field acceptance of the biological eye without relying on grazing-incidence reflection.

For example:

- Arrays of micromirrors can replicate the geometric tiling of a lobster eye to produce wide-angle panoramic cameras operating in the visible or infrared.
- Infrared metasurfaces can perform the same angular mapping by using subwavelength resonators tuned for specific wavelengths, effectively realizing a flat, compact compound-eye lens.
- These designs carry the same biomimetic logic, many small directional channels acting in parallel, but their operating principle is refractive or diffractive, not grazing reflective.

While it is physically possible to design a lobster-eye structure for infrared wavelengths, it would not operate under the same principles that make it successful for X-rays. At long wavelengths, the optical interactions shift from grazing-incidence reflection to ordinary reflection and refraction. Nevertheless, the biomimetic concept of the lobster eye—a distributed, wide-field sensing architecture—continues to inspire innovative IR systems, especially in panoramic imaging, metasurface design, and compact compound optics.

8.11 Conclusion

From coastal shallows to cosmic distances, the lobster's reflective eye—arrays of square microchannels operating at grazing incidence—has become a blueprint for wide-field X-ray imaging. By translating this structure into MCP-based optics, engineers have opened new possibilities across astronomy, medicine, security, and industry. The story is as much about perspective as technology: observation → abstraction → transfer, with constraint-aware validation yielding sustained insight.

Analogy-thinking wrap: Using the Chapter 1 vocabulary, this chapter has traced a full cycle: source domain → mapping → abstraction → transfer with careful constraint matching and iterative validation, culminating in sustained insight generation across domains.

References

[1] Land, M. F., et al., "Superposition Images are Formed by Reflection in the Eyes of Some Decapod Crustaceans," *Nature,* Vol. 263, No. 5580, 1976, pp. 764–765, https://doi.org/10.1038/263764a0.

[2] Angel, J. R. P. "Lobster Eyes as X-Ray Telescopes," in *Proceedings Volume 0184, Space Optics Imaging X-Ray Optics Workshop,* Society of Photo-Optical Instrumentation Engineers (SPIE), 1979, 1979 Huntsville Technical Symposium, https://doi.org/10.1117/12.957437.

[3] Peele, A. G., et al., "X-Ray Focusing with Lobster-Eye Optics: A Comparison of Theory with Experiment," *Applied Optics,* Vol. 35, No. 22, 1996, pp. 4420–4425, https://doi.org/10.1364/AO.35. 004420.

[4] Peele, A. G., et al., "Lobster-Eye X-Ray Optics: A Rapid Evaluation of the Image Distribution," *Applied Optics,* Vol. 37, No. 4, 1998, pp. 632–636, https://doi.org/10.1364/AO.37.000632.

[5] Irving, D. H., et al., "Optical Metrology for Analysis of Lobster-Eye X-Ray Optics," *Applied Optics,* Vol. 42, No. 14, 2003, pp. 2422–2426, https://doi.org/10.1364/AO.42.002422.

[6] Ouyang, M., et al., "Comprehensive Understanding of the Focal Property of Lobster-Eye Optics," *Applied Optics,* Vol. 59, No. 14, 2020, pp. 4263–4270, https://doi.org/10.1364/AO.390101.

[7] Peele, A, G., et al., "LIGA for Lobster: First Observation of Lobster-Eye Focusing from Microchannel Plates," *Review of Scientific Instruments,* Vol. 72, No. 3, 2001, pp. 1843–1848, https://doi.org/10.1063/1.1347377.

[8] Zhang, C., et al., "First Wide Field-of-View X-Ray Observations by a Lobster-Eye Focusing Telescope in Orbit," *The Astrophysical Journal Letters,* Vol. 941, No. 1, 2022, p. L2, https://doi.org/10.3847/2041-8213/aca32f.

[9] Schmidt, W. K. H., "A Proposed X-Ray Focusing Device with Wide Field of View for Use in X-Ray Astronomy," *Nuclear Instruments and Methods,* Vol. 127, No. 2, 1975, pp. 285–292, https://doi.org/10.1016/0029-554X(75)90501-7.

[10] Vukusic, P., et al., "Photonic Structures in Biology," *Nature,* Vol. 424, No. 6950, 2003, pp. 852–855, https://doi.org/10.1038/nature01941.

[11] Shastri, S. D., et al., "Focusing with Saw-Tooth Refractive Lenses at a High-Energy X-ray Beamline," *Journal of Synchrotron Radiation,* Vol. 27, 2020, pp. 590–598, https://doi.org/10.1107/ S1600577520003665.

[12] Lider, V. V., et al., "Kirkpatrick–Baez and Wolter X-Ray Focusing Optics (Review)," *Journal of Surface Investigation: X-ray, Synchrotron and Neutron Techniques,* Vol. 13, 2019, pp. 670–682, https://doi.org/10.1134/S102745101904027X.

[13] Roth, T., et al., "Materials for X-ray Refractive Lenses Minimizing Wavefront Distortions," *MRS Bulletin*, Vol. 42, No. 6, 2017, pp. 430–436, https://doi.org/10.1557/mrs.2017.117.

[14] Snigirev, A., et al., "Focusing High-Energy X-Rays by Compound Refractive Lenses," *Applied Optics*, Vol. 37, No. 4, 1998, pp. 653–662, https://doi.org/10.1364/AO.37.000653.

[15] Li, L., et al., "Stiffener and Staggered-Square Honeycomb Structure Design for Lobster Eye X Ray Micro Pore Optics," *Optics Letters*, Vol. 49, No. 15, 2024, pp. 4413–4416, https://doi.org/10.1364/OL.533211.

[16] Ma, S., et al., "Comprehensive Simulation and Analysis of Lobster Eye Micro-Pore Optics Imaging in the Hard X-Ray Band," *Optics Communications*, Vol. 560, 2024, p. 130457, https://doi.org/10.1016/j.optcom.2024.130457.

[17] Ling, Z., et al., "Study on the Optical Properties of Iridium-Coated Lobster Eye X-Ray Micro Pore Optics," *Nuclear Instruments & Methods in Physics Research A*, Vol. 1014, 2023, p. 166598, https://doi.org/10.1016/j.optmat.2023.114120.

[18] Hudec, R., et al., "Lobster Eye X-Ray Optics," *Handbook of X-Ray and Gamma-Ray Astrophysics*, 2024, pp. 137–175, https://doi.org/10.1007/978-981-19-6960-7 3.

[19] Kaaret, P. E., et al., "X-Ray Focusing Using Microchannel Plates," *Applied Optics*, Vol. 31, No. 34, 1992, pp. 7339–7343, https://doi.org/10.1364/AO.31.007339.

[20] Peele, A. G., "Investigation of Etched Silicon Wafers for Lobster-Eye Optics," *Review of Scientific Instruments*, Vol. 70, No. 2, 1999, pp. 1268–1273, https://doi.org/10.1063/1.1149584.

[21] Zhang, C., et al., "First Wide Field-of-View X-Ray Observations by a Lobster Eye Focusing Telescope in Orbit," *The Astrophysical Journal Letters*, Vol. 941, No. 2, 2022, p. L14, https://doi.org/10.3847/2041-8213/aca32f.

[22] Yuan, W., et al., "Einstein Probe–a Lobster-Eye Satellite for Time-Domain Astronomy," *Proceedings SPIE*, Vol. 10699, 2018, pp. 543–551, https://doi.org/10.1117/12.2313358.

[23] Saha, T. T., et al., "Transverse Ray Aberrations for Paraboloid–Hyperboloid Telescopes," *Applied Optics*, Vol. 24, No. 12, 1985, pp. 1856–1863, https://doi.org/10.1364/AO.24.001856.

[24] Aschenbach, B., "X-Ray Telescopes," *Reports on Progress in Physics*, Vol. 48, No. 5, 1985, pp. 579–629, https://doi.org/10.1088/0034-4885/48/5/001.

[25] Harvey, J. E., et al., "Grazing-Incidence Hyperboloid–Hyperboloid Designs for Wide-Field X-Ray Imaging Applications," *Applied Optics*, Vol. 40, No. 1, 2001, pp. 136–144, https://doi.org/10.1364/AO.40.000136.

[26] van Speybroeck, L. P., et al., "Design Parameters of Paraboloid–Hyperboloid Telescopes for X-Ray Astronomy," *Applied Optics*, Vol. 11, No. 2, 1972, pp. 440–445, https://doi.org/10.1364/AO.11.000440.

[27] Willingale, R., et al., "Gamma-Ray Bursts and Fast Transients," *Space Science Reviews*, Vol. 207, 2017, pp. 63–86, https://doi.org/10.1007/s11214-017-0366-4.

[28] Bhalerao, V., et al., "Transients and Time Domain Astrophysics," *Journal of Astrophysics and Astronomy*, Vol. 46, 2025, p. 40, https://doi.org/10.1007/s12036-025-10073-9.

[29] Sun, H., et al., "Extragalactic High-Energy Transients: Event Rate Densities and Luminosity Functions," *The Astrophysical Journal*, Vol. 812, No. 1, 2015, p. 33, https://doi.org/10.1088/ 0004-637X/812/1/33.

[30] Woellert, K., et al., "Cubesats: Cost-Effective Science and Technology Platforms for Emerging and Developing Nations," *Advances in Space Research*, Vol. 47, No. 4, 2011, pp. 663–684, https://doi.org/10.1016/j.asr.2010.10.009.

[31] Villela, T., et al., "Towards the Thousandth CubeSat: A Statistical Overview," *International Journal of Aerospace Engineering*, Vol. 1, 2019, pp. 1–13, https://doi.org/10.1155/2019/5063145.

[32] Lemmer, K., "Propulsion for CubeSats," *Acta Astronautica*, Vol. 134, 2017, pp. 231–243, https://doi.org/10.1016/j. actaastro.2017.01.048.

[33] Xue, Y., et al., "Small Satellite Remote Sensing and Applications–History, Current and Future," *International Journal of Remote Sensing*, Vol. 29, No. 15, 2008, pp. 4339–4372, https://doi.org/10.1080/01431160801914945.

[34] Serjeant, S., et al., "The Future of Astronomy with Small Satellites," *Nature Astronomy*, Vol. 4, 2020, pp. 1031–1038, https://doi.org/10.1038/s41550-020-1201-5.

[35] Fraser, G. W., et al., "The Operation of Microchannel Plates with Square Pores," *Nuclear Instruments and Methods in Physics*, Vol. 310, No. 1–2, 1991, pp. 292–298, https://doi.org/10.1016/0168-9002(91)91046-X.

[36] Brunton, A. N., et al., "Metrology and Modeling of Microchannel Plate X-Ray Optics," *Applied Optics*, Vol. 36, No. 22, 1997, pp. 5461–5470, https://doi.org/10.1364/AO.36.005461.

[37] Willingale, R., et al., "Hard X-Ray Imaging with Microchannel Plate Optics," *Experimental Astronomy*, Vol. 8, 1998, pp. 281–296, https://doi.org/10.1023/A:1008099315152.

[38] Land, M., F. Burton, and V. Meyer-Rochow, "The Optical Geometry of Euphausiid Eyes," *Journal of Comparative Physiology*, Vol. 130, 1979, pp. 49–62, https://doi.org/10.1007/BF02582973.

[39] Yuan, W., et al., "Einstein Probe: A Lobster-Eye Telescope for Monitoring the X-Ray Sky," in *Proc. SPIE* 10699, *Space Telescopes and Instrumentation 2018: Ultraviolet to Gamma Ray*, 2018, p. 1069925, https://doi.org/10.1117/12.2313358.

[40] Yuan, W., et al., "Science Objectives of the Einstein Probe Mission," *Sci. China Phys., Mech. Astron.*, Vol. 68, No. 1, 2025, https://doi.org/10.1007/s11433-024-2600-3.

[41] Wajid, A. M., et al., "X-Ray Backscattering: Design, Applications, and Prospects in Nondestructive Evaluation," *Journal of Radioanalytical and Nuclear Chemistry*, Vol. 334, 2025, pp. 3033–3054, https://doi.org/10.1007/s10967-025-10103-4.

[42] Su, L., et al., "Modeling the Focusing Efficiency of Lobster-Eye Optics for Image Shifting Depending on the Soft X-Ray Wavelength," *Applied Optics*, Vol. 56, No. 22, 2017, pp. 6267–6271, https://doi.org/10.1364/AO.56.006267.

[43] Xu, J., et al., "A Novel Lobster-Eye Imaging System Based on Schmidt-Type Objective for X-Ray-Backscattering Inspection," *Review of Scientific Instruments*, Vol. 87, No. 7, 2016 https://doi.org/10.1063/1.4955472.

[44] Ebrahimpour, Z., et al., "Characterization in the Extreme Ultraviolet Domain of Microchannel Plate Based Device Using Synchrotron Radiation," *Journal of Applied Physics*, Vol. 134, No. 6, 2023, p. 063102, https://doi.org/10.1063/5.0161701.

[45] Mazuritskiy, M. I., et al., "Wave Propagation and Focusing of Soft X-Rays by Spherically Bent Microchannel Plates," *Journal of Synchrotron Radiation*, Vol. 28, No. 2, 2021, pp. 383–391, https://doi.org/10.1107/S1600577520016458.

[46] Xu, J., et al., "A Novel Lobster-Eye Imaging System Based on Schmidt-Type Objective for X-Ray Backscattering Inspection," *Review of Scientific Instruments*, Vol. 87, No. 7, 2016, p. 073103, https://doi.org/10.1063/1.4955215.

[47] Peng, Songwu, et al., "Preliminary Geometric Parameters Optimization of Lobster-Eye-Type Wide Field of View Soft X-Ray Imager," *Optical Engineering*, Vol. 58, No. 9, 2019, p. 093101, https://doi.org/10.1117/1.OE.58.9.093101.

[48] Tichy, V., et al., "Effective Collecting Area of Lobster Eye Optics and Optimal Value of Effective Angle," *Experimental Astronomy*, Vol. 47, No. 2, 2019, pp. 161–175, https://doi.org/10.1007/ s10686-019-09622-2.

[49] Zhang, C., et al., "First Wide Field-of-View X-ray Observations by a Lobster-Eye Focusing Telescope in Orbit," *The Astrophysical Journal Letters*, Vol. 939, No. 1, 2022, p. L3, https://doi.org/10.3847/2041-8213/aca32f.

[50] Jia, P., et al., "Target Detection Framework for Lobster Eye X-Ray Telescopes with Machine-Learning Algorithms," *The Astrophysical Journal Supplement Series*, Vol. 264, No. 2, 2023, p. 43, https://doi.org/10.3847/1538-4365/acab02.

[51] Li, L., et al., "Comprehensive Characterization and Simulation of the Point Spread Function of Lobster Eye X-Ray Micro Pore Optics," *Optics Communications*, Vol. 569, October 2024, p. 130821, https://doi.org/10.1016/j.optcom.2024.130821.

9

Photonic Neural Networks: From Neural Analogies to Lightwave Intelligence

After reading this chapter you should be able to:

- Explain how photonic neural networks (PNNs) mimic biological neural systems and why photonics offers advantages in speed, energy efficiency, and parallelism over traditional electronics.
- Describe key architectural components of PNNs—including Mach-Zehnder interferometers (MZIs), optical ring resonators, and nonlinear optical elements—and how they enable neural network operations such as weighted summation and activation.
- Evaluate the current challenges and potential applications of PNNs, including the limitations in training and nonlinearity, and identify areas where PNNs can transform edge computing, quantum systems, and real-time signal processing.

9.1 Introduction

The dream of achieving intelligent computation using light has transitioned from theoretical abstraction to practical exploration, fueled by rapid advancements in integrated photonics and artificial intelligence (AI) [1–3]. At the center of this transformation is the photonic neural network (PNN), a class of computing architectures that aims to reproduce the functional capabilities of artificial neural networks (ANNs) using the physics of light rather than the flow of electrons [1, 3–5]. This innovation is not an arbitrary leap, it is deeply grounded in a biological analogy. The human brain, with its interconnected neurons and adaptive synapses, serves as both inspiration and blueprint.

As illustrated in Figure 9.1, the PNN architecture directly parallels the structure and operation of a biological neuron. In biological systems, dendrites function as the primary receptors of electrochemical signals, channeling them toward the cell body (soma), where they are integrated and transformed into an electrical impulse [6–8]. This impulse then propagates along the axon and is transmitted to other neurons via synaptic terminals. Analogously, in a PNN, each optical input ($x_1, x_2,..., x_n$) represents an incoming signal that is weighted ($w_1, w_2,..., w_n$) and summed to form a composite optical field. This field, modulated by a bias term (b), is passed through an optical activation function $f(\cdot)$ to produce an output (y), analogous to the action potential in neurons. The right-hand model in Figure 9.1 thus serves as a mathematical and physical abstraction of biological computation, where light waves replace electrical impulses, enabling ultrafast, low power, and inherently parallel information processing in photonic systems.

As shown in Figure 9.2, the underlying concept of ANNs is inspired by the biological neuron's information processing mechanism. The figure highlights four widely adopted activation functions—sigmoid, hyperbolic tangent (tanh), exponential linear unit (ELU), and rectified linear unit (ReLU)—each

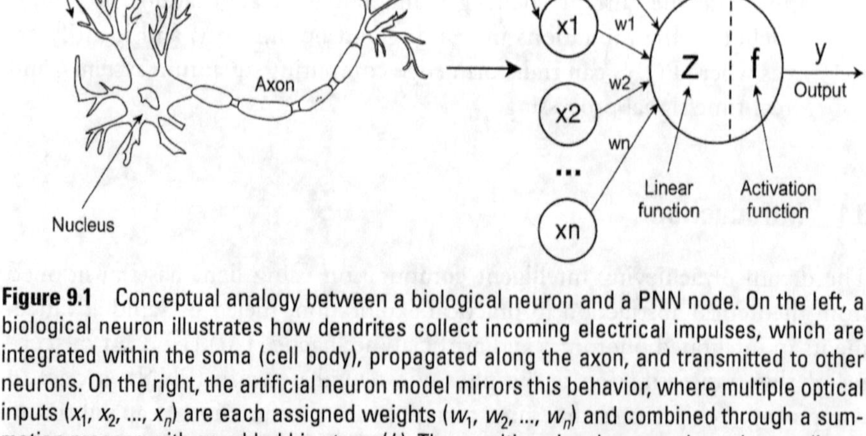

Figure 9.1 Conceptual analogy between a biological neuron and a PNN node. On the left, a biological neuron illustrates how dendrites collect incoming electrical impulses, which are integrated within the soma (cell body), propagated along the axon, and transmitted to other neurons. On the right, the artificial neuron model mirrors this behavior, where multiple optical inputs ($x_1, x_2, ..., x_n$) are each assigned weights ($w_1, w_2, ..., w_n$) and combined through a summation process with an added bias term (b). The resulting signal passes through a nonlinear activation function $f(\cdot)$, yielding an optical output y. This comparison demonstrates how photonic hardware can emulate neuronal computation, leveraging light–matter interactions to achieve parallel, high-speed, and energy-efficient information processing.

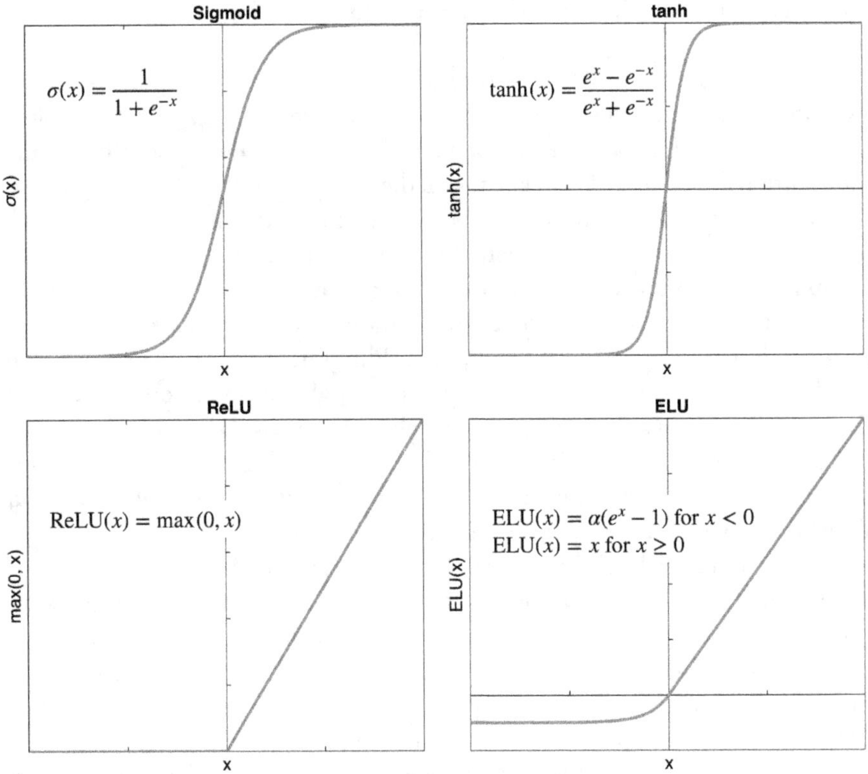

Figure 9.2 Illustration of the four commonly employed nonlinear activation functions (sigmoid, tanh, ELU, and ReLU), each introducing nonlinearity essential for enabling the artificial neural network to learn complex input–output mappings.

imparting distinct nonlinear properties [9]. While the sigmoid and tanh functions are smooth and bounded, making them suitable for probabilistic outputs, ReLU has become dominant in modern deep learning architectures due to its computational efficiency and ability to mitigate gradient vanishing problems [10–13]. Together, these components form the foundation for constructing powerful computational models capable of representing complex mappings and hierarchical feature abstractions, analogous to biological perception and cognition.

As described in Chapter 1, this reflects a structured analogy process: the biological brain serves as the source domain, and photonic hardware becomes the target domain through mapping, abstraction, and transfer of principles. While electronic hardware has successfully mimicked some aspects of this structure, it does so with significant energy and speed constraints [14–16]. Photonics offers an alternative medium, one that operates with unmatched velocity and energy efficiency, making it an ideal platform for emulating neural architectures at scale.

The emergence of ANNs has reshaped entire domains, from image recognition to natural language processing. But as network depth and complexity have increased, so too has the demand for faster, more efficient hardware. Conventional electronic processors, whether CPUs, GPUs, or even domain-specific AI accelerators, are reaching thermal and bandwidth bottlenecks [17, 18]. They rely on moving large volumes of data between separate memory and compute units, often consuming more energy in data transport than in actual computation. This architectural limitation, known as the von Neumann bottleneck, hampers scalability and speed. Photonic neural networks sidestep this limitation by enabling in-memory and in-flight computation [19–24]. In PNNs, light propagates through optical elements (modulators, phase shifters, and interferometers) while simultaneously undergoing the operations that constitute neural inference. This unification of data transport and transformation dramatically reduces energy overhead and latency.

The conceptual bridge between biological neurons and photonic devices is more than metaphorical. In biological systems, neurons process and transmit information using action potentials modulated by synaptic weights. In photonics, analogs emerge in the form of optical modulators, which alter the intensity or phase of light in response to input signals, and interferometric networks, such as those based on Mach-Zehnder interferometers, which combine multiple optical paths to perform weighted summation, mimicking the aggregation function of a neuron [25, 26]. Furthermore, photonic systems benefit from the capacity to multiplex across multiple degrees of freedom, including wavelength, phase, polarization, and spatial modes. This parallelism is a natural match to the brain's concurrent processing of diverse sensory inputs [27]. Optical waveguides, by guiding multiple wavelengths simultaneously, can process many data streams without interference, enabling scalable and reconfigurable connectivity among layers [28, 29].

One of the primary attractions of PNNs is their potential for ultralow energy consumption [1, 30]. Electronic switching in transistors requires charging and discharging capacitive nodes, an inherently dissipative process. In contrast, optical transmission through passive components, such as waveguides or beam splitters, can occur with minimal energy loss. Research has shown that energy-per-operation figures in the femtojoule and attojoule range are achievable for integrated photonic systems [31], offering orders of magnitude improvement over their electronic counterparts. These gains are particularly impactful in edge applications, where power budgets are constrained, and in high-throughput systems where total energy efficiency defines scalability.

Despite these advantages, challenges remain in realizing the full potential of PNNs [1]. Implementing nonlinear activation functions purely in optics is nontrivial, due to the inherently linear nature of most optical systems. Approaches based on saturable absorbers, Kerr effect, and phase-change materials are being investigated, but integration and efficiency trade-offs must be carefully managed. Additionally, training PNNs poses fundamental difficulties because gradient-based optimization methods require back-propagation—an operation difficult to perform directly in analog photonic hardware. To date, many implementations rely on offline training, with optical hardware serving primarily as a high-speed inference engine [32]. Nevertheless, the development of photonic-compatible training methods, including in situ update algorithms and local learning rules, is a vibrant area of research [33–38].

As the field matures, PNNs are poised to redefine the landscape of AI. Their capacity to operate at the speed of light, with unmatched energy efficiency, opens doors to applications ranging from real-time signal processing in embedded sensors to scalable AI acceleration in data centers. Perhaps most profoundly, PNNs illustrate how the convergence of physics, biology, and computation can yield new paradigms for processing information—paradigms that move beyond the limitations of traditional silicon electronics and toward a future where intelligence is not just electrical, but optical.

9.2 Biological Inspiration and Photonic Analogs

This section exemplifies the first stages of analogy thinking discussed in Chapter 1: identifying a well-understood source domain (biological neurons) and transferring its governing principles into a novel technological target domain (photonic circuits). The remarkable efficiency of biological neural networks lies in their distributed, massively parallel architecture [39, 40]. In the human brain, information is processed through networks of neurons that receive and integrate signals via synaptic connections. Each neuron functions as a decision-making unit, summing excitatory and inhibitory inputs until a threshold is reached, triggering an action potential that propagates to connected neurons. These processes occur in parallel across billions of interconnected neurons, with plasticity in synaptic strength allowing the network to adapt and learn over time. Synapses modulate the strength and timing of signal transmission through chemical neurotransmitters, encoding learning and memory in the dynamics of connectivity [41–43]. This architecture has inspired generations of AI researchers and engineers seeking to build brain-like computational systems in hardware.

As illustrated in Figure 9.3, the architecture of an ANN is conceptually inspired by the functional structure of a biological neuron. In a biological system, information is received through dendrites, processed within the soma, and transmitted along the axon toward the axon terminals, where synaptic

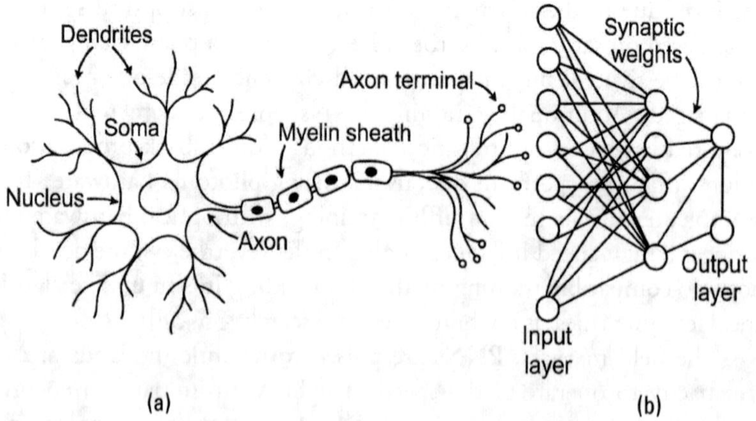

Figure 9.3 Schematics of (a) a biological neuron, and (b) a single-layer artificial neural network. The biological neuron consists of dendrites, soma, axon, and axon terminals, which transmit electrical signals via synapses. The artificial neural network mimics this structure computationally through interconnected nodes and weighted links, forming input, hidden, and output layers that collectively emulate neuronal signal processing.

connections enable communication with other neurons. Similarly, an ANN represents this information flow using layers of interconnected nodes. Each node corresponds to a simplified neuron model, and the synaptic connections are represented by adjustable weights that determine how input signals propagate and combine within the network. The hidden layer(s) in the ANN serve an analogous role to biological integration and processing, transforming input patterns into meaningful representations before generating the final output response. This structural analogy between biological and artificial systems underpins the motivation for developing neural computation models that emulate cognitive processes through algorithmic learning.

Translating these principles into the domain of photonics introduces both opportunities and challenges. The core idea is to map the fundamental components of neural computation—neurons, synapses, activation, and learning—onto physical optical devices. In PNNs, neurons can be represented by optical components capable of processing and modulating light. These include electro-optic modulators, optical resonators, and nonlinear optical devices that can implement thresholding or activation-like functions. Meanwhile, the role of synapses, which are responsible for scaling and integrating incoming signals, is assumed by tunable couplers, beam splitters, interferometers, and waveguide-based weighting elements, which determine how light from different inputs combines and propagates through the network.

A defining advantage of photonic systems is their intrinsic parallelism [44–46]. Unlike electrons, which suffer from mutual interference, heat dissipation, and resistance, photons can travel unimpeded through free space or within waveguides, even when multiplexed across different wavelengths. This allows photonic systems to perform many operations simultaneously, not just across spatially separated paths, but across frequency domains as well. WDM enables the encoding of multiple data streams on a single optical fiber or waveguide, allowing a single physical pathway to support multiple independent signals.

Furthermore, the physics of light itself lends naturally to core neural operations. Optical interference, which occurs when coherent light waves combine based on their relative phases, can perform the mathematical equivalent of weighted summation [47, 48]. This is the essence of synaptic integration in biological systems, replicated in photonics with beam splitters and interferometric networks. Each incoming optical signal is modulated by a weighting element, typically a phase shifter or tunable attenuator, and then combined with others in a controlled interference pattern. The output intensity corresponds to the sum of inputs, modulated by their respective weights, mimicking the aggregation that occurs in the soma of a biological neuron.

This ability to replicate neural behavior with passive or low-power optical elements is compelling for large-scale implementations. Systems based on MZI meshes and optical ring resonators are already demonstrating the feasibility of this mapping, achieving high-speed and high-throughput computations on integrated photonic platforms [31]. The use of nonlinear optical materials, such as Kerr media or saturable absorbers, adds the missing piece: the nonlinearity required to enable complex decision boundaries, just as biological neurons fire only when a threshold is crossed [49–51].

9.3 Architectures of Photonic Neural Networks

Following the analogy-mapping phase outlined in Chapter 1, the architectures discussed here represent concrete realizations of abstracted neural mechanisms translated into photonic components. The development of PNNs has progressed rapidly from early free-space demonstrations to integrated photonic platforms. Early systems relied on lenses, spatial light modulators (SLMs), and diffractive optical elements to emulate ANN behavior, but suffered from size, alignment sensitivity, and poor scalability.

The next generation shifted toward integrated photonic circuits, leveraging silicon photonics. Optical waveguides guide light across densely packed chips. A key architectural element is the MZI mesh. Each MZI consists of two beam splitters and two arms, one or both with phase shifters. By adjusting the phase difference, the interference at the output can be controlled to represent a weighting coefficient. In essence, each MZI acts as a tunable synapse, capable of complex multiplication on an optical signal. Networks of MZIs arranged in a mesh configuration can perform matrix–vector multiplications, the foundational operation of neural layers.

As illustrated in Figure 9.4, the MZI operates by dividing an incoming optical signal into two separate propagation paths using a Y-splitter. The light traveling through the upper and lower arms accumulates distinct optical phases, which can be modulated by external factors such as refractive index changes, temperature variations, or applied voltages. Upon recombination at the second Y-splitter, the two beams interfere constructively or destructively depending on their relative phase difference. This controllable interference enables the MZI to function as a fundamental optical component in photonic circuits, serving roles in modulation, switching, and photonic neural network architectures.

The power of the MZI mesh lies in its reconfigurability and speed. Once programmed, the mesh processes optical signals at line rate, enabling ultra-fast forward propagation. Coherent light allows constructive and destructive interference to encode results within the physics of light. MZI meshes can be optimized to represent arbitrary unitary matrices and, with architectural modifications, more general matrix forms, enabling fully connected neural networks on-chip.

Complementing MZI meshes, optical ring resonators confine light in closed loops and support resonance at specific wavelengths, making them sensitive to both amplitude and phase of the incoming signals. They can serve as memory elements and as nonlinear activation units, depending on how their transmission characteristics are modulated [31]. Their compact footprint and wavelength-selective behavior make them suitable for dense, multiplexed architectures.

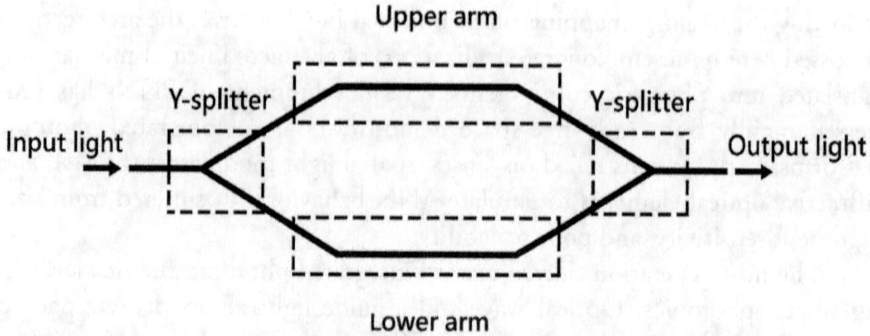

Figure 9.4 Schematic diagram of a Mach-Zehnder interferometer showing the splitting and recombination of light waves. The input light is divided by a Y-splitter into two optical paths, an upper arm and a lower arm, that later recombine through a second Y-splitter to produce the output light. The interference at the output depends on the relative optical phase difference accumulated along the two arms.

9.4 Optical Nonlinearities and Activation Functions

These nonlinear processes illustrate how analogy thinking moves from structural similarity to functional correspondence, a key form of mapping and abstraction described in Chapter 1. Nonlinearity is a cornerstone of neural network computation. In the photonic domain, however, the implementation of nonlinearity is significantly more difficult. Linear optics can perform efficient matrix-vector multiplications, but without nonlinearity, the model is limited.

Researchers have explored several methods. Saturable absorbers exhibit decreased absorption at higher intensities, effectively creating an intensity-dependent transmission function [52, 53]. Kerr media exploit an intensity-dependent refractive index to achieve phase shifts that can be harnessed in interferometric setups [31]. Two-photon absorption introduces a sharp intensity threshold [54]. These effects typically require high intensities or long interaction lengths, posing integration challenges for compact, low-power circuits.

As a practical compromise, many systems adopt hybrid optoelectronic approaches: linear computation occurs optically; photodetectors convert signals for electronic activation before re-encoding into light. While this introduces conversion overhead, it preserves photonic speed and parallelism for the most intensive operations.

Dynamic materials and programmable photonic elements combine optical and electronic control. Phase-change materials (PCMs), already used for weight storage, can exhibit nonlinear switching behavior, potentially serving as memory and activation elements.

9.5 Training Photonic Neural Networks

In analogy terms, this stage echoes the constraint and refinement phase from Chapter 1, where ideas transferred from the source domain are iteratively adapted to the realities of the target system. Training PNNs is challenging. Traditional ANNs rely on backpropagation with precise arithmetic and accessible memory. In photonic hardware designed for all-optical operation, direct gradient computation and backpropagation are complex. Many elements are static once fabricated, lacking simple reconfiguration unless augmented with materials or electronic controls. Consequently, most implementations use offline training in simulation and then program the photonic hardware with the resulting weights [32].

To move beyond this, researchers are developing tunable or trainable photonic components. One approach uses phase-change materials such as Ge2Sb2Te5 (GST), which can reversibly switch between amorphous and crystalline states with distinct refractive indices. Changes can be induced optically or thermally, allowing weights to be encoded directly into waveguides or resonators. Once programmed, GST holds its state without constant power, providing nonvolatile, analog weight storage. Systems incorporating PCMs have demonstrated dynamic weight updates, enabling local learning rules or hardware-level gradient estimation [31].

Another strategy employs thermo-optic phase shifters or electro-optic modulators embedded within MZI meshes. These can be driven electrically to adjust phase shifts and interference patterns, effectively modifying the weight matrix. In this hybrid model, the photonic core handles high-speed matrix-vector products, while electronics perform slower yet precise weight updates. Some in situ training algorithms approximate gradients from local measurements or feedback signals, allowing partial on-chip training, though challenges in scalability and stability remain.

The vision of fully optical learning—where both inference and training occur optically—would eliminate conversion overhead and enable autonomous, self-adaptive photonic systems. Progress will require advances in materials, device engineering, and algorithms aligned with light-based computation (e.g., perturbation-based training and forward-mode differentiation).

9.6 Technological Implications and Applications

PNNs promise drastic reductions in power consumption and increases in throughput. Photons propagate with minimal loss and no intrinsic electrical resistance, enabling computations at rates beyond digital electronics while consuming orders of magnitude less energy.

This makes PNNs appealing for edge computing, where local, fast, efficient processing of sensory data is essential for autonomous vehicles, drones, or wearable medical sensors. The ability to process high-bandwidth data (e.g., video, biosignals) in real time without cloud dependence is crucial. PNNs, with light-speed operation and low thermal footprint, offer a pathway to such systems.

Real-time signal processing (e.g., telecommunications decoding, noise filtering, modulation classification) demands enormous throughput and very low latency. In optical communication systems, the bottleneck is often conversion of optical signals to electronic form. PNNs can receive, process, and act on information entirely in the optical domain, preserving bandwidth and reducing complexity.

These optical architectures also hold promise in quantum information science, which requires precise manipulation of single or entangled photons.

Perhaps most significantly, PNNs sidestep the von Neumann bottleneck (Figure 9.5). Photonic systems, especially those with integrated memory

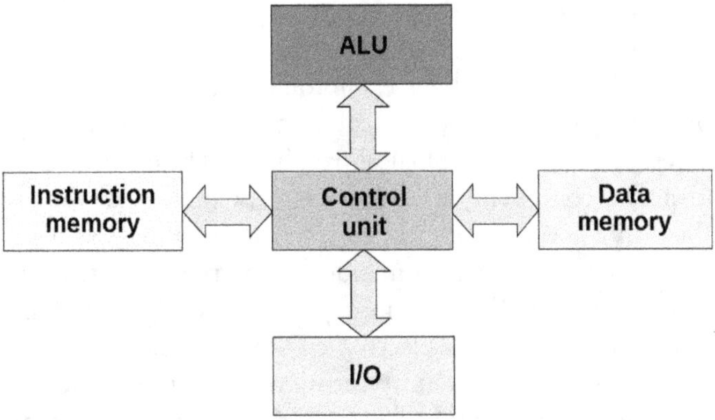

Figure 9.5 Schematic diagram of the classical von Neumann architecture, illustrating the separation between processing and memory units. The architecture consists of an ALU, a control unit, instruction memory, data memory, and input/output (I/O) modules interconnected through a shared communication bus. This structural division leads to the well-known von Neumann bottleneck, where data transfer between memory and processing units becomes a major constraint on computational speed and energy efficiency, particularly in data-intensive workloads.

elements like PCMs or nonlinear resonators, allow computation to occur in situ, as light moves through the network—in-memory, in-flight processing.

As depicted in Figure 9.5, the von Neumann architecture represents the foundational model of conventional electronic computing systems. In this design, the computational process is organized around a CPU that is composed of a control unit and an arithmetic logic unit (ALU). The control unit orchestrates the execution of instructions retrieved sequentially from instruction memory, while the ALU performs arithmetic and logical operations on data obtained from data memory. Both instruction and data streams share the same communication pathway, known as the system bus, which links the CPU to memory and input/output (I/O) modules. This shared bus design enforces a serial transfer of data and instructions, resulting in the characteristic von Neumann bottleneck. The bottleneck manifests as a throughput limitation, where the CPU must idle while waiting for data to be fetched or written back to memory, an inefficiency that scales poorly with increasing data size and computational complexity. Although this architecture underpins nearly all classical digital systems, it constrains parallel execution and contributes to significant energy overhead during data movement. These limitations have spurred the exploration of non–von Neumann paradigms, such as photonic, neuromorphic, and in-memory computing, which aim to colocate computation and storage to achieve greater efficiency and scalability.

Both academic and commercial groups are investing in photonic AI accelerators, including chip-scale coprocessors, optical cores in supercomputers, and intelligent photonic nodes in distributed sensor networks [55].

9.7 Conclusion

Viewed through the lens of Chapter 1, photonic neural networks stand as a culmination of analogy thinking—demonstrating how abstraction, mapping, and transfer can reshape entire technological paradigms. The convergence of optical physics and neural computation is a growing discipline with the potential to transform how intelligence is embedded into machines. PNNs offer a fundamentally new way to process information by marrying the structure of biological neural networks with the intrinsic properties of light. In doing so, they sidestep limitations of traditional electronics—thermal inefficiency, bandwidth bottlenecks, and latency—while introducing a computing paradigm built on speed, parallelism, and energy efficiency. As the chapter of this work has explored in detail, the story of PNNs is one of conceptual elegance, technological ingenuity, and cross-disciplinary collaboration.

From their roots in biological analogy, PNNs draw not just inspiration, but an operational model that directly informs design. The neuron, with its weighted summation and threshold activation, finds a counterpart in optical modulators and interferometric circuits. The synapse—adaptable connectivity in the brain—is mirrored in tunable optical couplers and phase shifters. These mappings translate biological function into scalable photonic circuitry, enabled by decades of advancement in silicon photonics, waveguide design, and nanofabrication. Unlike biological systems limited by biochemical signaling speeds, PNNs leverage near-instantaneous propagation of light, operating at rates suited to time-sensitive applications.

Architectural innovations—MZI meshes, optical ring resonators, and waveguide pathways—encode dense matrix multiplications and, when paired with nonlinear optical materials, emulate activation functions essential for learning and decision-making. The significance goes beyond performance metrics: computation can be reframed through the medium of light, inheriting advantages of optics: low latency, high bandwidth, and minimal energy dissipation.

While PNNs face challenges, such as training methodologies, nonlinear integration, scalability, the momentum suggests these are engineering frontiers. Hybrid optoelectronic training, in situ adjustment via phase-change materials, and algorithmic adaptation for optical hardware are emerging rapidly. These advances are driven by practical demands: the rise of edge AI, the proliferation of Internet of Things sensors, and the push toward real-time, low-power inference at the network edge.

Applications extend beyond classical machine learning. Natural compatibility with optical signal processing suits telecommunications, real-time imaging, and security. In quantum information, where photons carry entanglement and coherence, PNNs could provide intelligent feedback and error correction. In biomedical diagnostics needing real-time spectral and spatial analysis, PNN speed and sensitivity enable new capabilities. Even in astrophysics, where massive observational data streams demand rapid filtering and feature extraction, PNNs could accelerate discovery of transient phenomena.

Perhaps the most profound implication is a redefinition of computing substrates. For decades, computation has been tied to electronic materials and transistor scaling. PNNs introduce a post-Moore paradigm based on new physics, materials, and architectures. In this view, intelligence can be encoded in phase shifts, splitting ratios, and resonance frequencies—fundamentally different quantities that nonetheless support the same mathematics.

As photonic technologies mature and integrate with digital systems, heterogeneous platforms will exploit the best of both domains. The future of AI may lie not solely in silicon, but in waveguides, resonators, and modulators operating at terahertz speeds. In this vision, light becomes more than a messenger, it becomes a medium of cognition.

In conclusion, PNNs embody a new frontier in computing—one that bridges disciplines, challenges conventions, and opens the door to intelligent systems limited not by electrons, but by imagination. Their development underscores the power of analogies between nature and technology and suggests that solutions to pressing computational problems may be found in the fundamental behavior of light itself.

References

[1] Liao, K., et al., "Integrated Photonic Neural Networks: Opportunities and Challenges," *ACS Photonics*, Vol. 10, No. 7, 2023, pp. 2001–2010, https://doi.org/10.1021/acsphotonics.2c01516.

[2] Xu, X.- Y., et al., "Integrated Photonic Computing Beyond the von Neumann Architecture," *ACS Photonics*, Vol. 10, No. 4, 2023, pp. 1027–1036, https://doi.org/10.1021/acsphotonics.2c01543.

[3] Tsakyridis, A., et al., "Photonic Neural Networks and Optics-Informed Deep Learning Fundamentals," *APL Photonics*, Vol. 9, No. 1, 2024, p. 011102, https://doi.org/10.1063/5.0169810.

[4] Xiang, S. Y., et al., "A Review: Photonics Devices, Architectures, and Algorithms for Optical Neural Computing," *Journal of Semiconductors*, Vol. 42, No. 2, 2021, p. 023105, http://doi.org/10.1088/1674-4926/ 42/2/023105.

[5] Khonina, S. N., et al., "Exploring Types of Photonic Neural Networks for Imaging and Computing—A Review," *Nanomaterials*, Vol. 14, No. 8, 2024, p. 697, https://doi.org/10.3390/ nano14080697.

[6] Spruston, N., et al., "Pyramidal Neurons: Dendritic Structure and Synaptic Integration," *Nature Reviews Neuroscience*, Vol. 9, 2008, pp. 206–221, https://doi.org/10.1038/nrn2286.

[7] Hausser, M., et al., "Diversity and Dynamics of Dendritic Signaling," *Science*, Vol. 290, No. 5492, 2000, p. 739–744, https://doi.org/10.1126/science.290.5492.739.

[8] Magee, J. C., et al., "Dendritic Integration of Excitatory Synaptic Input," *Nature Reviews Neuroscience*, Vol. 1, No. 3, 2000, pp. 181–190, https://doi.org/10.1038/35044552.

[9] Lederer, J., "Activation Functions in Artificial Neural Networks: A Systematic Overview," *arXiv preprint arXiv:2101.09957*, 2021, https://arxiv.org/abs/2101.09957.

[10] Dubey, S. R., et al., "Activation Functions in Deep Learning: A Comprehensive Survey and Benchmark," *Neurocomputing*, Vol. 503, 2022, pp. 92–108, https://doi.org/10.1016/j.neucom.2022.06.111.

[11] Krizhevsky, A., et al., "ImageNet Classification with Deep Convolutional Neural Networks," *Communications of the ACM*, Vol. 60, No. 6, 2017, pp. 84–90, https://doi.org/10.1145/3065386.

[12] Schmidhuber, J., "Deep Learning in Neural Networks: An Overview," *Neural Networks*, Vol. 61, 2015, pp. 85–117, https://doi.org/10.1016/j.neunet.2014.09.003.

[13] LeCun, Y., et al., "Deep Learning," *Nature*, Vol. 521, No. 7553, 2015, pp. 436–444, https://doi.org/10.1038/nature14539.

[14] Lo, K. H., et al., "Broadband Optical Engine System Integration by Wafer Level Process in HPC/AI Era," in *2025 IEEE 75th Electronic Components and Technology Conference (ECTC)*, San Diego, CA, May 27–30, 2025, https://doi.org/10.1109/ECTC51687.2025.00284.

[15] Chen, M. F., et al., "Optical and Electrical Characterization of a Compact Universal Photonic Engine," in *2025 IEEE 75th Electronic Components and Technology Conference (ECTC)*, San Diego, CA, May 27–30, 2025, https://doi.org/10.1109/ECTC51687.2025.00015.

[16] Hsia, H., et al., "EPIC-BOE: An Electronic-Photonic Chiplet Integration Technology with IC Processes for Broadband Optical Engine Applications," in *2024 IEEE International Electron Devices Meeting (IEDM)*, San Francisco, CA, December 7–11, 2024, https://doi.org/10.1109/IEDM50854.2024.10873589.

[17] Gholami, A., et al., "AI and Memory Wall," *IEEE Micro*, Vol. 44, No. 3, 2024, pp. 33–39, https://doi.org/10.1109/MM.2024. 3373763.

[18] Sun, Z., et al., "A Full Spectrum of Computing-in-Memory Technologies," *Nature Electronics*, Vol. 6, No. 11, 2023, pp. 901–915, https://doi.org/10.1038/s41928-023-01053-4.

[19] Shen, Y., et al., "Deep Learning with Coherent Nanophotonic Circuits," *Nature Photonics*, Vol. 11, 2017, pp. 441–446, https://doi.org/10.1038/nphoton.2017.93.

[20] Feldmann, J., et al., "Parallel Convolutional Processing Using an Integrated Photonic Tensor Core," *Nature*, Vol. 589, 2021, pp. 52–58, https://doi.org/10.1038/s41586-020-03070-1.

[21] Carlos Rios, C., et al., "In-Memory Computing on a Photonic Platform," *Science Advances*, Vol. 5, No. 2, 2019, p. eaau5759, https://doi.org/10.1126/sciadv.aau5759.

[22] Wetzstein, G., et al., "Inference in Artificial Intelligence with Deep Optics and Photonics," *Nature*, Vol. 588, 2020, pp. 39–47, https://doi.org/10.1038/s41586-020-2973-6.

[23] Chen, Z., et al., "Deep Learning with Coherent VCSEL Neural Networks," *Nature Photonics*, Vol. 17, 2023, pp. 723–730, https://doi.org/10.1038/s41566-023-01233-w.

[24] Feldmann, J., et al., "All-Optical Spiking Neurosynaptic Networks with Self-Learning Capabilities," *Nature*, Vol. 569, 2019, pp. 208–214, https://doi.org/10.1038/s41586-019-1157-8.

[25] Zappacosta, S., et al., "General Differential Hebbian Learning: Capturing Temporal Relations Between Events in Neural Networks and the Brain," *PLOS Computational Biology*, Vol. 14, No. 8, 2018, p. e1006227, https://doi.org/10.1371/journal.pcbi.1006227.

[26] Cem, A., et al., "Data-Driven Modeling of Mach-Zehnder Interferometer-Based Optical Matrix Multipliers," *Journal of Lightwave Technology*, Vol. 41, No. 16, 2023, pp. 5425–5436, https://doi.org/10.1109/JLT.2023.3263235.

[27] Yu, X., et al., "Parallel Optical Computing Capable of 100-Wavelength Multiplexing," *eLight*, Vol. 5, 2025, p. 10, https://doi.org/10.1186/s43593-025-00088-8.

[28] Nelson, E., et al., "Wavelength Division Multiplexed Optical Interconnect Using Short Pulses," *IEEE Journal of Selected Topics in Quantum Electronics*, Vol. 9, No. 2, 2003, pp. 486–491, https://doi.org/10.1109/JSTQE.2003.813322.

[29] Tan, Y., et al., "Silicon-Based Hybrid Demultiplexer for Wavelength-and Mode-Division Multiplexing," *Optics Letters*, Vol. 43, No. 9, 2018, pp. 1962–1965, https://doi.org/10.1364/OL.43.001962.

[30] Xiao, X., et al., "Large-Scale and Energy-Efficient Tensorized Optical Neural Networks on III-V-on-Silicon MOSCAP Platform," *APL Photonics*, Vol. 6, No. 12, 2021, p. 126107, https://doi.org/10.1063/5.0070913.

[31] Miller, D. A. B., "Attojoule Optoelectronics for Low-Energy Information Processing and Communications," *Journal of Lightwave Technology*, Vol. 35, No. 3, 2017, pp. 346–396, https://doi.org/10.1109/JLT.2017.2647779.

[32] Tait, A., et al., "Neuromorphic Photonic Networks using Silicon Photonic Weight Banks," *Scientific Reports*, Vol. 7, No. 1, 2017, p. 7430, 2017. https://doi.org/10.1038/s41598-017-07754-z. https://doi.org/10.1038/s41598-017-07754-z.

[33] Cheng, Y., et al., "Photonic Neuromorphic Architecture for Tens-of-Task Lifelong Learning," L*ight: Science & Applications*, Vol. 13, 2024, p. 56, https://doi.org/10.1038/s41377-024-01395-4.

[34] Chen, H., et al., "Effective Methods and Framework for Energy-Based Local Learning of Deep Neural Networks," *Frontiers in Artificial Intelligence*, Vol. 8, 2025, p. 1605706, https://doi.org/10.3389/frai.2025.1605706.

[35] Xu, T., et al., "Control-Free and Efficient Integrated Photonic Neural Networks via Hardware-Aware Training and Pruning," *Optica*, Vol. 11, No. 8, 2024, pp. 1039–1049, https://doi.org/10.1364/OPTICA.523225.

[36] Rubeling, P., et al., "In Situ Training in Programmable Photonic Frequency Circuits," *Nanophotonics*, Vol. 14, No. 16, 2025, https://doi.org/10.1515/nanoph-2025-0125.

[37] Lin, Z., et al., "120 GOPS Photonic Tensor Core in Thin-Film Lithium Niobate for In Situ Training," *Nature Communications*, Vol. 15, 2024, p. 9081, https://doi.org/10.1038/s41467-024-53261-x.

[38] Bandyopadhyay, S., et al., "Single-Chip Photonic Deep Neural Network with Forward-only Training," *Nature Photonics*, Vol. 18, 2024, pp. 1335–1343, https://doi.org/10.1038/s41566-024-01567-z.

[39] Jeon, I. et al., "Distinctive Properties of Biological Neural Networks and Recent Advances in Bottom-Up Approaches Toward a Better Biologically Plausible Neural Network," *Frontiers in Computational Neuroscience*, Vol. 17, 2023, https://doi.org/10.3389/fncom.2023.1092185.

[40] Morrison, A., et al., "Advancing the Boundaries of High-Connectivity Network Simulation with Distributed Computing," *Neural Computation*, Vol. 17, No. 8, 2005, pp. 1776–1801, https://doi.org/10.1162/0899766054026648.

[41] Ma, S., et al., "Synaptic Modifications in Learning and Memory–A Dendritic Spine Story," *Seminars in Cell and Developmental Biology*, Vol. 125, 2022, pp. 84–90, https://doi.org/10.1016/j. semcdb.2021.05.015.

[42] Abraham, W. C., et al., "Is Plasticity of Synapses the Mechanism of Long-Term Memory Storage?" *NPJ Science of Learning*, Vol. 4, 2019, p. 9, https://doi.org/10.1038/s41539-019-0048-y.

[43] Mansvelder, H. D., et al., "Synaptic Plasticity in Human Cortical Circuits: Cellular Mechanisms of Learning and Memory in the Human Brain?" *Current Opinion in Neurobiology*, Vol. 54, 2019, pp. 186–193, https://doi.org/10.1016/j.conb.2018.06.013.

[44] Shen, Y., et al., "Silicon Photonics for Extreme Scale Systems," *Journal of Lightwave Technology*, Vol. 37, No. 2, 2019, pp. 245–259, https://doi.org/10.1364/JLT.37.000245.

[45] Dong, B., et al., "Partial Coherence Enhances Parallelized Photonic Computing," *Nature*, Vol. 632, 2024, pp. 55–62, https://doi.org/10.1038/s41586-024-07590-y.

[46] Yu, X., et al., "Parallel Optical Computing Capable of 100-Wavelength Multiplexing," *eLight*, Vol. 5, p. 10, https://doi.org/10.1186/s43593-025-00088-8.

[47] Wu, J., et al., "Analog Optical Computing for Artificial Intelligence," *Engineering*, Vol. 10, 2022, pp. 133–145, https: //doi.org/10.1016/j.eng.2021.06.021.

[48] Dan, D., et al., "Optoelectronic Integrated Circuits for Analog Optical Computing," *Frontiers in Physics*, Vol. 10, 2022, https://doi.org/10.3389/fphy.2022.1064693.

[49] Zuo, Y., et al., "All-Optical Neural Network with Nonlinear Activation Functions," *Optica*, Vol. 6, No. 9, 2019, pp. 1132–1137, https://doi.org/10.1364/OPTICA.6.001132.

[50] Xu, Z., et al., "Reconfigurable Nonlinear Photonic Activation Function for Photonic Neural Network Based on Non-Volatile Opto-Resistive RAM Switch," *Light: Science and Applications*, Vol. 11, 2022 p. 288, https://doi.org/10.1038/s41377-022-00976-5.

[51] Pankov, A. V., et al., "Optical Neural Network Based on Synthetic Nonlinear Photonic Lattices," *Physical Review Applied*, Vol. 17, 2022, p. 024011, https://doi.org/10.1103/PhysRevApplied.17. 024011.

[52] Haiml, M., et al., "Optical Characterization of Semiconductor Saturable Absorbers," *Applied Physics B*, Vol. 79, 2004, pp. 331–339, https://doi.org/10.1007/s00340-004-1535-1.

[53] Marini, A., et al., "Theory of Graphene Saturable Absorption," *Physical Review B*, Vol. 95, 2017, p. 125408, https://doi.org/10.1103/PhysRevB.95.125408.

[54] Sánchez Muñoz, C., et al., "Quantum Metrology of Two-Photon Absorption," *Physical Review Research*, Vol. 3, 2021, p. 033250, https://doi.org/10.1103/PhysRevResearch.3.033250

[55] Chrostowski, L., and M. Hochberg, *Silicon Photonics Design: From Devices to Systems*, Cambridge, UK: Cambridge University Press, 2015.

10

Seeing the Invisible Thread

Between empirical observation and theoretical abstraction lies a domain where discovery originates. This book has explored that domain—not as a collection of devices or equations, but as an examination of the cognitive structures that enable innovation. Central to this exploration has been the role of analogical reasoning. As demonstrated throughout the preceding chapters, analogy is not a peripheral linguistic device; it is a generative mechanism that enables the transfer of insight across conceptual boundaries. In photonics, a field that routinely contends with phenomena beyond direct human perception, analogical reasoning provides a foundational framework for both conceptualization and design.

The purpose of this final chapter is to reconsider not the technologies themselves, but the form of reasoning that produced them. The case studies presented—from the periodic architectures of photonic crystals to the bioinspired microtextures of antireflective coatings, from optical communication networks to the geometrically informed design of X-ray optics—illustrate that innovation in photonics frequently arises from structural and functional analogies. The most transformative developments in the field were not solely the result of incremental technological advances, but of a reframing of how light and its interactions were understood. Progress followed when researchers began to perceive light through alternative conceptual models, mapping its properties onto other physical or biological systems.

This cognitive shift has strong historical precedent. Biological morphology has long inspired optical design; mathematical frameworks originating in acoustics or crystallography have provided representational tools for photonics; and conceptual metaphors, such as channels, bottlenecks, or mirrors, have guided the physical implementation of communication and sensing systems. These are not decorative correspondences. They alter the structure of problems themselves, enabling researchers to apply the logic of one system to another. Such transpositions expand the design space and generate new theoretical and practical opportunities.

Contemporary photonics extends across multiple sectors: data infrastructure, defense, biomedical imaging, and consumer electronics. In all these domains, the path from concept to prototype often begins with an act of analogical abstraction. For example, the development of photonic neural networks requires the alignment of two distinct systems—the propagation of electromagnetic waves and the architecture of neural computation. Achieving this synthesis demands not only multidisciplinary expertise, but a facility for metaphorical reasoning: an ability to see correspondences between optical interference and information processing, between laser dynamics and synaptic activation. The capacity to perform such mappings, rather than mere technical proficiency, enables genuine innovation.

A key argument advanced throughout this book is that analogical reasoning is a teachable cognitive skill. It begins with the abstraction of a problem beyond disciplinary vocabulary: asking what must flow, what must resonate, what must be guided or gated. When problems are expressed at this structural level, insights from disparate domains—biology, mechanics, computation—become transferable. The relevance of knowledge is no longer dictated by subject matter but by relational structure.

Nevertheless, analogy alone does not constitute invention. It represents an initiating condition for innovation, not its culmination. Realizing an analogical insight requires iterative experimentation, validation, and refinement. Many conceptual breakthroughs in photonics, such as photonic crystals or hollow-core fibers, were articulated decades before fabrication technologies permitted their realization. Their persistence in theoretical discourse underscores the resilience of analogy as a cognitive scaffold: it establishes a conceptual map long before technical execution becomes feasible.

In this context, the present work functions less as a retrospective and more as a methodological guide. The intent is to cultivate in the reader a disciplined capacity for analogical perception—an approach that integrates technical precision with conceptual breadth. Whether in the design of waveguides, imaging systems, quantum detectors, or display technologies, progress depends upon the ability to translate insights across domains. The most significant advances often emerge in this intermediate zone, where depth of expertise intersects with lateral vision.

The continued evolution of photonics will depend on this capacity. As research advances toward quantum and neuromorphic regimes—requiring simultaneous reasoning in probabilistic, wave-based, and circuit-level frameworks—analogical thinking will remain indispensable. It facilitates communication across scales, from single-photon interactions to integrated optical systems, and across paradigms, bridging physical, mathematical, and biological perspectives.

Even seemingly mature subfields illustrate this dynamic. The graded-index principle underlying modern antireflective coatings, for instance, originated from the microstructures of moth eyes but now informs metasurface engineering and AI-driven optical optimization. These developments exemplify a feedback process in which nature and technology continually inform one another through structural correspondence.

The conclusion of this book, therefore, is not an endpoint but an invitation—to adopt analogy not as ornamentation but as analytical practice. When approaching a design challenge, the most productive question may not be "What equation is missing?" but "What system elsewhere behaves in a comparable manner?" When encountering unfamiliar scientific literature, the analogical thinker searches for structural resonance rather than topical similarity. In communicating complex research, effective analogies serve not merely to simplify, but to reveal latent conceptual symmetry.

Throughout these chapters, analogy has been presented as a mode of seeing and reasoning, which is integral to research, pedagogy, collaboration, and communication. The strongest proposals, explanations, and innovations in photonics have all depended on this capacity to map the known onto the unknown. Mastering this practice equips researchers to transform uncertainty into discovery.

The light that photonics seeks to manipulate is, ultimately, only part of the story. The other part lies in how we conceptualize it—how we imagine its structure, its analogs, and its potential futures. That capacity for structured imagination will continue to determine not only what we build, but how we understand the invisible threads that connect ideas across disciplines.

As illustrated in Figure 10.1, progress is modeled as motion along a diagonal corridor where abstraction and observation advance in concert. The corridor's width encodes the permissible mismatch between what a model asserts and what experiments report; moving outside the band quantifies imbalance. Points above the band (empirical drift) mark regimes in which measurements accumulate faster than the governing abstractions are refined, often signaling an opportunity to extract structure, formalize invariants, or simplify the hypothesis space. Points below the band (theoretical overreach) mark regimes in which conjecture outstrips verification, suggesting the need for targeted assays, higher fidelity instrumentation, or relaxed assumptions. In practice, the distance from a point to the band can be read as a corrective priority: reduce empirical–theoretical residuals via data collection when below, or via model consolidation when above. The figure thus provides a compact operational heuristic for managing research trajectories and maintaining productive coupling between ideas and evidence.

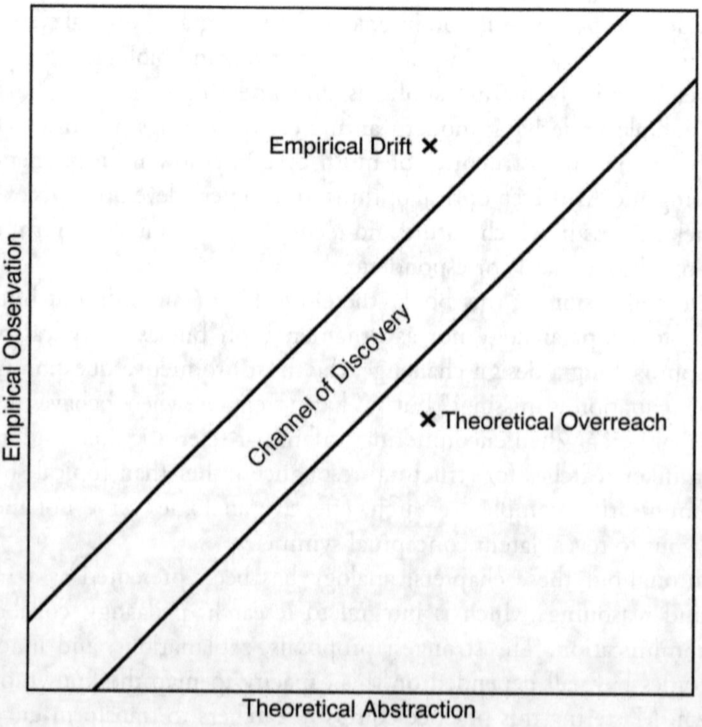

Figure 10.1 Channel of Discovery. The axes denote theoretical abstraction (x, increasing to the right) and empirical observation (y, increasing upward). The two parallel lines at 45 degrees, define a diagonal band in which theory and evidence coevolve; the band's thickness represents acceptable tolerance between prediction and measurement. Points above the band (e.g., empirical drift) indicate observation outpacing conceptual framing, whereas points below the band (e.g., theoretical overreach) indicate speculative theory insufficiently supported by data.

About the Author

Khaled Ahmed is a pioneering technologist and prolific inventor whose career spans more than three decades across microelectronics, semiconductor processing, advanced displays, and photonic systems. He earned his BSc and MSc degrees in electrical engineering from Ain Shams University, Egypt, in 1991 and 1994, respectively, before moving to the United States to pursue doctoral studies. In December 1998, he received his PhD in microelectronics from North Carolina State University.

Dr. Ahmed's industrial journey began at AMD, where he played a critical role in the productization of the K6 microprocessor—AMD's first non-Intel-clone CPU. He then contributed to the advancement of SiGe BiCMOS technologies at Conexant Systems, targeting high-performance communication applications. In 2001, Dr. Ahmed joined Applied Materials, where his work on plasma nitridation became foundational for CMOS fabrication across the semiconductor industry throughout the 2000s.

His career continued to evolve through strategic roles that bridged advanced manufacturing and emerging technologies. At Applied's SunFab initiative, he led SubFab productization for solar manufacturing. He then joined Intermolecular, driving R&D on novel materials for thin-film transistors in next-generation displays.

In 2015, Dr. Ahmed joined Intel's Global Supply Chain organization, where he led the development of a microLED ecosystem tailored to Intel's vision for future laptop displays. His focus shifted in 2021 toward leveraging microLEDs for high-speed optical communication, laying the groundwork for ultra-efficient interconnects. In 2024, he joined the Extreme Scale Computing group in Intel's Office of the CTO, where he continues to drive innovation at the intersection of microLED photonics and hyperspectral imaging.

Dr. Ahmed holds approximately 200 issued and pending patents spanning semiconductor devices, fabrication equipment, microscopic assembly, display systems, and optical technologies. He was named Intel Inventor of the Year for four consecutive years, a rare distinction reflecting both breadth and impact. In August 2024, he authored the first original book on microLEDs, cementing his role as a thought leader in the field.

With a career defined by bold transitions and deep technical contributions, Dr. Ahmed stands at the forefront of innovation in electronics, displays, and photonic systems.

Index

Artech House
Applied Photonics Library

Brian Culshaw and Marc Wuilpart, Series Editors

The ABCs of Fiber Optic Communication, Sudhir Warier

Acousto-Optic Signal Processing: Fundamentals and Applications, Pankaj K. Das and Casimer De Cusatis

Advanced Optical Communication Systems and Networks, Milorad Cvijetic and Ivan B. Djordjevic

Analogy Thinking for Photonics Innovation, Khaled Ahmed

Applications of Modern RF Photonics, Preetpaul Singh Devgan

Bistabilities and Nonlinearities in Laser Diodes, Hitoshi Kawaguchi

Chemical and Biochemical Sensing with Optical Fibers and Waveguides, Gilbert Boisdé and Alan Harmer

Coherent and Nonlinear Lightwave Communications, Milorad Cvijetic

Coherent Doppler Wind Lidars in a Turbulent Atmosphere, Viktor Banakh and Igor Smalikho

Coherent Lightwave Communication Systems, Shiro Ryu

Digital Optical Measurement Techniques and Applications, Pramond Rastogi, editor

DWDM Fundamentals, Components, and Applications, Jean-Pierre Laude

Electromagnetic and Photonic Simulation for the Beginner: Finite-Difference Frequency-Domain in MATLAB®, Raymond C. Rumpf

Electro-Optical Systems Performance Modeling, Gary Waldman

Elliptical Fiber Waveguides, R. B. Dyott

Engineering Optical Networks, Sudhir Warier

Fiber Bragg Gratings: Fundamentals and Applications in Telecommunications and Sensing, Andrea Othonos and Kyriacos Kalli

Fiber Optics Communications, Henry F. Taylor

The Fiber-Optic Gyroscope, Third Edition, Hervé C. Lefèvre

Fiber-Optic Sensors for Biomedical Applications, Daniele Tosi and Guido Perrone

Field Theory of Acousto-Optic Signal Processing Devices, Craig R. Scott

Finite Element Modeling Methods for Photonics, B. M. Azizur Rahman and Arti Agrawal

Frequency Stabilization of Semiconductor Laser Diodes, Tetsuhiko Ikegami, Shoichi Sudo, and Yoshihisa Sakai

Fundamentals of Multiaccess Optical Fiber Networks, Dennis Mestdagh

Germanate Glasses: Structure Spectroscopy and Properties, Alfred Margaryan and Michael Piliavin

Handbook of Distributed Feedback Laser Diodes, Second Edition, Geert Morthier and Patrick Vankwikelberge

Highly Coherent Semiconductor Lasers, Motoichi Ohtsu

High-Power Optically Activated Solid-State Switches, Arye Rosen, Fred Zutavern

Integrated Optics: Design and Modeling, Reinhard Marz

Introduction to Electro-Optical Imaging and Tracking Systems, Shaheen A. Hovanessian and Khalil Seyrafi

Introduction to Glass Integrated Optics, S. Iraj Najafi

Introduction to Infrared and Electro-Optical Systems, Third Edition, Ronald G. Driggers, Melvin H. Friedman, John W. Devitt, Orges Furxhi and Anjali Singh

Introduction to Radiometry and Photometry, Second Edition, William Ross McCluney

Introduction to Semiconductor Integrated Optics, Hans P. Zappe

Laser Communications in Space, Stephen Lambert

Liquid Crystal Devices: Physics and Applications, Vladimir G. Chigrinov

Lithium Niobate Photonics, James E. Toney

MicroLED Devices and Systems, Khaled Ahmed

Military Laser Technology and Systems, David H. Titterton

Natural Photonics and Bioinspiration, Sébastien R. Mouchet and Olivier Deparis

New Photonics Technologies for the Information Age: The Dream of Ubiquitous Services, Shoichi Sudo and Katsunari Okamoto, editors

Optical Document Security, Third Edition, Rudolf L. van Renesse

Optical FDM Network Technologies, Kiyoshi Nosu

Optical Fiber Amplifiers: Materials, Devices, and Applications, Shoichi Sudo, editor

Optical Fiber Communication Systems, Leonid Kazovsky, Sergio Benedetto and Alan Willner

Optical Fiber Sensors, Volume One: Principles and Components, Brian Culshaw and John P. Dakin

Optical Fiber Sensors, Volume Two: Systems and Applications, Brian Culshaw and John P. Dakin

Optical Fiber Sensors, Volume Three: Components and Subsystems, John Dakin and Brian Culshaw, editors

Optical Fiber Sensors, Volume Four: Applications, Analysis, and Future Trends, John Dakin and Brian Culshaw, editors

Optical Interconnection: Foundations and Applications, John Caulfied and Christopher S. Tocci

Optical Measurement Techniques and Applications, Pramod Rastogi

Optical Network Theory, Yitzhak Weissman

Optical Transmission for the Subscriber Loop, Norio Kashima

Optical Transmission Systems Engineering, Milorad Cvijetic

Optoelectronics for Low-Intensity Conflicts and Homeland Security, Anil K. Maini

Optoelectronic Techniques for Microwave and Millimeter-Wave Engineering, William M. Robertson

Passive Optical Components for Optical Fiber Transmission, Norio Kashima

Photonic Applications for Radio Systems and Networks, Fabio Cavaliere and Antonio D'Errico

Plasmonic Optical Fiber Biosensors, Christophe Caucheteur and Médéric Loyez

Polarization in Optical Fibers, Alan Rogers

Principles of Modern Optical Systems, Volume 1, Deepak Uttamchandani and Ivan Andonovic, editors

Principles of Modern Optical Systems, Volume 2, Deepak Uttamchandani and Ivan Andonovic, editors

Reliability and Degradation of III-V Optical Devices, Osamu Ueda

Securing Information and Communications Systems, Stephen Furnell, Sokratis Katsikas, Javier Lopez, and Ahmed Patel

Semiconductor Raman Lasers, Ken Suto and Jun-Ichi Nishizawa

Semiconductors for Solar Cells, Hans Joachim Moller

Semiconductor Quantum Wells and Superlattices for Long-Wavelength Infrared Detectors, M. Omar Manasreh

Signal Processing and Performance Analysis for Imaging Systems, S. Susan Young, Ronald G. Driggers, and Eddie L. Jacobs

Single-Mode Optical Fiber Measurement: Characterization and Sensing, Giovanni Cancellieri and Franco Chiaraluce

Smart Structures and Materials, Brian Culshaw

Substrate Surface Preparation Handbook, Max Robertson

Surveillance and Reconnaissance Imaging Systems: Modeling and Performance Prediction, Jon C. Leachtenauer and Ronald G. Driggers

Tunable Laser Diodes, Markus-Christian Amann and Jens Buus

Ultrafast Diode Lasers: Fundamentals and Applications, Peter Vasilev

Understanding Optical Fiber Communications, Alan Rogers

Wavelength Division Multiple Access Optical Networks, Andrea Borella, Giovanni Cancellieri and Franco Chiaraluce

Wired and Wireless Seamless Access Systems for Public Infrastructure, Tetsuya Kawanishi

For further information on these and other Artech House titles, including previously considered out-of-print books now available through our In-Print-Forever® (IPF®) program, contact:

Artech House
685 Canton Street
Norwood, MA 02062
Phone: 781-769-9750
Fax: 781-769-6334
e-mail: artech@artechhouse.com

Artech House
16 Sussex Street
London SW1V 4RW UK
Phone: +44 (0)20 7596-8750
Fax: +44 (0)20 7630-0166
e-mail: artech-uk@artechhouse.com

Find us on the World Wide Web at: www.artechhouse.com

www.ingramcontent.com/pod-product-compliance
Lightning Source LLC
LaVergne TN
LVHW011524200926
848484LV00001B/142
9781685691073